# 油气井工程设计规范与法规
（双语版）

索彧 孔翠龙 赵万春 王彪 主编

石油工业出版社

## 内容提要

本书基于编写人员长期从事油气井工程领域的理论研究及现场经验，整合了非常规油气资源及新能源的相关热点及现场案例，整理了油气井工程领域相关的设计规范及法律法规，可培养学生参与油气井开发全过程的能力。

本书可作为石油高等院校相关专业的指导教材，也可供从事油气井工程领域工作的技术人员及研究人员参考使用。

### 图书在版编目（CIP）数据

油气井工程设计规范与法规：汉、英 / 索彧等主编 . -- 北京：石油工业出版社，2025.6. -- ISBN 978-7-5183-7077-1

Ⅰ. TE64-37

中国国家版本馆 CIP 数据核字第 2025VS9787 号

---

出版发行：石油工业出版社
　　　　　（北京安定门外安华里 2 区 1 号　100011）
　　　网　　址：www.petropub.com
　　　编辑部：（010）64523708
　　　图书营销中心：（010）64523633
经　　销：全国新华书店
印　　刷：北京九州迅驰传媒文化有限公司

2025 年 6 月第 1 版　2025 年 6 月第 1 次印刷
787×1092 毫米　开本：1/16　印张：22.75
字数：550 千字

定价：80.00 元
（如出现印装质量问题，我社图书营销中心负责调换）
**版权所有，翻印必究**

# 《油气井工程设计规范与法规（双语版）》编写人员

| | | | | |
|---|---|---|---|---|
| 索 彧 | 孔翠龙 | 赵万春 | 王 彪 | 孙晓峰 |
| 胡超洋 | 孙士慧 | 王 鑫 | 张 旭 | 陈建国 |
| 王桂全 | 赵思聪 | 汪道兵 | 李玉伟 | 张 军 |
| 魏显航 | 杨 楠 | 孔铉文 | 马宇泽 | 贾振甲 |
| 彭东喆 | 黄刘科 | 肖诚成 | 赵延杰 | 李子豪 |
| 李芬芬 | 滕 阳 | 张锐铎 | 张程晨 | 杨稳晶 |
| 关文燕 | 武 超 | 王晓光 | 周凌陟 | 董牧宇 |

# 前言 Preface

鉴于能源行业的极端重要性及战略意义，石油和天然气工程学科始终是能源类高校的核心学科。随着我国常规油气资源开采难度增大，非常规油气资源及新能源作为保证我国油气高质量发展的重要接替资源越来越受到关注和重视。随着非常规资源和新能源的开发，相关的油气井工程理论、工艺及技术也在快速发展，油气井工程相关教材的内容应及时更新，对新知识、新技术进行有效整合。

本书以新工科建设理念为指导、以培养学生的创新能力和实践能力为出发点、以培养德才兼备的石油人才为目标，编写人员基于长期从事油气井工程相关理论研究和工程项目的丰富经验，整理了学生应掌握的油气井工程领域的相关知识和技术，以及油气井工程领域的设计规范及涉及的法律法规，在案例分析部分重点考虑油气井工程工艺流程的基本操作规范、HSE 相关的现场操作规范等，以期培养学生具备参与油气井开发全流程的能力。本书为中英双语版，可进一步加强读者对油气井工程领域相关的石油专业词汇的熟悉度；本书可以作为高等石油院校相关专业的指导教材，可供从事钻井工程相关工作的科研人员参考使用，也可作为油气行业技术人员的培训用书。

本书整理了石油与天然气行业中的实际操作规范、行业标准和法律法规，充分融入目前非常规油气钻井工程、固井工程等领域的成熟理论、工艺、技术，辅以现场案例分析，有助于提升学生的专业素质和实践能力，更好地适应当前油气行业的人才需求。本书兼具理论性、知识性和实用性，还可有效支撑石油与天然气工程学科中的学科交叉互融建设。同时，本书还深入挖掘石油行业蕴含的丰富思政要素，通过典型思政案例实现专业传授与德育引导的有机统一，培育学生坚定爱国主义信念及树立能源报国的理想。

本书第一章、第二章、第三章分别为钻井、完井、固井的相关理论知识及技术介绍，第四章为 HSE 相关知识简介，第五章为典型事故及处理措施，第六章、第七章为油气井工程相关的法律条款及规范标准。全书由索彧等编写人员共同完成。在本书的编写过程中，大庆油田有限责任公司采油工艺研究院孔翠龙、中国石油集团大庆钻探工程有限公司齐悦、吉林油田非常规资源开发公司叶勤友、吉林油田公司钻井工艺研究院何军、吉林油田公司油气工程研究院段永伟对本书的现场案例进行了审核。

本书得到了黑龙江省研究生课程思政高质量建设项目、东北石油大学研究生教材建设项目及东北石油大学石油与天然气工程学科培优专项建设项目的资助。中国石油大学（北京）冯永存、西南石油大学李皋、长江大学张菲菲、中国石油大学（华东）黄贤斌等专家为本书提出了宝贵意见；本书在撰写过程中引用和参考了相关文献与资料，在此特向资料提供者和文献作者一并表示感谢。

由于笔者水平有限，书中难免存在疏漏之处，敬请读者批评指正。

# 目录 Contents

Chapter 1　Drilling ·········· 1
　1.1　Basic Knowledge of Drilling ········ 1
　1.2　The Basic Theory of Drilling Engineering Design ············ 4
　1.3　Drill Bit ················ 13
　Ideological and Political Case Studies ··· 19

Chapter 2　Completion ················ 27
　2.1　Overview of the Development of Modern Drilling and Completion Technologies ············ 27
　2.2　Development of Drilling and Completion Technology ············ 32
　2.3　Completion Wellbore Structure ··· 35
　2.4　The Principle of Well Completion Design ············ 40
　Ideological and Political Case Studies ··· 51

Chapter 3　Well Cementing Design ··· 62
　3.1　Characteristics of Cementing ······ 62
　3.2　Casing String Design ············ 69
　3.3　Cementing Mud and Injection Design ············ 80
　Ideological and Political Case Studies ··· 95

Chapter 4　HSE Knowledge Related to the Oil and Gas Industry ··· 103
　4.1　Introduction to HSE-Related Knowledge ············ 103

第一章　钻井 ············ 1
　1.1　钻井的基本知识 ············ 1
　1.2　钻井工程设计的基本理论 ········ 4
　1.3　钻头 ················ 13
　思政案例 ············ 19

第二章　完井 ············ 27
　2.1　现代钻井与完井技术的发展概况 27
　2.2　钻完井技术的发展 ············ 32
　2.3　完井井底结构 ············ 35
　2.4　完井设计原则 ············ 40
　思政案例 ············ 51

第三章　固井工程设计 ············ 62
　3.1　固井 ················ 62
　3.2　套管串设计 ············ 69
　3.3　固井水泥浆及注替设计 ········ 80
　思政案例 ············ 95

第四章　油气行业 HSE 相关知识 ······ 103
　4.1　HSE 相关知识简介 ············ 103

| | |
|---|---|
| 4.2 Master HSE Risk Identification for Drilling Operations ............ 104 | 4.2 掌握钻井作业 HSE 风险识别 ... 104 |
| 4.3 On Site Medical Emergency ...... 112 | 4.3 现场医疗急救 .................. 112 |
| Chapter 5 Typical Accidents and Handling Measures .......... 118 | 第五章 典型事故及处理措施............ 118 |
| 5.1 Drill Bit Accidents ............... 118 | 5.1 钻头事故 ..................... 118 |
| 5.2 On-site Accident Cases ........... 139 | 5.2 现场事故案例 .................. 139 |
| Chapter 6 Specifications and Standards Related to Oil and Gas Well Engineering .................. 178 | 第六章 油气井工程相关的规范与标准............ 178 |
| 6.1 Safety Technical Regulations for Drilling Well Site Equipment Operation (SY/T 5974-2020) (Part) ........................... 178 | 6.1 《钻井井场设备作业安全技术规程》（SY/T 5974—2020）（部分）...... 178 |
| 6.2 Safety Regulations for Underground Operations (SY 5727-2014) (Part) ........................... 210 | 6.2 《井下作业安全规程》（SY 5727—2014）（部分） ...... 210 |
| 6.3 Special Safety Signs for Oil and Gas Production (SY/T 6355-2017) (Part) ........................... 224 | 6.3 《石油天然气生产专用安全标志》（SY/T 6355—2017）（部分）...... 232 |
| Chapter 7 Laws and Regulations Related to Oil and Gas Well Engineering .................. 267 | 第七章 油气井工程相关的法律法规... 267 |
| 7.1 Work Safety Law of the People's Republic of China (Excerpt) ...... 268 | 7.1 《中华人民共和国安全生产法》（节选）...................... 268 |
| 7.2 Unified Regulations on the Rational and Comprehensive Utilization of Mineral Resources in Kazakhstan (Excerpt) ........................ 285 | 7.2 哈萨克斯坦矿产资源合理和综合利用统一规定（节选） ............. 285 |
| 7.3 Russian Regulations on the Development of Oil and Gas Fields (Excerpt) ... 341 | 7.3 俄罗斯的油气田开发法规（节选） .................... 341 |

# Chapter 1 Drilling

## 1.1 Basic Knowledge of Drilling

### 1.1.1 The Basic Meaning of Drilling

Drilling refers to the use of certain equipment, tools and technical means to form a hole from the surface to a certain depth underground with different trajectory shapes. It refers to the process of drilling holes in rocks, soil or other formations underground or underwater by using drilling tools, drills and other equipment. This process is often used to obtain geological information, extract mineral resources, build a wellbore, or develop groundwater.

The purpose and significance of drilling are geological evaluation, discovery and development of oil and gas reservoirs. Types of drilling include manual excavation, dry drilling and rotary drilling. Drilling is the "locomotive" of the Petroleum industry.

### 1.1.2 Rotary Drilling Technology

Rotary drilling technology development including 4 parts.

Conceptual period (1901-1920). Including the combination of drilling and washing, the use of cone bits, cement cementing, etc.

# 第一章 钻井

## 1.1 钻井的基本知识

### 1.1.1 钻井的基本含义

钻井指利用一定的设备、工具和技术手段形成一个从地表到地下某一深度处具有不同轨迹形状的孔道。即指通过使用钻具、钻头等设备，在地下或水下的岩石、土壤或其他地层中钻出孔洞的过程。这个过程通常是为了获取地质信息、开采矿产资源、建造井筒或进行地下水开发等目的。

钻井的目的和意义在于地质评价、发现油气藏和开发油气藏。钻井的种类包括人工挖掘、顿钻和旋转钻井。钻井是石油工业的"火车头"。

### 1.1.2 旋转钻井技术

旋转钻井技术发展包括以下四个阶段。
概念时期（1901—1920年）。包括钻井和洗井相结合、牙轮钻头的使用、水泥固井等。

Development period (1920-1948). Technologies such as cone bit, cementing process and drilling fluid have been further developed, and drilling related equipments have been further improved.

Scientific drilling period (1948-1969). At this stage, a large amount of scientific research was carried out, which enabled the rapid development of drilling technology. Outstanding technical achievements including: high-pressure jet drilling, efficient cone bits (insert teeth, sliding seal bearing bits), high-quality drilling fluid (low solid phase, no solid phase non-dispersed system drilling fluid), optimal parameter drilling (optimal weight on bit, rotational speed and hydraulic parameters), formation pressure detection technology, oil and gas well pressure control technology, drilling fluid solid phase control technology, balanced pressure drilling technology.

Automated drilling period (1969-). It mainly embodies the application of electronic instrument, automatic measurement and computer in drilling engineering. Measurement while drilling, integrated logging, automated drilling rig, geological and automatic guidance, vertical drilling system, expansion tube technology, small hole, underbalanced pressure drilling, extended reach horizontal and branch Wells, continuous pipe drilling, ultra-high pressure jet drilling technology, etc.

Rotary drilling method including: rotary table drilling, downhole power rotary drilling and top drive rotary drilling.

## 1.1.3 Drilling Technology Development

Domestic drilling technology development level as follows:

In 1976, China completed the first ultra-

发展时期（1920—1948年）。牙轮钻头、固井工艺、钻井液等技术得到进一步发展，钻井相关设备进一步改进。

科学化钻井时期（1948—1969年）。这个阶段开展了大量的科学研究，使钻井技术得以迅速发展。突出的技术成就有：高压喷射钻井、高效牙轮钻头（镶齿、滑动密封轴承钻头）、优质钻井液（低固相、无固相不分散体系钻井液）、优选参数钻井（优选钻压、转速和水力参数）、地层压力检测技术、油气井压力控制技术、钻井液固相控制技术、平衡压力钻井技术。

自动化钻井时期（1969年至今）。主要体现电子仪表、自动测量和计算机在钻井工程中的应用。随钻测量、综合录井、自动化钻机、地质和自动导向、垂直钻井系统、膨胀管技术、小井眼、欠平衡压力钻井、大位移水平井和分支井、连续管钻井、超高压喷射钻井技术等。

旋转钻井方法包括：转盘旋转钻井、井下动力旋转钻井和顶部驱动旋转钻井。

## 1.1.3 钻井技术发展情况

国内钻井技术发展水平如下。

1976年我国在四川地区完成了第一口6011m的超深井。

deep well of 6011m in Sichuan.

Between 1976 and 1985, 170 deep Wells and 10 ultra-deep Wells were completed. The Guanji well in Sichuan is 7175m; Gu2 Well 7002m in Xinjiang.

From 1986 to 1997, a total of 688 deep and ultra-deep Wells were completed in China, including 34 ultra-deep Wells. Tacan 1 Well in Tarim is currently the deepest well in China, with a depth of 7200m.

Foreign drilling technology development level as follows:

The United States: in 1938, the world's first 4573m deep well was drilled, and in 1949, the ultra-deep well was drilled into 6255m.

Soviet Union: The world record for ultra-deep Wells of 12200m was set in 1984.

Germany: An ultra-deep well of 9107m was drilled in 1994.

## 1.1.4 Development Trend and Restriction Factors of Drilling Technology

Development trend:

(1) Drilling is closely connected with the exploration and development of oil and gas fields;

(2) Diversification of well trajectories (special process Wells);

(3) Improve drilling rate (especially deep Wells), underbalanced pressure drilling, ultra-high pressure drilling, power drilling + rotary drilling +PDC bit (or other special drill);

(4) Automatic guidance drilling, automatic control of well trajectory;

(5)Automated drilling: data acquisition, surface (integrated logging instrument), underground, MWD, LWD, FEWD, data transmission, (storage, analysis, processing, decision-making) feedback to the wellsite

1976—1985 年，完成了 170 口深井和 10 口超深井。其中四川的关基井 7175m；新疆的固 2 井 7002m。

1986—1997 年，我国共完成深井超深井 688 口，其中超深井 34 口。塔里木盆地的塔参 1 井是我国当时最深的井，井深达 7200m。

国外钻井技术发展水平如下：

美国：1938 年钻成世界上第一口 4573m 的深井，1949 年钻成 6255m 的超深井。

苏联：1984 年创造了 12200m 的世界特超深井的纪录。

德国：1994 年钻成一口 9107m 的超深井。

## 1.1.4 钻井技术的发展趋势及制约因素

发展趋势：

（1）钻井与油气田的勘探开发紧密相连；

（2）井眼轨迹的多样化（特殊工艺井）；

（3）提高机械钻速（特别是深井超深井），欠平衡压力钻井、超高压钻井，动力钻+旋转钻+PDC 钻头（或其他特种钻头）；

（4）自动导向钻井，井眼轨迹的自动控制；

（5）自动化钻井：数据采集，地面（综合录井仪），地下，MWD、LWD、FEWD，数据传输，（存储、分析处理、决策）反馈到井场执行机构，形成优化钻井的闭环系统。

actuator, forming a closed-loop system for optimizing drilling.

Typical new drilling technologies are: horizontal well, extended reach well, underbalanced pressure drilling, small hole well, ultra-deep well, branch well, multi-bottom well, guided drilling, etc.

典型的钻井新技术有：水平井、大位移井、欠平衡压力钻井、小井眼井、超深井、分支井、多底井、导向钻井等。

## 1.2 The Basic Theory of Drilling Engineering Design

## 1.2 钻井工程设计的基本理论

### 1.2.1 Analysis of Geological Tasks and Drilling Difficulties

### 1.2.1 地质任务与钻井难点分析

#### 1.2.1.1 Geological Tasks and Quality Requirements

#### 1.2.1.1 地质任务与质量要求

The primary goal of drilling construction design is to complete the geological task, and the construction design should analyze the special requirements of the drilling geological task and take targeted technical measures in order to complete the drilling task with high quality. Usually, the special requirements of geological tasks can include.

(1) Requirements for data recording. Data logging is one of the main tasks of exploratory wells. Geological tasks that have a greater impact on drilling construction are mainly large coring scales, frequent geological circuits, and midway tests, which lead to longer drilling time and thus more difficulty in stabilizing the borehole. Special attention needs to be paid to the drilling fluid system, formulation and the ability of the drilling fluid to stabilize the borehole.

As large boreholes have a certain impact on the logging series and the quality of logging data, in order to obtain the logging data of the

钻井施工设计的首要目标是完成地质任务，施工设计应分析钻井地质任务的特殊需求，采取针对性的技术措施，才能高质量地完成钻井任务。通常地质任务的特殊要求有：

（1）对资料录取的要求。资料录取是探井的主要任务之一。对钻井施工影响较大的地质任务主要是取心进尺多、地质循环频繁、中途测试等，这些工作导致钻井时间的延长，由此导致井眼稳定难度加大。需在钻井液体系、配方及钻井液稳定井壁能力方面特别引起注意。

由于大井眼对测井系列与测井资料质量有一定影响，为获取上部兼探层位的测井资料，有时需要先钻常规尺寸井眼，完成测井后再进行扩眼，这时需要制订出安全扩眼的技术措施。

upper part of the partially explored layer, it is sometimes necessary to drill a regular-sized borehole first, and then expand the borehole after completing the logging, and then it is necessary to formulate technical measures for the safe expansion of the borehole.

(2) Special requirements on the quality of well casing. Ultra-deep wells due to the long drilling cycle, the drilling tools rotate for a long time will produce serious wear on the casing, and the casing of ultra-deep wells is subjected to a large load, the safety factor is low, so the protection work for casing is very important, in addition to the measures that must be taken to protect the drilling rod from abrasion and other measures to control a more stringent well inclination is also one of the most important measures, which requires the formulation of the corresponding control of the inclination of the well of the technical measures.

Control of well quality also includes higher requirements for borehole expansion rates to ensure cementing quality and logging data quality.

(3) Special requirements for cementing quality. The problem of wellhead pressurization during the production of high-pressure gas wells is a difficult problem that has not yet been completely solved in the world. In order to delay the time of wellhead pressure and reduce the severity of wellhead pressure, it is necessary to put forward more stringent requirements on the quality of cementing and wellbore integrity.

(4) Special requirements for drilling fluids. Drilling fluid density control is required for oil and gas discovery, and when drilling under certain narrow density window conditions, a more refined density control

（2）对井身质量的特殊要求。超深井由于钻井周期长，钻具长时间的旋转对套管将产生严重的磨损，而超深井套管所受载荷大，安全系数低，因此对套管的保护工作非常重要，除必须采取的钻杆防磨保护等措施外，对井斜的严格控制也是非常重要的措施之一，由此需要制订相应的控制井斜的技术措施。

对井身质量的控制也包括为保证固井质量、测井资料品质而对井眼扩大率提出更高的要求。

（3）对固井质量的特殊要求。高压气井生产过程中的井口带压问题是目前世界上尚没有完全解决的难题。为延缓井口带压时间，减少井口带压严重程度，需要对固井质量及井筒完整性提出更为严格的要求。

（4）对钻井液的特殊要求。钻井液密度控制是油气钻井的需要，而在某些窄密度窗口情况下钻井时，从保证井眼稳定性出发，需要采取更为精细的密度控制方法，才能保证不喷、不漏、不垮、不塌、不阻、不卡。

method is required from the point of view of ensuring the stability of the borehole, in order to ensure that there is no blowout, leakage, collapse, collapse, blockage, or jamming.

### 1.2.1.2 Analysis of Drilling Difficulties

The analysis of drilling difficulties includes a full analysis of the special requirements of the geological task and a detailed analysis of the neighboring well data. Although Party A has the obligation to make a comprehensive and in-depth analysis of the data of neighboring wells, it cannot replace the construction unit to make a more in-depth analysis of the data of neighboring wells, so as to analyze the difficulties of drilling at a deeper level. If Party A's design does not fully analyze the data of the neighboring wells, the construction unit should analyze the drilling situation of the neighboring wells comprehensively and sort out the difficulties of drilling this well.

### 1.2.2 Drilling Technology Response

Technical countermeasures for drilling are formulated on the basis of an analysis of the special needs of geological tasks for drilling wells and the difficulties of drilling wells. The technical countermeasures highlight the concretization of the recommended measures in the drilling engineering design. They include the following.

### 1.2.2.1 Construction Design of Drill Bits and Drilling Parameters

If there is no specific drill bit model in the design of the drilling project, the construction design must form a specific plan for the selection and use of the drill bit. If there is a specific drill bit model in the design of drilling project, it should be refined into a drill bit

### 1.2.1.2 钻井难点分析

钻井难点分析包括充分分析地质任务的特殊要求和对邻井资料进行细致的分析。虽然甲方钻井工程设计有义务对邻井资料进行全面深入的分析，但这并不能代替施工单位对邻井资料进行更为深入的分析，从而在更深层次对钻井难点进行分析。如果甲方设计没有对邻井资料进行充分的分析，则施工单位更要在全面分析邻井钻井情况基础上，梳理出本井钻井的难点。

### 1.2.2 钻井技术对策

在分析地质任务对钻井的特殊需求与钻井难点基础上，制订钻井技术对策。技术对策突出钻井工程设计中推荐措施的具体化。包括以下内容：

### 1.2.2.1 钻头与钻井参数施工设计

钻井工程设计中如果没有明确具体钻头型号，施工设计中就必须形成钻头的具体选型与使用的方案。如果钻井工程设计中已有了具体的钻头型号，则应针对钻头使用方案，细化成为使用的钻头序列方案，包括钻头使用异常后的应对措施。施工设

sequence program for the use of the drill bit, including the countermeasures after abnormal use of the drill bit. The construction design should emphasize the consideration of details, and reduce the occurrence of accidents during the drilling construction process through careful design, so that the drilling production can be carried out continuously and smoothly.

The construction design should be refined to determine whether a single bit will be used to drill the entire footage in the first pass, and whether old bits can be used. For the second drill, the cement plug and substructure can be drilled with a more economical steel-toothed bit, and thereafter how many feet each bit may drill, and according to the wear grading of the bit after it is lifted out, the type of the next bit will be determined, and so on, until the drill bit combinations for the whole well are designed. In addition, for complex deep wells that may have variable conditions, several sets of drill programs should be prepared in order to avoid the situation where no drill bit is available for use.

As the construction unit generally has a good supply relationship with the drill bit manufacturer, the design of the drill bit and drilling parameters in the construction design can also give full play to the role of the drill bit manufacturer. Let the drill bit manufacturer according to the neighboring well logging data, the neighboring well drill bit use, put forward the drill bit design scheme, and signed according to the effect of the use of the drill bit to pay the cost of the drill bit agreement, so that the drill bit manufacturer from the design. This allows the drill bit manufacturer to follow up the whole process from design to use, thus improving the effect of the drill bit. If any abnormality occurs during the use, the

计中重点要突出对细节的考虑，通过周密的设计，减少钻井施工过程中意外情况的发生，使钻井生产得以连续顺利地进行。

施工设计应细化到一开是否采用一只钻头钻完全部进尺，能否使用旧钻头。二开钻水泥塞与下部结构可以使用较为经济的钢齿钻头，此后每只钻头可能钻多少进尺，根据钻头起出后磨损分级情况，确定下只钻头的型号，如此直到设计出全井的钻头组合。另外，针对复杂深井可能存在井况多变的情况，还应有备选的钻头方案，以避免钻头使用中出现没有钻头可用的情况发生。

由于施工单位一般都与钻头厂商建立了良好的供货关系，施工设计中钻头与钻井参数设计还可以充分发挥钻头厂商的作用。让钻头厂商根据邻井测井资料、邻井钻头使用情况，提出钻头设计方案，并签订根据钻头使用效果支付钻头费用的协议，让钻头厂商从设计到使用，全程跟踪，以提高钻头使用效果。如果使用中出现异常还可以及时修改钻头设计，以保证钻头的使用效果。

design of the drill bit can be modified in time to ensure the effectiveness of the drill bit.

The drilling parameter design should be designed for each drill bit to enter the well modeling, grinding and other specific parameters and process measures. For the drilling process may appear gravel layer and other abnormal conditions should be designed specific countermeasures to ensure that the drilling construction process is safe, fast, economical and efficient.

1.2.2.2 Construction Design of Drilling Tools and Antio-slope

The design of drilling tools combination is as conducive as possible to strengthening the drilling parameters and reducing the occurrence of accidents and complications while meeting the anti-slope requirements, and at the same time is conducive to dealing with accidents and complications under the well. Construction design should analyze the experience and effect of the use of drilling tools combinations in neighboring wells drilling engineering design of drilling tools combinations design principles, the characteristics of the recommended drilling tool combinations etc., to refine the drilling tool combinations for each well section, and put forward the use of supporting technical measures, such as well-slope monitoring measures, short-range starting and lowering of the drilling and drilling and eye marking measures, drilling parameters to use, and so on. Each time the drilling is started to drill the cement plug and casing substructure, a drilling tool combination that is conducive to protecting the casing should be designed.

In the drilling of wells for the monitoring of well inclination and the complexity of the well, the design of adjustable drilling

钻井参数设计应对每只钻头入井造型、磨合等具体参数与工艺措施等进行具体设计。对钻井过程中可能出现的砾石层等异常情况都要设计具体的应对措施，保证钻井施工过程安全、快速、经济、高效。

1.2.2.2 钻具组合与防斜施工设计

钻具组合设计是在满足防斜要求的情况下，尽可能强化钻井参数，减少事故及井下复杂情况的发生，同时有利于处理事故与井下复杂情况。施工设计应分析邻井钻具组合使用的经验与效果钻井工程设计中钻具组合的设计原则，推荐的钻具组合的特点等，细化每一井段的钻具组合，并提出使用中的配套技术措施，如井斜监测措施、短程起下钻与划眼措施、钻井参数使用等。每次开钻钻水泥塞和套管下部结构，要设计有利于保护套管的钻具组合。

钻井中针对井斜监测情况及井下复杂情况，要设计可供调整的钻具组合，以便于出现各种新的情况后及时采取有效的措施，避免出现井身质量问题，也有利于安全顺利钻进。

tool combinations, in order to facilitate the emergence of a variety of new situations can be taken in a timely manner to take effective measures to avoid the emergence of the quality of the well, but also conducive to safety and smoothly drilling.

### 1.2.2.3 Construction Design for Casing Protection and Anti-friction and Drag Reduction

For complex deep wells, specific technical measures need to be carefully designed for the protection of casing, including the weighting agent of drilling fluid, the addition of extreme pressure lubricant, the design of anti-wear protection belt for drill pipe, the use of anti-wear sleeve for casing head, and so on, and if necessary, the installation of anti-wear and damping joints of drill pipe should be added, so as to further minimize the abrasion of the casing. When designing the above measures, it is necessary to put forward specific requirements for specific operations during drilling, such as the inspection and replacement of casing head anti-wear sleeve, and the inspection and repair welding of drill pipe wear-resistant belt.

Reducing wear on the casing requires strict control of well slope changes during the last drilling, proposing higher-than-standard well slope control indicators, and designing anti-slope drilling tool combinations and well slope monitoring and control measures.

### 1.2.2.4 Construction Design for Special Processes

Special process wells in the construction technology measures and the use of equipment, instruments, tools are closely related to the need for clear construction requirements, give full consideration to the technical parameters

### 1.2.2.3 套管保护与防磨减阻施工设计

对于复杂深井来说，需要针对套管的保护精心设计具体的技术措施，包括钻井液的加重剂、极压润滑剂的添加、钻杆防磨保护带设计、套管头防磨套使用等，必要时要加装钻杆防磨减阻接头，以进一步减少套管的磨损。设计以上措施时，需对钻井期间的具体操作提出具体要求，如套管头防磨套的检查与更换、钻杆耐磨带的检查与补焊等。

减少对套管的磨损需要严格控制上次开钻期间的井斜变化，提出高于标准规定的井斜控制指标，并有针对性地设计防斜钻具组合与井斜监测及控制措施。

### 1.2.2.4 特殊工艺施工设计

特殊工艺井中，施工技术措施与使用的设备、仪器、工具密切相关，需要在明确施工要求的情况下，充分考虑自己的装备技术参数，设计出可行的施工工艺措施。

of their own equipment, design a feasible construction technology measures.

1.2.2.4.1 Directional and Horizontal Well Trajectory Control and Guidance Construction Design

The directional drilling instrument determines the directional and guiding process. In view of different formation inclination characteristics, it is necessary to carefully select the type and parameter of inclination-forming and inclination-enhancing drilling tools. After drilling to the reservoir, it is necessary to design the guided drilling method according to the characteristics of the reservoir to ensure the encounter rate of the reservoir and make the trajectory extend in the most favorable position of the reservoir, so as to obtain a better development effect.

In long horizontal section horizontal wells and large displacement wells, the trajectory of the borehole determines the range of displacement that can be drilled, and in order to break through the technical barriers as much as possible and extend the length of the horizontal section, it is necessary to fully optimize the profile type of the horizontal well.

1.2.2.4.2 Underbalanced/gas Drilling Construction Design

Underbalanced/gas drilling construction design requires the development of detailed countermeasures for various complex situations based on specific equipment and tool performance to ensure safe and smooth construction.

The main contents of underbalance construction design include underbalance equipment matching, installation requirements, underbalance process measures, underbalance control parameters, underbalance HSE requirements, underbalance work team and

1.2.2.4.1 定向井、水平井轨迹控制与导向施工设计

定向井的随钻测量仪器决定了定向与导向工艺。针对不同的地层造斜特性，需要精心选择造斜与增斜钻具类型及参数。在钻达储层后，要针对储层的特点设计导向钻井方式，确保储层钻遇率，并使轨迹在储层最有利位置延伸，从而获取更好的开发效果。

在长水平段水平井与大位移井中，井眼轨迹决定了可以钻达的位移范围。为尽可能突破技术的障碍，延伸水平段长度，需要充分优化水平井的剖面类型。

1.2.2.4.2 欠平衡/气体钻井施工设计

欠平衡/气体钻井施工设计要求根据具体的设备与工具性能，制订详细的各种复杂情况的应对措施，保证施工的安全顺利进行。

欠平衡施工设计主要内容包括欠平衡设备配套、安装要求，欠平衡工艺措施，欠平衡控制参数，欠平衡 HSE 要求，欠平衡作业队伍与井队的配合内容，欠平衡施工的组织指挥等内容。

well team cooperation content, underbalance construction organization and command.

1.2.2.4.3　Directional Well Defense Design

The general drilling design has given the directional well anti-collision calculation of the normal surface scanning map, the task of the construction design is for the specific construction process, to develop a perfect anti-collision measures. During the construction of a series of wells, the trajectory of the drilled borehole becomes the basis of the construction design for the subsequent wells. The construction unit must always determine the anti-collision measures based on the trajectory of the specific drilled borehole.

The focus of the anti-collision construction design is the tracking design of the borehole trajectory, to determine a reasonable trajectory control program, and to reduce the chances of borehole collision. The requirements of anti-collision should be fully considered in the use of drilling tools combination.

1.2.2.4.4　Drilling Fluid Design

Drilling fluid design is based on the system, formula and performance design of Party A's drilling engineering design, further refinement of performance control indexes, optimization of specific manufacturers of drilling fluid materials, and preference for the fit products to achieve the best performance of the design. Further refinement of drilling fluid formulation and maintenance treatment. For example, the treatment measures before adding various treating agents, the order of addition, the speed of addition, etc., to form the specific measures that can be operated on site.

1.2.2.4.5　Key Construction Process Measures for Each Well Section

Clearly define the key construction

1.2.2.4.3　定向井防碰设计

一般钻井设计中已给出了定向井防碰计算的法面扫描图，施工设计的任务是针对具体施工过程，制订出完善的防碰措施。在丛式井施工过程中，已钻的井眼轨迹对后续井而言是施工设计的依据。施工单位必须随时根据具体已钻成井的轨迹，确定防碰措施。

防碰施工设计的重点是井眼轨迹的跟踪设计，确定合理的轨迹控制方案，减少发生井眼相碰的概率。在钻具组合使用上应充分考虑防碰的要求。

1.2.2.4.4　钻井液设计

钻井液设计是在甲方钻井工程设计的体系、配方、性能设计的基础上，进一步细化性能控制指标，优化钻井液材料具体生产厂家，优选最合适的产品以达到设计的最佳性能。对钻井液配制、维护处理进行进一步的细化，如各种处理剂加入前的处理措施、加入顺序、加入速度等，形成现场可操作的具体措施。

1.2.2.4.5　各井段重点施工工艺措施

明确各井段施工的重点措施，针对可

measures for each well section, and prepare construction measures plans for possible complexities and accidents under well to reduce their occurrence. For accidents and complexities that are expected to occur, treatment plans should be prepared, and the necessary tools and materials should be ready to minimize downtime caused by accidents and complexities, as well as to prevent them from becoming more complex due to lack of timely handling.

Drilling engineering design is the process of designing the wellbore to ensure the safe and efficient completion of drilling operations. The following are the basic theories of drilling engineering design:

(1) Well design objectives: The primary objective of drilling engineering design is to ensure well integrity and wellhead safety, and to achieve these objectives while maximizing drilling efficiency and reducing costs.

(2) Wellbore design: Wellbore design includes determining parameters such as well depth, diameter and wall stability. This takes into account formation properties, wellhead and bottom hole conditions, and drilling fluids. The borehole depth and diameter will affect subsequent borehole stability, borehole quality and drilling speed.

(3) Drilling fluid (Mud) design: Drilling fluid plays an important role in the drilling process, including cooling and lubricating the bit, controlling well pressure, and cleaning the hole. Drilling fluid design should take into account wellbore stability, formation protection, drilling efficiency and environmental factors.

(4) Drill tool design: drill tool includes drill bit, drill pipe, drill collar and other tools and equipment. The design of drilling tools should take into account the factors such as

能出现的复杂与事故，做好施工措施预案，减少事故与井下复杂情况的发生。对于预计可能发生的事故与井下复杂情况，应做好处理预案，准备好相关的工具与材料，减少因事故与复杂而引起的停工，也避免事故与复杂因得不到及时处理而变得更加复杂化。

钻井工程设计是针对井筒的设计过程，旨在确保安全高效地完成钻井作业。以下是钻井工程设计的基本理论。

（1）井设计目标：钻井工程设计的首要目标是确保井孔的完整性和井口的安全性，并在达到这些目标的同时，尽可能提高钻井效率和降低成本。

（2）井筒设计：井筒设计包括确定井深、直径和井壁稳定性等参数。这需要考虑地层特性、井口和井底条件及钻井液。井深和直径将影响后续井筒稳定性、井筒质量和钻速。

（3）钻井液（泥浆）设计：钻井液在钻井过程中扮演着重要角色，包括冷却和润滑钻头、控制井口压力和清洗井眼。钻井液设计应考虑井筒稳定性、地层保护、钻井效率和环境因素。

（4）钻具设计：钻具包括钻头、钻杆、钻铤等工具和设备。钻具设计要考虑到井深、地层性质、井眼尺寸和钻井液性质等因素。钻具的选择和设计要确保其能够满足钻井作业的要求，并具有足够的强度和耐用性。

well depth, formation properties, hole size and drilling fluid properties. Drilling tools are selected and designed to meet the requirements of drilling operations and have sufficient strength and durability.

(5) Drilling parameters design: drilling parameters include drilling rate, weight on bit, rotational speed, feed rate, etc. The selection of these parameters should comprehensively consider the factors such as wellbore stability, formation properties, drilling tool performance and drilling fluid performance to ensure the safety and efficiency of drilling.

(6) Safety design: Safety is one of the most important considerations in drilling engineering design. Safety design includes well control measures, accident prevention and emergency response. In the design of drilling engineering, it is necessary to take into account the safety measures of wellhead and bottom, as well as the measures to prevent blowout, well collapse, well kick and other accidents.

## 1.3  Drill Bit

Drill bit is the main tool to break rock to form a borehole. At present, the drill bits used in oil industry drilling are categorized into scraper bits, tooth wheel bits and diamond bits.

### 1.3.1  Scraper Bits

Scraper bits consist of an upper bit body, a lower bit body, a blade and a carbide nozzle.

Scraper bits mainly break formation rock by cutting, shearing and squeezing. Since these types of rock breaking methods mainly overcome the shear strength of the rock, easier than the rock breaking methods that overcome the compressive strength of the rock.

（5）钻井参数设计：钻井参数包括钻速、钻压、转速、进给速度等。这些参数的选择要综合考虑井筒稳定性、地层性质、钻具性能和钻井液性能等因素，以确保钻井的安全和高效。

（6）安全设计：钻井工程设计中安全是最重要的考虑因素之一。安全设计包括井控措施、事故预防和应急响应等方面。在钻井工程设计中，要考虑到井口和井底的安全措施，以及防止井喷、井塌、井涌等事故的措施。

## 1.3  钻头

钻头是破碎岩石形成井眼的主要工具。目前，石油行业钻井使用的钻头分为刮刀钻头、牙轮钻头和金刚石钻头。

### 1.3.1  刮刀钻头

刮刀钻头由上钻头体、下钻头体、刀片和硬质合金喷嘴组成。

刮刀钻头主要以切削、剪切和挤压方式破碎地层岩石。这几种破岩方式主要是克服岩石的抗剪强度，因此比克服岩石的抗压强度的破岩方式容易。

Scraper bits HSE Requirements.

(1) Scraper bits must be transported in boxes, and cannot be thrown at random when loading and unloading, so as to prevent touching and damaging the scraper blades;

(2) Check the welds of scraper bits, and those with unqualified quality are not allowed to enter the well;

(3) Scraper bit whole drill is a normal phenomenon, but to prevent serious whole drill, in order to prevent the whole off the scraper blade;

(4) When drilling with a scraper bit, do not pressurize the turntable to start it;

(5) When the drilling speed slows down or the bit contacts the bottom of the well with elevated pump pressure, the drilling fluid should be circulated to start the drill.

### 1.3.2 Tooth-wheel Drill Bits

According to the number of tines, the tooth-wheel drill bits are single-tine bits, two-tine bits, three-tine bits and four-tine bits. Currently, three-tine bits are commonly used (Figure 1.1). When rotating the three-tine bits, it has the effect of impacting, crushing and shearing to break the rock; the cutting teeth contact the bottom of the well alternately, and the breaking torque is small; the contact area between the cutting teeth and the bottom of the well is small, and the specific pressure is high, so it is easy to eat into the stratum; the total length of the cutting teeth is large, and thus reduces the wear and tear. The three-tine tooth-wheel drill bit is able to adapt to soft, hard and other properties of the formation, and has become the most widely used type of bit for oil drilling.

刮刀钻头 HSE 要求：

（1）刮刀钻头必须装箱运输，装卸时不能随意抛掷，防止碰坏刮刀片；

（2）检查刮刀钻头焊缝，质量不合格者不能入井；

（3）刮刀钻头整钻是正常现象，但要防止严重整钻，以防整掉刮刀片；

（4）用刮刀钻头钻进时，严防加压启动转盘；

（5）钻速变慢或钻头接触井底泵压升高时，应循环钻井液起钻。

### 1.3.2 牙轮钻头

牙轮钻头按牙轮数量分为单牙轮钻头、两牙轮钻头、三牙轮钻头和四牙轮钻头。目前普遍使用的是三牙轮钻头（图1.1）。三牙轮钻头旋转时具有冲击、压碎和剪切破碎岩石的作用；切削齿交替接触井底，破岩扭矩小；切削齿与井底接触面积小，比压高，容易吃入地层；切削齿总长度大，因而减少了磨损。三牙轮钻头能够适应软、硬等多种性质的地层，是石油钻井使用最广泛的一种钻头。

Figure 1.1  Three-tine Tooth-wheel Drill Bit

图 1.1  三牙轮钻头

The three-tine tooth-wheel drill bit is composed of 3 tooth wheel cones symmetrically distributed at 120° and complexly machined together with other parts. Its structure can be divided into five basic parts: drill body, slap, tooth wheel, bearing and water eye. Sealed jet drill bits have oil storage and sealing system in addition to the above basic parts.

The body of the drill is the main body of the drill, the upper part is threaded to connect with the drilling tool and support the tooth wheel. The palm is connected with the tooth wheel shaft, and there is a journal on it, which is used to support the tooth wheel. The outside of the wheel has milled teeth or inlaid teeth. The bearing is an important factor in determining the life of the drill bit. The oil reservoir sealing system not only ensures lubrication of the bearings, but also effectively prevents the drilling fluid from entering the bearings of the drill bit, which greatly improves the life of the bearings as well as the drill bit. Water holes are the channels through which the drilling fluid flows out of the bit, and there are generally three water holes in a

三牙轮钻头是 3 个牙轮锥体按 120°对称分布并与其他部分一起经复杂加工组成。其结构可分为钻头体、巴掌、牙轮、轴承和水眼 5 个基础部分。密封喷射式钻头除上述基础部分外，还有储油密封系统。

钻头体是钻头的本体，上部车有螺纹与钻具相连，起支撑牙轮的作用。巴掌与牙轮轴相连，上面有轴颈，用于支撑牙轮。牙轮外面车有铣齿或镶入镶齿。牙轮内装有轴承和储油密封系统，轴承是决定钻头寿命的一个重要因素。储油密封系统既能保证轴承得到润滑，又可以有效地防止钻井液进入钻头的轴承内，大幅度提高了轴承及钻头的寿命。水眼是钻井液流出的通道，一般三牙轮钻头上有 3 个水眼。普通钻头水眼是在钻头体的适当位置开孔并焊上水眼套。

three-tine bits. In general, there are three water holes in a three-tine bits tooth-wheel drill. The water hole of an ordinary drill bit is made in the appropriate position of the drill bit body and welded on the water hole sleeve.

Tooth-wheel drill bits HSE Requirements:

(1) Tooth-wheel drill bits must be boxed for transportation, and cannot be thrown at random during loading and unloading to prevent the tooth wheel from being touched and damaged;

(2) Check the weld of the drill bit, the quality is not qualified to enter the well;

(3) The drill bit contacts the bottom of the well, adopts low speed and light pressure to grind the tooth wheel for about 0.5h, and then gradually increases to the design drilling pressure to drill;

(4) When the drill bit has jump drilling phenomenon, adjust the drilling pressure speed appropriately, and then resume normal drilling after it becomes normal;

(5) When the drill bit has the phenomenon of complete drilling, we should analyze the reason and deal with it in time. If the drill bit is used in the late stage of the whole drill, the drilling operation should be stopped immediately, the drilling fluid should be circulated well, and the drilling should be organized to start drilling and change the drill bit;

(6) When there are phenomena such as large load on the rotary table, uneven rotation, backing up, and obvious decrease in drilling speed at the later stage of drilling, the drilling should be stopped, the drilling fluid should be circulated, and the drilling should be organized to start drilling and change the drill bit, and the test drilling should not be carried out again.

牙轮钻头 HSE 要求：

（1）牙轮钻头必须装箱运输，装卸时不能随意抛掷，防止碰坏牙轮；

（2）检查钻头焊缝，质量不合格者不能入井；

（3）钻头接触井底，采用低速、轻压磨合牙轮 0.5h 左右，再逐渐加至设计钻压钻进；

（4）钻头有跳钻现象时，要适当调整钻压转速，待正常后，再恢复正常钻进；

（5）钻头有整钻现象时，要认真分析原因，及时进行处理。钻头用到后期如发生整钻，要立即停止钻进作业，循环好钻井液，组织起钻、换钻头；

（6）钻头用到后期出现转盘负荷大、转动不均匀、打倒车、钻速明显下降等现象时，应停钻、循环好钻井液，并组织起钻、换钻头，不可再行试钻。

### 1.3.3 Diamond Drill Bits

Drill bits that use diamond material as the cutting edge are called diamond drill bits. Depending on the material used to manufacture the cutting teeth, diamond bits are categorized into three types: natural diamond drill bits, PDC drill bits and TSP drill bits (barlas bits). Currently, the type of diamond drill bit widely used in oil mines is the PDC drill bit (Figure 1.2).

Figure 1.2  PDC Drill Bits

The full name of PDC drill bit is Polycrystalline Diamond Composite (PDC) drill bit. It is an integral drill bit with no moving parts. On the erosion-resistant tungsten carbide carcass in the crown of the drill bit, a thin layer of synthetic diamond is welded with a thickness of less than 1mm, which serves as the working layer when cutting rock; Polycrystalline diamond compound, composite piece is based on diamond powder as raw material, adding binder sintered at high temperature and high pressure into the cutting tooth shape for the round slice. The organic combination between the two makes PDC drill bits have both the hardness and wear resistance of diamond and the structural strength and

### 1.3.3 金刚石钻头

用金刚石材料作为切削刃的钻头称为金刚石钻头。根据切削齿制造材料的不同，金刚石钻头分为天然金刚石钻头、PDC钻头和TSP钻头（巴拉斯钻头）三种类型。目前，石油矿场上广泛使用的金刚石钻头类型是PDC钻头（图1.2）。

图1.2  PDC钻头

PDC钻头全称聚晶金刚石复合片钻头。它是一种没有活动零件的整体钻头。在钻头冠部耐冲蚀的碳化钨胎体上，焊有薄层人造金刚石，厚度小于1mm，切削岩石时作为工作层；聚晶金刚石复合片以金刚石粉为原料，加入黏结剂在高温高压下烧结而成切削齿形状为圆片状。两者之间的有机结合，使PDC钻头既具有金刚石的硬度和耐磨性，又具有碳化钨的结构强度和抗冲击能力。在低钻压、高转速钻进过程中，PDC钻头破岩能力强。加之喷嘴安装在钻头的冠部，形成的水力破岩作用比三牙轮钻头大，因而在软到中硬地层中钻进时能取得较高的机械钻速。同时，切削齿在钻进过程中具有自锐作用，钻头的纯钻井时间也增加。PDC钻头在水眼处装有硬质合金喷嘴，喷嘴是可拆卸的，钻头使用后喷嘴还可卸下重复使用。

impact resistance of tungsten carbide. In the process of low drilling pressure and high speed drilling, PDC drill bit has strong rock breaking ability. In addition, the nozzle is installed in the crown of the bit, forming a hydraulic rock-breaking effect is larger than that of the three-tine bits tooth-wheel drill bit, so it can achieve higher mechanical drilling speed when drilling in soft to medium-hard strata. At the same time, the cutting teeth have self-sharpening effect in the drilling process, and the pure drilling time of the bit is also increased. The PDC bit is equipped with a carbide nozzle at the water hole, and the nozzle is removable, and the nozzle can be unloaded and reused after the bit is used.

PDC drill bits HSE Requirements:

(1) PDC drill bits must be transported in boxes, and must not be thrown during loading and unloading to prevent damage to the carcass and polycrystalline diamond composite piece;

(2) Trial drilling: at the beginning of drilling, it should be slow rotation, light pressure (about 20kN), and gradually increase the drilling pressure and rotational speed after the formation of the new well wall, at the same time selecting the drilling pressure and rotational speed of drilling under the maximum mechanical drilling speed;

(3) Avoid scratching eyes with PDC drill bits;

(4) Do not start the turntable under pressure;

(5) In the later stages of drilling, it is possible to increase the drilling pressure in order to maintain a high mechanical drilling speed;

(6) Drilling tools are within the bare eye, and taps are not allowed to sit on the turntable for repairs;

(7) Dry drilling and downhole drops are

PDC 钻头 HSE 要求：

（1）PDC 钻头必须装箱运输，装卸时不能随意抛掷，防止碰坏胎体和聚晶金刚石复合片；

（2）试钻：开始钻进时，应慢转、轻压（20kN 左右），待新井壁形成后逐渐加大钻压和转速，同时选择在最大机械钻速下的钻压和转速钻进；

（3）避免用 PDC 钻头划眼；

（4）严禁加压启动转盘；

（5）钻头用到后期可以适当提高钻压，以保持高机械钻速；

（6）钻具在裸眼以内，严禁水龙头坐在转盘上修理；

（7）严禁干钻和井下落物；

strictly prohibited;

(8) If there is a significant decrease in mechanical drilling speed, an increase in riser pressure, or a heavier load on the rotary table during drilling, the drilling should be considered to be started.

# Ideological and Political Case Studies

Academician Shen Zhonghou is a leading figure in the field of high-pressure water jet technology in China's petroleum industry. He is an outstanding expert in oil and gas well engineering technology and water jet technology, and the founder of China's oil and gas well engineering discipline. He is known as the "Father of Drill Bits" in China. He mainly engaged in research on the theory and technology of oil and gas well drilling engineering, as well as water jet theory and technology. He has won the National Science and Technology Progress Second Prize twice, and the provincial and ministerial level Science and Technology Progress First Prize four times. Additionally, he has been honored with titles such as "National Energy Industry Outstanding Model Worker" and "Expert with Outstanding Contributions to the Petroleum Industry."

Shen Zhonghou was born on February 13, 1928, in a mountain village in Dazhu County, Sichuan Province. The people there are simple and honest, and his parents named him "Zhonghou", which means "faithful and sincere". In 1947, Shen Zhonghou was admitted to Chongqing University. When asked why he chose the mining department, He replied, "Our country is too poor, we can't develop without

（8）如钻进中出现机械钻速显著下降、立管压力增高、转盘负荷加重等情况时，应考虑起钻。

# 思政案例

沈忠厚院士是中国石油工业高压水射流领域的领军人物。沈忠厚院士是杰出的油气井工程技术专家、水射流专家，我国油气井工程学科的奠基人，被誉为中国"钻头之父"。他主要从事油气井钻井工程理论与技术和水射流理论与技术等研究工作，曾获国家科技进步二等奖2次，省部级科技进步一等奖4次，并荣获"全国能源工业特等劳模""石油工业有突出贡献专家"等荣誉称号。

沈忠厚1928年2月13日出生在四川省大竹县一个山寨里，山里人纯朴憨厚，父母为他起了"忠厚"这个名字。沈忠厚1947年考上了重庆大学，有人问他为什么选择采矿系时，沈忠厚回答："我们中国太穷了，没有矿产不行；找到丰富的矿藏，中国才有可能走向富裕。"

minerals. Only by discovering rich mineral deposits can China have a chance to become prosperous."

In 1951, after graduating from university, Shen Zhonghou was sent to the Yumen Oilfield for a one-year internship. After returning to the school, he took up the teaching of petroleum drilling engineering. In 1955, he was transferred to the Beijing Petroleum Institute (the predecessor of China University of Petroleum). China's petroleum industry truly developed after the founding of the People's Republic of China. At the beginning of the establishment of the People's Republic, the entire drilling and oil extraction system of the country had only about 800 employees, 3 drilling rigs, and produced a total of only 120000t of oil per year. Petroleum is stored in complex and mysterious strata thousands of meters underground. The cost of petroleum drilling is high, with each well costing several million yuan or even more. To reduce drilling costs, it is necessary to improve drilling efficiency. Drilling speed mainly depends on the drilling bit used. To increase oil production, efforts must be made at the core of drilling—the bit. Improving the efficiency of the drill bit became the subject of Shen Zhonghou's years of research.

In the 1960s, a Japanese jetliner encountered a light rain while flying and found numerous small holes on its body after landing. Some scientists concluded the theory of water jet based on this incident — water can generate tremendous force at extremely high speeds. In the early 1970s, water jet technology emerged internationally, with countries like the United States, the Soviet Union and Germany starting to research how to apply water jet technology in industry. By chance, Shen

1951年，沈忠厚大学毕业后被派往玉门油田实习一年，回校后担任石油钻井工程课的教学，1955年调入北京石油学院（中国石油大学前身）。中国的石油工业是新中国成立后才真正发展起来的。建国之初，全国整个钻井采油系统加在一起只有约800名职工、3部钻机，一年总共才产油12万吨。石油蕴藏在地下几千米复杂而神秘的岩层中，石油钻井付出的代价是高昂的，每打一口井要耗资几百万甚至更多。要想降低钻井成本，就必须提高钻井效率。钻井速度主要取决于钻井用的钻头，要想提高石油产量，必须在钻井的核心部位——钻头上下功夫。提高钻头的效率，成为沈忠厚多年钻研的课题。

20世纪60年代，日本一架喷气式飞机在空中飞行时遇到一场小雨，降落后发现机身上都是小孔。一些科学家由此得出水射流理论——水在速度极快的情况下能产生巨大的力量。20世纪70年代初，国外出现水射流技术，美国、苏联、德国等开始研究如何把水射流技术用在工业上。一个偶然的机会，沈忠厚到成都飞机厂参观，看到了水射流切割机。那小巧玲珑的切割机切起合金钢材就像切豆腐一样游刃有余。他脑海里灵光一闪，困惑他多年的钻井难题终于有了解决的方案。

Zhonghou visited the Chengdu Aircraft Factory and saw a water jet cutter. The dexterous cutter sliced through alloy steel as easily as cutting through tofu. A light bulb went off in his mind, and the drilling problem that had puzzled him for years finally had a solution.

Shen Zhonghou resolutely chose this world-class challenge—to utilize water jet rock-breaking technology to improve drilling efficiency, and he began to harness the seemingly weak force of water to confront the hard rock. A wise man delights in water, and Shen Zhonghou's research spanned 20 years. Over those two decades, Shen Zhonghou systematically revealed the laws of flow and dynamics of submerged non-free jets, as well as the laws of pressure and water power attenuation, establishing theoretical calculation models. For the first time, he proposed a new method and model for optimizing hydraulic parameters with the objective function of maximize water power at the bottomhole surface in drilling engineering, establishing a new hydraulic design theory and solving important theoretical problems that had long plagued drilling engineering. He invented a new type of extended nozzle roller bit, under the same conditions, can improved the average mechanical drilling speed by 25% to 30% compared to conventional roller bits, achieving direct economic benefits of 165 million yuan. He also developed the theory of self-vibrating cavitation jet, established basic relational expressions and structural mathematical models for two types of resonant cavities, and invented a self-vibrating cavitation jet bit.

Shen Zhonghou has been engaged in the research of high-pressure water jet theory and technology. He combined jet technology with petroleum engineering, making creative contributions to drilling engineering and

沈忠厚坚定地选择了这个世界级难题——利用水射流破岩技术提高钻井效率。他开始利用看起来柔弱的水向坚硬的岩石宣战。智者乐水，沈忠厚的研究一做就是20年。20年间，沈忠厚系统揭示了淹没非自由射流流动及动力学规律以及压力和水功率衰减规律，建立了理论计算模型；首次提出了钻井工程以井底岩面获最大水功率为目标函数，优选水力参数的新方法和模型，建立了新的水力设计理论，解决了钻井工程长期没有解决的重要理论问题；发明了新型加长喷嘴牙轮钻头，该新型钻头在相同条件下与普通牙轮钻头相比，平均机械钻速提高25%至30%，获直接经济效益1.65亿元；发展了自振空化射流理论，建立了两种谐振腔的基本关系式和结构数学模型，发明了自振空化射流钻头。

沈忠厚一直从事高压水射流理论与技术的研究工作，他将射流技术与石油工程相结合，对钻井工程做出创造性成果和贡献，对石油工程做出了卓著的贡献，其主要成果有：

distinguished contributions to the petroleum industry. His main achievements include:

(1) In the development of high-efficiency drill bits, he made significant breakthroughs and developments in the dynamic laws of submerged non-free jets, the theory of self-vibrating cavitation jets, and the theory of mechanical and hydraulic combined rock breaking. Based on this, he invented three types of high-efficiency drill bits: a new type of extended nozzle roller bit, a self-vibrating cavitation jet bit, and a mechanical and hydraulic combined rock-breaking bit. These three types of bits have significantly increased the mechanical drilling speed in oilfield applications, with more than 3500 bits used in over 10 oilfields and regions, yielding direct economic benefits of over one billion yuan.

(2) In the field of drilling engineering both domestically and internationally, he pioneered a new hydraulic design method with the objective function of maximize water power at the bottomhole surface, solving an important theoretical problem that had long been in need of resolution.

(3) He was the first in China to conduct research on radial horizontal drilling technology, filling a research gap in the country.

(4) He created a new technology for treating near-wellbore formations and deblocking in oil wells using self-vibrating cavitation rotating jets, which has been applied to more than 400 wells in 11 oilfields and regions, creating economic benefits of over one billion yuan.

Shen Zhonghou's research achievements have won the National Science and Technology Progress Second Prize once, the National Invention Third Prize once, the provincial and ministerial level Science and Technology

（1）结合研制高效钻头，对淹没非自由射流动力学规律、自振空化射流理论和机械及水力联合破岩理论有重要突破和发展。在此基础上又发明了新型加长喷嘴牙轮钻头、自振空化射流钻头和机械及水力联合破岩钻头等3种高效钻头。3种钻头在油田应用中大幅度提高机械钻速，在10余个油田和地区应用3500只，获直接经济效益过亿元。

（2）在国内外钻井工程领域，首次创立了以井底岩面获最大水功率为目标函数的水力设计新方法，解决了长期以来亟待解决的重要理论问题。

（3）在国内首次开展径向水平钻井技术研究，填补了国内研究的空白。

（4）首创自振空化旋转射流处理油井近井地层及解堵新技术，在11个油田和地区400余口井上应用，创经济效益过亿元。

沈忠厚的研究成果先后获国家科技进步二等奖1次、国家发明三等奖1次、省部级科技进步一等奖2次和二等奖3次，获中外专利13项，在中外刊物发表论文70余篇，出版英文专著1部、中文专著2部，并荣获"全国能源工业特等劳动模范"等6项省部级以上荣誉称号和奖励。

Progress First Prize twice and Second Prize three times. He has obtained 13 patents in China and abroad, published more than 70 papers in Chinese and international journals, published one English monograph and two Chinese monographs, and been honored with six provincial and ministerial level titles and awards, including "National Energy Industry Outstanding Model Worker".

The wise man uses his strategies to the fullest, and the benevolent spreads his kindness. Shen Zhonghou is both a wise and benevolent man, and even more so, a loyal man. He is loyal to the country he loves and to the academic field in which he has always been passionate. The strategy for sustainable development is a grand plan for national development established by the Party and the state. In May 2003, the State Council proposed the topic of "Strategic Research on China's Sustainable Development of Oil and Gas Resources" to promote strategic development in the energy sector. As a member of the advisory committee for the topic, Shen Zhonghou deeply felt the enormity of the task and the weight of responsibility. Despite being over 70 years old, he dedicated himself wholeheartedly to understanding the current situation of oil and gas resources, carefully studying the domestic and international situations, and conducting in-depth investigations and interdisciplinary, interdepartmental, and inter-industry research on the key and critical issues of sustainable development of China's oil and gas resources, providing the research group with a wealth of reference opinions. "I hope my opinions can make even a little contribution to this work", Shen Zhonghou said very seriously. "A single sentence, a single piece of data, might influence

智者尽其谋，仁者播其惠。沈忠厚是位智者、仁者，更是一位忠者。他忠于他热爱的祖国和一直乐此不疲的学术领域。可持续发展战略是党和国家制定的民族发展大计。2003年5月，国务院提出了"中国可持续发展油气资源战略研究"课题，以促进能源领域的战略发展。作为课题咨询委员会成员，沈忠厚深感此项任务的艰巨和责任的重大。他不顾70余岁高龄，全身心深入了解油气资源现状、认真研究国内外形势，围绕我国油气资源可持续发展的重点和关键问题，进行了深入的调查研究和跨学科、跨部门、跨行业的论证工作，为课题组提供大量参考意见。"我希望我的意见能够为这项工作做哪怕是一点点贡献。"沈忠厚十分认真地说，"一句话，一个数据，可能会影响我国石油战略未来几十年的发展方向。"

the direction of China's oil strategy for decades to come."

Participating in national energy strategy research and formulating a blueprint for the direction of the national petroleum industry, devoting himself to the country's development, is a joy for Shen Zhonghou, because "working is happiness". Shen Zhonghou says that as long as the country needs him, he is always ready to be dispatched.

Reaching the pinnacle of one's career is akin to the peak of a mountain. Shen Zhonghou's academic achievements have reached the forefront and are among the best in the field of water jet technology both domestically and internationally. However, even higher than his academic level is his profound virtue and the tangible manifestation of this virtue in educating and benefiting others. In dealing with accidents such as well blowouts, Shen Zhonghou does not simply stop at solving the problem but applies the cutting-edge knowledge and practical experience from the production frontlines to guide the entire drilling discipline, water jet field, and even petroleum education.

In early 2007, a well blowout occurred at a certain gas field, and it took several attempts to successfully control it. No hydrogen sulfide was detected upon inspection. Lured by the temptation of over 26 million cubic meters of gas per day, the production team was eager to proceed. When Shen Zhonghou arrived, he found that they had already connected the pipelines and were ready for production. "We absolutely cannot take risks. Just because there's no hydrogen sulfide now doesn't mean it won't appear in the future," warned Shen Zhonghou, nearly 80 years old at the

通过参与国家能源战略研究、制定国家石油工业方向性的蓝图，为国家的发展尽心尽力，这对于沈忠厚是一种享受，因为"工作着是快乐的"。沈忠厚说，只要国家需要，他随时听从调遣。

人到山顶德为峰。沈忠厚的学术成果在国内外水射流领域达到了前沿和一流，而比他的学术水平更高的是他深厚的德泽，以及这种德行在育人利物中的具体体现。在处理类似井喷等事故上，沈忠厚并不停留在解决问题上，而是把学科前沿、生产一线的实践经验用在指导整个钻井学科、水射流领域乃至石油教育中。

2007年初，某气田一探井发生了井喷，经过多次压井才取得成功。经检查并未发现硫化氢，面对每天2600多万立方米气的诱惑，生产方又动了心。当沈忠厚赶到的时候，发现他们把管线都接好了，准备生产。"坚决不能冒险。现在没有硫化氢，能保证将来没有吗？"年近80的沈忠厚告诫大家，不要被眼前利益迷惑，气已经被压住，资源早晚跑不了。他反复强调，现在的设备都不防硫，一旦硫化氢泄露出来，其威力比原子弹还要厉害。生产方按照沈忠厚的建议停了工。事后证明：气体中含有高浓度的硫化氢。

time. He cautioned against being blinded by short-term interests, as the gas was already under control, and the resources weren't going anywhere. He repeatedly emphasized that the current equipment was not designed to handle sulfur, and if hydrogen sulfide were to leak, its potency would be even more devastating than an atomic bomb. The production team followed Shen Zhonghou's advice and halted operations. It was later confirmed that the gas contained high concentrations of hydrogen sulfide.

While receiving praise for his judgment, Shen Zhonghou was considering what lessons the repeated well blowouts should impart to the school's research on "three-high" gas fields and teaching. After each national strategic research effort and accident response, he would offer suggestions for the development of the discipline and talent cultivation. "Such tragic accidents could be avoided if all our engineers had rich experience," Shen Zhonghou advised his students and team. "Having practical experience is half the skill in engineering." "As a national key university, we should have leading figures; as the leader in the petroleum industry, we should cultivate a group of leading figures in the field."

Over the decades, Shen Zhonghou has gathered a team of professionals of different ages, capable of tackling tough challenges, under his leadership. His team include academicians, professors, doctoral supervisors, academic leaders, senior engineers, and some have even taken leadership positions in world-renowned petroleum companies. Shen Zhonghou says, "I want to step back and be your advisor." "In life, if you think less of yourself, you'll be much happier" he adds. As

正当大家为他的判断力交口称赞的时候，沈忠厚却在考虑：多次井喷事故应该给学校研究"三高"气田和教学带来哪些思考？每次参与国家层面的战略研究和事故处理后，他都会为学科的发展以及人才培养提出建议。"如果我们的工程师都有很丰富的经验，这样惨痛的事故是可以避免的。沈忠厚告诫他的学生和他带领的团队，"搞工程没有实践经验，有本事也只有一半。""作为一所国家重点高校，就要有一些领军人物；作为石油行业的领头雁，就要培养一批石油界的领军人物。"

几十年来，沈忠厚以高尚的人格魅力和深厚的学术造诣凝聚起一支老中青相结合、能打硬仗的队伍。在他的麾下，有院士、教授、博士生导师，有学术带头人、高级工程师，还有人走上世界著名石油企业的领导岗位。沈忠厚说："我要退到后面，做你们的顾问。""人生在世，把自己想得少一点，心情就会愉快很多。"他说。作为智者，他是学术界的领军人物，站在学术的前沿瞭望；作为师者，他是学术界的宗师，培养了一批又一批的杰出学者。尽管已是耄耋之年，但沈忠厚院士依然为我国石油工程领域的发展殚精竭虑。他以自己丰硕的学术成果和行为世范，实现了自己为中国石油工业奋斗终生的初心和誓言。

a wise man, he is a leading figure in academia, standing at the forefront of scholarship; as a teacher, he is a master in academia, cultivating batch after batch of outstanding scholars. Despite his advanced age, Academician Shen Zhonghou still worries about the development of China's petroleum engineering field. With his rich academic achievements and exemplary conduct, he has realized his lifelong commitment and oath to fight for China's petroleum industry.

From his youthful ambitions to his relentless pursuit in later years, he has always been devoted to the development of China's petroleum industry, dedicated to national strategic needs, disciplinary construction, and the cultivation of high-level talent. With his life's blood, he has exemplified the meaning of "loyalty" and "thickness" in his name—being loyal to the country and the university, and nurturing others with his profound virtue.

This excerpt is from "Loyal Character for the Country, Thick Virtue for the Benefit of People— Remembering Academician Shen Zhonghou of the Chinese Academy of Engineering, Professor of China University of Petroleum (East China)" by Liu Jishun, published in Shandong Education (Higher Education) Issue 4, 2019.

从青年时期的壮志豪情,到年事已高时的孜孜以求,他始终心系我国石油工业的发展,一心致力于国家战略需求、学科建设发展和高层次人才培养,以毕生的心血诠释了他名字中"忠""厚"二字的内涵———以"忠"实品格为国为校,以"厚"重德泽育人利物。

本部分节选自刘积舜发表于《山东教育(高教)》2019年第4期的《"忠"实品格为国"厚"德载物育人——记中国工程院院士、中国石油大学(华东)教授沈忠厚》。

# Chapter 2 Completion

# 第二章 完 井

## 2.1 Overview of the Development of Modern Drilling and Completion Technologies

## 2.1 现代钻井与完井技术的发展概况

With the development of modern science and technology, drilling technology and completion technology has also developed rapidly, which is characterized by the following: the development of drilling from early experience to scientific drilling; the development of shallow wells, medium-depth wells to deep wells and ultra-deep wells; the development of straight wells (vertical wells) and directional wells to directional wells with large slopes, clusters of wells, and horizontal wells; and the development of drilling from onshore wells to offshore wells and deep-sea wells.

Before the early 1960s, the traditional rotary drilling technology was still dominant in foreign countries. During this period, jet theory and technology were introduced, the "three-in-one" tooth-wheel drill bit was developed, jet drilling appeared, and the mechanical drilling speed was greatly increased. On the basis of jet drilling, the optimization parameters were expanded from hydraulic parameters to mechanical parameters, and the goal of the optimization technology was to reduce drilling

随着现代科学技术的发展，钻井技术与完井技术也得到迅速发展，其特点是：从早期的经验钻井发展到科学化钻井；从浅井、中深井发展到深井、超深井，从钻直井（垂直井）、定向井发展到大斜度定向井、丛式井、水平井；从陆地钻井发展到近海钻井和深海钻井。

在20世纪60年代初期以前，传统的转盘钻井技术仍在国外占主导地位。在这一时期引入了射流理论和技术，开发了"三合一"牙轮钻头，出现了喷射钻井，机械钻速大幅提高。在喷射钻井的基础上，将优化的参数从水力参数扩大到机械参数，而优化技术的目标则在于降低钻井成本。这两项技术的研发与应用，标志着传统的转盘钻井已趋于成熟。

costs. The development and application of these two technologies marked the maturity of traditional rotary drilling.

In the middle and late 1960s, screw drilling tools and photographic single-point and multi-point inclinometers were successively developed, coupled with the development of borehole track design methods and bottom drilling tool combination force and deformation analysis methods, which laid a good foundation for the development and wide application of directional wells and clumped wells drilling technology in the 1970s. Since then, the proportion of downhole power drilling has gradually increased and the proportion of rotary drilling has gradually decreased in drilling work.

In the 1970s, the introduction of computer technology and the development of wireless follow-drilling measurement technology was a new milestone in the development of drilling technology. As an efficient computing tool, computer promoted drilling mathematical modeling and quantitative analysis, and accelerated the development of scientific drilling: as an advanced tool for receiving, processing and storing information, computer became the broad controller of drilling control system. The development of Measurement While Drilling (MWD) is a result of the introduction of telemetry and telecommunication technology into drilling, followed by the development of wireless inclinometers for drilling, and the expansion of the measurement content to two types of parameters, namely, engineering and geology. The successful development of polycrystalline diamond composite (PDC) bits is the result of the introduction of new hard wear-resistant

20世纪60年代中后期，相继开发出螺杆钻具和照相式单点、多点测斜仪，再加上井眼轨道设计方法和底部钻具组合受力与变形分析方法的发展，为20世纪70年代定向井、丛式井钻井技术的发展与广泛应用打下了良好基础。从此，在钻井工作中，井下动力钻井所占的比重逐步上升，转盘钻井所占的比重逐步下降。

20世纪70年代，计算机技术的引入和无线随钻测量技术的研发，是钻井技术发展的一个新的里程碑。计算机作为一种高效计算工具，推动了钻井数学建模与定量分析，加快了科学化钻井的发展；计算机作为接收、处理、存储信息的先进工具，又成为钻井控制系统的广义控制器。随钻测量（MWD）的研发，是遥测遥传技术引入钻井的结果，之后开发出无线随钻测斜仪，在测量内容方面向工程和地质两类参数扩展。聚晶金刚石复合片（PDC）钻头的研发成功，是硬质耐磨新材料和烧结新工艺引入钻井的结果，延长了钻头的使用寿命，提高了机械钻速。以上新技术的出现，不仅进一步提高了定向井和直井的钻井效率，也为20世纪80年代水平钻井技术的发展创造了条件。

materials and new sintering technology into drilling, which prolonged the service life of the bits and increased the mechanical drilling speed. The emergence of the above new technologies, in addition to further improving the drilling efficiency of directional wells and straight wells, created conditions for the development of horizontal drilling technology in the 1980s.

In 1980s, the guided screw drilling tools replaced the straight screw drilling tools and elbow joints, and the combined application of guided screw drilling tools and MWD, coupled with the development of the theory of borehole trajectory control and the method of calculating downhole resistance and torque, successfully realized the geometric guidance of horizontal well drilling, and the development of the first generation of logging-with-drilling (LWD) was successfully carried out in 1989, which enabled the development of the drilling technology from geometric to geologic guidance, and further ensured that the advantage of "fewer wells with higher production" could be brought into play. This further ensured the realization of the advantage of "fewer wells, higher production" of horizontal wells. In addition, the early 1980s was the peak period of deep well drilling, only the United States in 1982 completed 1289 deep wells to ensure the safety and quality of wells, is the key to deep wells and ultra-deep well drilling. The outstanding progress of deep well ultra-deep well drilling technology is the result of the development of vertical drilling system (VDS), marking the beginning of the well slope control technology to automation direction has a breakthrough development.

20世纪80年代，导向螺杆钻具替代了直螺杆钻具和弯接头，导向螺杆钻具与MWD结合应用，再加上井眼轨迹控制理论和井下摩阻、扭矩计算方法的发展，成功地实现了水平井钻井的几何导向。1989年第一代随钻测井（LWD）开发成功，使水平井钻井技术由几何导向发展到地质导向，进一步保证了水平井"少井高产"优势的发挥。另外，20世纪80年代前期是深井钻井的高峰期，仅美国在1982年就完成深井超深井1289口。确保井下安全与质量，是深井超深井钻井的关键。深井超深井钻井技术的突出进展，是研发出垂直钻井系统（VDS）的结果，标志着井斜控制技术在自动化方向有了突破性发展。

At the end of the 1980s, the rapid development of technology to protect oil and gas reservoirs mainly relied on the improvement of understanding, that the damage to oil and gas reservoirs is mainly caused by the pressure of the liquid column in the well being greater than the formation pressure, and the greater the pressure difference is, the greater the harm to the reservoirs, rather than the greater density of drilling fluids that cause greater damage to the reservoirs. In fact, the damage to the reservoir is caused throughout the entire construction process, including drilling, logging, cementing, perforating, acidizing, fracturing, well washing, water injection, well workover, and other operations and construction processes can cause damage to the reservoir. Another improvement in awareness is that different drilling fluids, perforating fluids, and acidizing fluids should be used for different types of reservoir formation characteristics. These working fluids should be matched with the lithology of the reservoir in order to minimize the damage caused to the reservoir, instead of aimlessly reducing the water loss, decreasing the solid-phase content, and so on. Using reasonable drilling fluid density and ensuring balanced pressure drilling is only a prerequisite for protecting oil and gas reservoirs. Foreign countries have formed a series of supporting reservoir evaluation and engineering evaluation test technologies, developed and applied some drilling fluids and completion fluids that cause little damage to the formation, and applied them in the field. Based on the introduction of international advanced technology, China has formed a complete theory and method of protecting oil and gas reservoirs, including lithology determination and analysis, reservoir sensitivity evaluation

在20世纪80年代末，保护油气层技术的快速发展主要依靠认识上的提高，即认为对油气层伤害主要是井内液柱压力大于地层压力造成的，这个压力差越大，对油气藏造成的伤害越大，而不是钻井液密度大对油气层造成的伤害大。实际上对油气层造成的伤害贯穿于整个施工过程，包括钻井、测井、固井、射孔、酸化、压裂、洗井、注水、修井等作业施工过程均会对油气层造成伤害。认识上的另一个提高就是应该针对不同类型油气藏岩层特点使用不同的钻井液、射孔液、酸化液。这些工作液要和油气层岩性配伍，才能减少对油气层造成的伤害，而不是无目的地降低失水量、减少固相含量等。使用合理的钻井液密度并确保平衡钻井，仅是保护油气层的前提条件。国外形成了系列配套的储层评价、工程评价试验技术，发展和应用了一些对地层伤害小的钻井液、完井液，并在现场应用。我国在引入国际先进技术的基础上，形成了完整的保护油气层理论和方法，包括岩性测定与分析、储层敏感性评价技术、油气层伤害机理研究、矿场油气层损害评价技术、保护油气层的钻井液完井液技术、保护油气层的固井技术、负压射孔技术和保护油气层的酸化压裂投产技术，使得我国的油气层保护技术自20世纪90年代以来，总体上处于国际先进水平。

technology, research on the mechanism of oil and gas reservoir damage, mine site oil and gas reservoir damage evaluation technology, drilling and completion fluid technology for protecting oil and gas reservoirs, cementing technology for protecting oil and gas reservoirs, negative pressure perforation technology, and acid fracturing and production technology for protecting oil and gas reservoirs, this has put China's overall technology for protecting oil and gas reservoirs at the international advanced level since the 1990s.

In the 1990s, the development of drilling technology has advanced by leaps and bounds. New technology (directional wells, horizontal wells, branching wells, large displacement wells, underbalanced drilling, gas drilling, fishbone wells, etc.) and new tools and equipments (PDC drill bits, MWD, LWD, geosteering drilling system, etc.) have brought unprecedented splendor to the oil and gas industry. Developed countries such as Europe and the United States have drilled ultra-deep wells of more than 10000m as well as difficult wells with large displacements.

After entering the 21st century, drilling and completion technology began to develop in the direction of automation, informationization and intelligence, and certain achievements have also been made. There are mainly the following aspects; (1) Ground drilling automation, mainly including AC frequency conversion electric drive drilling rig, top drive device, automatic pipe draining equipment, automatic wellhead equipment, automatic drill pipe handling device, automatic pressure control system, integrated driller control room, multi-parameter measuring instrument and comprehensive logging instrument, etc.; (2) Downhole automation,

20世纪90年代，钻井技术发展突飞猛进。新的工艺技术（定向井、水平井、分支井大位移井、欠平衡钻井、气体钻井、鱼骨井等）和新的工具设备（PDC钻头、MWD、LWD、地质导向钻井系统等）为石油天然气工业带来了前所未有的辉煌。部分发达国家已钻成逾万米的超深井及高难度大位移井。

进入21世纪后，钻井与完井技术开始朝着自动化、信息化、智能化的方向发展，也取得了一定的成就。主要有以下几方面；（1）地面钻井自动化，主要包括交流变频电驱动钻机、顶部驱动装置、自动排管设备、全自动井口设备、全自动钻杆处理装置、自动控压系统、一体化司钻控制室、多参数测量仪及综合录井仪等；（2）井下自动化，主要包括MWD、LWD、近钻头地质导向仪（井下"眼睛"）、随钻地震（井下"望远镜"）、自动垂直钻井系统及旋转导向钻井系统（井下"方向盘"）、智能完井技术；（3）钻井信息化，主要体现在远程实时作业中心，实现钻前方案优化、钻中决策支持、钻后效益评估功能。中国的钻井企业近年来开展了卓有成效的信息化建设工作，国内各大石油公司都建立了自己的数据中心，在"十三五"期间也基本实现了钻井动态数据的实时采集、传输和远程监控。

mainly including MWD, LWD, near-bit geosteering instrument (downhole "eye"), drilling seismic (downhole "telescope"), automatic vertical drilling system and rotary guided drilling system (downhole "steering wheel"), intelligent completion technology; (3) Drilling informationization, which is mainly reflected in the remote real-time operation center, realizing the functions of pre-drilling plan optimization, drilling decision-making support, post-drilling benefit evaluation. China's drilling enterprises have carried out fruitful informationization construction work in recent years, and major domestic oil companies have established their own data centers, and basically realized real-time collection, transmission and remote monitoring of drilling dynamic data during the "13th Five-Year Plan" period.

With the rapid development of big data, artificial intelligence, cloud computing and other technologies, the fourth industrial revolution represented by artificial intelligence has come. Driven by the tide of automation, intelligence and unmanned technology, oil drilling technology is also experiencing a stage of development from automated drilling to intelligent drilling. Intelligent drilling technology will be an important field for the development of drilling technology for a long time.

## 2.2　Development of Drilling and Completion Technology

The concept of drilling originated from the successful Zhuojian wells drilled by Ton in ancient China (1041—1054 AD), and the rotary drilling technology was developed in the early 19th century. Early drilling technology relied on people's accumulated experience to guide

随着大数据、人工智能、云计算等技术的快速发展,以人工智能为代表的第四次工业革命已经来到。在自动化、智能化和无人化技术大潮的推动下,石油钻井技术也正经历由自动化钻井转向智能化钻井发展的阶段。智能钻井技术将是未来钻井技术发展的重要领域。

## 2.2　钻完井技术的发展

钻井的概念起源于我国古代顿钻成功的卓简井(公元1041—1054年),19世纪初发展成为旋转钻井技术。早期的钻井技术是靠人们积累下来的经验指导钻井。随着科学技术的发展,钻井技术也步入科学化发展阶段。回顾历史,世界钻井技术发展

drilling. With the development of science and technology drilling technology also stepped into the stage of scientific development. Looking back at history, the development of drilling technology in the world has gone through about four stages.

(1) Conceptual period (1900-1920): The two processes of drilling and well washing began to be combined, and the use of tooth-wheel drill bit and cementing of casing began.

(2) Development period (1920-1948): improvement of tooth-wheel drill bit, development of cementing process and drilling fluid technology, and the use of high-power drilling rigs.

(3) Scientific drilling stage (1948-1968): new types of tooth-wheel drill bit (studded, sliding bearing and sealed rolling bearing bits) were developed, low-solid phase/no-solid phase non-dispersive drilling fluid system was developed, formation pressure detection, well control and cementing technology was applied, jet drilling was developed, and the theory of balanced drilling and the drilling and completing method of protecting oil and gas formations were proposed.

(4) The period of automated drilling (1968 to present): the application of automated drilling rigs and wellhead automation tools began, automatic measurement of drilling parameters and application of computers in drilling engineering were realized, the theory of system engineering analysis (optimal drilling technology), downhole closed-loop drilling and automated/geologically guided drilling systems were established, horizontal well drilling technology and underbalanced drilling technology were vigorously developed.

People's republic of China's oil drilling and completion business has achieved great

大约经历了四个阶段。

（1）概念时期（1900—1920年）：开始把钻井和洗井两个过程结合起来，开始使用牙轮钻头和用水泥封固套管。

（2）发展时期（1920—1948年）：牙轮钻头有了改进，固井工艺和钻井液技术得以发展，大功率钻机开始投入使用。

（3）科学化钻井阶段（1948—1968年）：开发了新型牙轮钻头（镶齿、滑动轴承、密封滚动轴承钻头），研制了低固相/无固相不分散钻井液体系，应用了地层压力检测、井控和固控技术，发展了喷射钻井，提出了平衡钻井理论和保护油气层的钻井、完井方法。

（4）自动化钻井时期（1968年至今）：开始应用自动化钻机和井口自动化工具：实现了钻井参数自动测量和计算机在钻井工程中应用，建立了系统工程分析理论（最优化钻井技术）、井下闭环钻井和自动/地质导向钻井系统，大力发展了水平井钻井技术和欠平衡钻井技术。

中华人民共和国成立以来的石油钻井完井事业，取得了巨大发展，经历了组建、发展和提高三个时期。

development and gone through three periods of formation, development and improvement.

(1) The period of oil drilling establishment (1949—1957): the number of drilling teams grew from 8 to 205, the number of drilling rigs reached 319, the total drilling footage reached $52.9 \times 10^4$m, and the crude oil production reached $145 \times 10^4$t.

(2) Oil drilling development period (1958—1977): the number of drilling teams grew to 570, the number of drilling rigs reached 942, and the total drilling footage reached $352.3 \times 10^4$m. The drilling speed was improved, with the 1205 drilling team of Daqing Oilfield setting a record of the highest annual footage of $12.7 \times 10^4$m (in 1971), and a single team drilling 56 wells in a month; the 3235 drilling team of Shengli Oil field setting the highest annual footage of $15.1 \times 10^4$m (in 1973), and the annual footage of a single rig reaching 18000m, which was close to the international level. During this period, there were 31 deep wells drilled over 4000m, including Dagang Xingang 57 wells drilled to a depth of 5127m, Shengli Dongfeng 2 wells drilled to a depth of 5006m, Jianghan Wang 2 wells drilled to a depth of 5163m, and Sichuan Nvji wells drilled to a depth of 6011.6m, but the drilling speed was very slow. Drilling technology has been developed, a set of drilling methods to deal with oil and gas reservoirs such as fractured carbonate rocks has been worked out, and new types of drilling bits have appeared. Drilling equipment has been improved, electric drilling rigs have been developed, and 5000m drilling rigs, hydraulic and universal blowout preventers have been successfully developed.

（1）石油钻井组建时期（1949—1957年）：钻井队由8个发展到205个，钻机数量达到了319台，钻井总进尺量达到52.9×10⁴m，原油产量达到145×10⁴t。

（2）石油钻井发展时期（1958—1977年）：钻井队发展到570个，钻机数量达到了942台，钻井总进尺量达到352.3×10⁴m。提高了钻井速度，大庆油田1205钻井队创造年进尺12.7×10⁴m的最高纪录（1971年），1个队1个月可钻56口井；胜利油田3235钻井队创造年进尺15.1×10⁴m的最高纪录（1973年），单台钻机年进尺达到18000m，接近国际先进水平。这个时期共钻超过4000m的深井31口，其中大港新港57井钻深5127m、胜利东风2井钻深5006m、江汉王2井钻深5163m、四川女基井钻深6011.6m，但钻速很慢。钻井技术有所发展，摸索出一套对付裂缝性碳酸盐岩等油气藏的钻井方法，出现了新型钻头。钻井装备有所改善，发展了电动钻机，研制成功了5000m钻机、液压和万能防喷器。

(3) The period of oil drilling improvement (1978 to the present): foreign drilling equipments, tools and instruments were introduced. Major drilling technologies reached the international level: the "5th Five-Year Plan" national key research projects implemented jet drilling technology, and the "6th Five-Year Plan" national key research projects implemented balanced pressure drilling and well control technology; In the "7th Five-Year Plan", the national key research projects successively implemented directional well/cluster well drilling technology and reservoir protection technology; in the "8th Five-Year Plan", the national key research projects successively implemented horizontal well/underbalanced drilling technology; in the "9th Five-Year Plan", the deep well/underbalanced drilling technology has been developed so far. Since then, it has developed deep/ultra-deep well drilling technology and started to carry out exploration and development drilling in shallow sea areas/thick oil reservoirs, which has achieved remarkable results and established basic drilling theories (including drilling fluid mechanics, drilling rock fragmentation, drilling tubular column mechanics, oilfield chemistry and drilling control engineering, etc.).

## 2.3 Completion Wellbore Structure

The selection of the wellbore structure for a well completion is the most important step in the completion process. Once realized, the wellbore structure of a well completion is essentially unchangeable. The wellbore structure of a well completion has a great influence on

（3）石油钻井提高时期（1978年至今）：开始引进国外钻井装备、工具和仪表。重大钻井技术达到了国际水平："五五"国家重点攻关项目实施了喷射钻井技术，"六五"国家重点攻关项目实施了平衡压力钻井和井控技术；"七五"国家重点攻关项目先后实施了定向井/丛式井钻井技术和油气层保护技术；"八五"国家重点攻关项目先后实施了水平井/欠平衡钻井技术；"九五"至今发展了深井/超深井钻井技术，开始对浅海地区/稠油油藏进行勘探开发钻井，取得了显著成果，建立了钻井基础理论（包括钻井流体力学、钻井岩石破碎学、钻井管柱力学、油田化学和钻井控制工程学等）。

## 2.3 完井井底结构

在完井过程中，选择完井的井底结构是最重要的一步。完井的井底结构一旦实现，基本上是不可更改的。完井的井底结构对井的钻进有极大的影响，对井的生产方式、井的产量、井的修复和增产有制约作用。在决定完井的井底结构时，首先考

the drilling of the well, and has a restraining effect on the production method of the well, the production rate of the well, the rehabilitation of the well and the increase of production. When deciding the bottoming structure of a completed well, the nature of the reservoir rock and the conditions of the oil recovery process are the two priority factors to be considered.

Factors to be considered in selecting the bottoming structure of well completion include: type of reservoir, lithology of formation, oil and gas content of reservoir, distribution of oil and gas, rock stability of completion section, presence of high-pressure and complex layers near the production layer, presence of bottom water or gas top of reservoir, porosity and permeability of the production layer, and technological requirements of oil production. For example, for homogeneous hard formation, barehole completion can be used, while for non-homogeneous hard formation, perforation completion is used; for non-stable formation, non-solidified screen pipe completion is used; for poor cementation of producing formation and sanding problem, sand-proof screen pipe completion should be used; and wells to be fractured and acidized should be cemented with casing.

### 2.3.1 Perforated Completion

Perforated completion is one of the most widely used completion methods at home and abroad, which can be adopted in straight wells, directional wells and horizontal wells. Perforated completion includes casing perforation completion and tail pipe perforation completion.

Casing perforated completion is to use the same size drill bit to drill through the oil formation to the designed depth, then lower the

虑储层岩石性质、采油工艺条件两个因素。

选择完井井底结构要考虑的诸因素有：储层类型、地层岩性、储层含油气情况、油气分布情况、完井层段的岩石稳定程度、生产层附近有无高压层和复杂层、储层有无底水或气顶、生产层的孔隙度和渗透率、采油生产的工艺要求等。例如，对于均质硬地层可采用裸眼完井，而非均质硬地层则采用射孔完井；对非稳定地层采用非固井式筛管完井；产层胶结性差，存在出砂问题时，则应采用防砂筛管完井；要进行压裂、酸化的井应下套管固井。

### 2.3.1 射孔完井

射孔完井是国内外使用最为广泛的一种完井方法，在直井、定向井、水平井中都可采用。射孔完井包括套管射孔完井和尾管射孔完井。

套管射孔完井是用同一尺寸的钻头钻穿油层直至设计井深，然后下油层套管至油层底部并注水泥固井，最后射孔。射孔弹射穿油层套管、水泥环并穿透油层一定深度，从而建立起油气流的通道。

oil casing to the bottom of the oil formation and cement the well, and finally shoot the hole. The shot hole projectile is shot through the oil casing and cement ring and penetrates the oil formation to a certain depth, thus establishing a channel for oil and gas flow.

Tail pipe perforated completion is to inject cement into the technical casing to cement the well after the drill bit drills to the top boundary of the oil formation, and then drill through the oil formation to the design depth with a smaller drill bit, and then use the drilling tools to send down the tail pipe and hang it on the technical casing. The overlapping section of the tail pipe and technical casing is generally not less than 50m, before cementing the tail pipe and then shooting the hole.

### 2.3.2 Openhole Completion

Openhole completion means that the borehole is completely exposed without any tubulars in the well. There are two types of completion procedures for openhole completion: first, after the drill bit near the top boundary of the oil formation, cement is injected into the technical casing to cement the well. After the cement slurry is returned to a predetermined design height, a smaller diameter drill bit is lowered from the technical casing, drilled through the cement plug, and drilled through the oil formation to the design well depth to complete the well. This is the first stage of barehole completion. The other process is to directly drill through the oil formation to the designed well depth without changing the drill bit, then lower the technical casing to the top boundary of the oil formation and inject cement to cement the well. Openhole completion can be used in straight wells, directional wells, and horizontal wells.

尾管射孔完井是在钻头钻至油层顶界后，下技术套管注水泥固井，然后用小一级的钻头钻穿油层至设计井深，用钻具将尾管送下并悬挂在技术套管上。尾管和技术套管的重合段一般不小于50m，再对尾管注水泥固井，然后射孔。

### 2.3.2 裸眼完井

裸眼完井就是井眼完全裸露，井内不下任何管柱。裸眼完井有两种完井工序：一是钻头钻至油层顶界附近后，下技术套管注水泥固井。水泥浆上返至预定的设计高度后，再从技术套管中下入直径较小的钻头，钻穿水泥塞，钻开油层至设计井深完井。另一种工序是不更换钻头，直接钻穿油层至设计井深，然后下技术套管至油层顶界附近，注水泥固井。裸眼完井在直井、定向井、水平井中都可采用。

### 2.3.3 Slotted Liner Completion

Slotted liner completion based on openhole completion in which a cut seam liner is placed inside the barehole well and can be used in straight, directional and horizontal wells. In contrast to barehole completions, there are two types of completions in the cuttings liner completion method. The first one is that after the drill bit drills to the top boundary of the oil formation, the technical casing is first injected with cement to cement the well, and then the drill bit with a smaller diameter is lowered from the technical casing to drill through the oil formation to the designed well depth. Finally, a pre-cut liner pipe is lowered in the oil formation part, relying on the liner pipe suspender (kava packer) on the top of the liner pipe to suspend the liner pipe on the technical casing and seal the annular space between the liner pipe and the casing, so that the oil and gas can flow into the wellbore through the cut of the liner pipe The second is to use the same size drill bit to drill through the oil formation, the lower end of the casing column connects the liner pipe to the lower part of the oil formation, and the annular space above the top boundary of the oil formation is sealed by the outer pipe packer and cementing joints cementing the well.

Slotted liner is in the liner wall along the axis of the parallel direction or vertical direction cut into a number of slits. The function of the slit is: on the one hand, to allow a certain number and size of crude oil can be carried to the ground "fine sand" through; the other hand, the larger particles of sand blocked in the liner outside. In this way, large sand particles in the liner pipe outside the formation of "sand

### 2.3.3 割缝衬管完井

割缝衬管完井是在裸眼完井的基础上，在裸眼井内下入割缝衬管，在直井、定向井、水平井中都可采用。与裸眼完井相对应，割缝衬管完井方法也有两种完井工序。第一种是钻头钻至油层顶界后，先下技术套管注水泥固井，再从技术套管中下入直径小一级的钻头钻穿油层至设计井深。最后在油层部位下入预先割缝的衬管，依靠衬管顶部的衬管悬挂器（卡瓦封隔器），将衬管悬挂在技术套管上，并密封衬管和套管之间的环形空间，使油气通过衬管的割缝流入井筒。第二种是用同一尺寸钻头钻穿油层后，套管柱下端连接衬管下入油层部位，通过管外封隔器和注水泥接头固井封隔油层顶界以上的环形空间。

割缝衬管就是在衬管壁上沿着轴线的平行方向或垂直方向割成多条缝眼。缝眼的功能是：一方面允许一定数量和大小的能被原油携带至地面的"细砂"通过；另一方面能把较大颗粒的砂子阻挡在衬管外面。这样，大砂粒就在衬管外形成"砂桥"或"砂拱"。"砂桥"中没有小砂粒，因为生产时此处流速很高，把小砂粒都带入井内了。"砂桥"的这种自然分选，使它具有良好的通过能力，同时起到保护井壁的作用。

bridge" or "sand arch". There is no small sand in the "sand bridge", because the production of high flow rate here, the small sand particles are brought into the well. "Sand bridge" of this natural sorting, so that it has a good ability to pass, while playing a role in protecting the well wall.

### 2.3.4 Gravel-fill Completions

For formations with serious loose cementation and sand outgrowth, the gravel-fill completion method should be generally adopted. It is the first wireline screen tube into the oil formation part of the well, and then use the filling liquid to pump the pre-selected gravel (gravel can be quartz sand, glass beads, resin-coated sand or ceramics) to the annular space between the wireline screen tube and the borehole or between the wireline screen tube and the casing, constituting a gravel-filled layer to block the oil formation sand from flowing into the wellbore, to achieve the purpose of protecting the wall of the well and preventing the sand from flowing into the well. Gravel-filled completions generally use stainless steel wire-wound screen tubing without cutting the liner. Reasons for this are:

(1) The slit width of slit liner pipe is limited by the strength of processing cutter, the minimum is 0.25~0.50mm, so the slit liner pipe is only suitable for medium and coarse sand grain oil layers and the slit width of wire-wound screen pipe can be as small as 0.12mm, so its scope of application is much larger;

(2) Filament-wound screen pipe is formed by winding filaments into a continuous seam, which has a much larger flow area than a cut-and-sewn liner pipe, and there is virtually no pressure drop when the fluid passes through

### 2.3.4 砾石充填完井

对于胶结疏松出砂严重的地层，一般应采用砾石充填完井方法。它是先将绕丝筛管下入井内油层部位，然后用充填液将在地面上预先选好的砾石（砾石可以是石英砂、玻璃珠、树脂涂层砂或陶粒）泵送至绕丝筛管与井眼或绕丝筛管与套管之间的环形空间内，构成一个砾石充填层，以阻挡油层砂流入井筒，达到保护井壁、防砂入井的目的。砾石充填完井一般都使用不锈钢绕丝筛管而不用割缝衬管。其原因有：

（1）割缝衬管的缝口宽度由于受加工割刀强度的限制，最小为 0.25~0.50mm。因此割缝衬管只适用于中（粗）砂粒油层。而绕丝筛管的缝隙宽度最小可达 0.12mm，故其适用范围要大得多；

（2）绕丝筛管是由绕丝形成一种连续缝隙，它的流通面积要比割缝衬管大得多，流体通过筛管时几乎没有压力降；

the screen pipe;

(3) wire screen tube with stainless steel wire as raw material, its strong corrosion resistance, long service life, high comprehensive economic efficiency. Gravel-fill complete and in the straight wells and directional wells can be used. But in the horizontal wells should be careful, because it is easy to make a bad sand card, so that the gravel filling failure, can not achieve the purpose of effective sand prevention. In order to adapt to the needs of different reservoir characteristics, both barehole completion and perforation completion can be filled with gravel, which are called barehole gravel filling and casing gravel filling respectively.

## 2.4 The Principle of Well Completion Design

### 2.4.1 Selection of Well Completion Method

The selection of well completion method is a complex systematic project, which requires comprehensive consideration of many factors, which mainly include whether the borehole is stable during the production process, whether the formation is sandy during the production process, geological and reservoir engineering characteristics, the size of the completion and production capacity, the economic benefits of drilling and completing the well and the requirements of oil recovery engineering, etc. Generally speaking, the reasonable completion method should try to meet the following requirements. Generally speaking, a reasonable completion method should try to meet the following requirements:

(1) Optimum connectivity should be maintained between the formation and the

（3）绕丝筛管以不锈钢丝为原料，其耐腐蚀性强，使用寿命长，综合经济效益高。砾石充填完井在直井、定向井中都可使用。但在水平井中应慎重，因为易发生砂卡，从而使砾石充填失败，达不到有效防砂的目的。为了适应不同油层特性的需要，裸眼完井和射孔完井都可以充填砾石，分别称为裸眼砾石充填和套管砾石充填。

## 2.4 完井设计原则

### 2.4.1 完井方法的选择

完井方法的选择是一项复杂的系统工程，需要综合考虑的因素很多，这些因素主要有生产过程中井眼是否稳定、生产过程中地层是否出砂、地质和油藏工程特性、完井产能大小、钻井完井的经济效益、采油工程要求等。合理的完井方法应该力求满足以下几点要求：

（1）油气层和井筒之间应保持最佳的连通条件，油气层所受的伤害最小；

wellbore, with minimum damage to the formation;

(2) There should be the largest possible seepage area between the oil and gas reservoir and the wellbore, and the resistance of the oil and gas well should be minimized;

(3) It should be able to effectively seal off the oil, gas and water layers, prevent gas or water flushing, and prevent mutual interference between the layers;

(4) It should be able to effectively control the sand out of the oil formation, prevent the well wall from collapsing, and ensure the long-term production of the well;

(5) Conditions should be available for carrying out measures such as layered water injection, gas injection, layered fracturing and acidizing, as well as for facilitating artificial lifting and downhole operations;

(6) In the case of a thickened oil field, the thick oil extraction is capable of meeting the requirements for thermal recovery (major steam throughput and steam drive);

(7) Conditions for side-drilling directional and horizontal wells are available at a later stage of oilfield development;

(8) The construction process is as simple as possible and the cost is as low as possible.

The choice of completion method should be based on production and development requirements, technological feasibility and economy to determine whether to use casing completion or barehole completion. If openhole completion is used, it is necessary to further determine the stability of the well wall during production. If the well wall is stable, then openhole unsupported completion can be considered. If the well wall is not stable, barehole perforated/cut pipe supported completion can

（2）油气层和井筒之间应具有尽可能大的渗流面积，油气入井的阻力最小；

（3）应能有效地封隔油气水层，防止气窜或水窜，防止层间的相互干扰；

（4）应能有效地控制油层出砂，防止井壁垮塌，确保油井长期生产；

（5）应具备进行分层注水、注气、分层压裂、酸化等措施，以及便于人工举升和井下作业等的工作条件；

（6）如为稠油油田，则稠油开采能达到热采（主要蒸汽吞吐和蒸汽驱）的要求；

（7）油田开发后期具备侧钻定向并及水平井的条件；

（8）施工工艺尽可能简便，成本尽可能低。

完井方式选择应在生产开发要求、工艺可行性及经济性的基础上，确定采用套管完井还是裸眼完井。若采用裸眼完井，则需要进一步判断生产过程中的井壁稳定情况。如果井壁稳定，则可考虑采用裸眼无支撑完井。如果井壁不稳定，对于石灰岩储层则可考虑裸眼打孔管/割缝管支撑完井；对于砂岩类储层则需要考虑防砂方式，并根据防砂方式设计，选择采用裸眼独立筛管完井或裸眼筛管充填完井。若采用套管完井，对于不需要防砂的井，可选用套管射孔完井；对于需要防砂的井，应根据防砂方式设计，选择采用套管独立筛管完井或套管筛管充填完井。

be considered for limestone reservoirs; for sandstone reservoirs, sand control methods need to be considered, and according to the design of sand control methods, barehole independent screen pipe completion or barehole screen pipe filling completion can be selected. If casing completion is used, for wells that do not need sand prevention, casing shot hole completion can be selected; for wells that need sand prevention, casing independent screen tube completion or casing screen tube filling completion should be selected according to the design of sand prevention method.

#### 2.4.1.1 Openhole Completion Method

Openhole completions include advanced openhole completions and later openhole completions. The selection principles are:

(1) Carbonate or sandstone reservoirs with hard and dense lithology, developed natural fractures, and stable well walls that do not collapse;

(2) Reservoirs with no air tops, no bottom water, no water-bearing interlayers and easily collapsible interlayers;

(3) A single reservoir, or a multilayer reservoir with essentially the same pressure and lithology;

(4) Reservoirs that are not prepared to implement segregated segments and selective treatment.

The main advantages of openhole completions are:

(1) Low cost;

(2) The reservoir is not harmed by the cement slurry;

(3) The use of expandable double packers allows for the implementation of production control and production enhancement operations in segregated segments.

#### 2.4.1.1 裸眼完井方式

裸眼完井包括先期裸眼完井和后期裸眼完井。其选择原则为：

（1）岩性坚硬致密、天然裂缝发育、井壁稳定不坍塌的碳酸盐岩或砂岩储层；

（2）无气顶、无底水、无含水夹层及易坍塌夹层的储层；

（3）单一储层，或压力、岩性基本一致的多层储层；

（4）不准备实施分隔层段及选择性处理的储层。

裸眼完井的主要优点是：

（1）成本低；

（2）储层不受水泥浆伤害；

（3）使用可膨胀式双封隔器，可以实施生产控制和分隔层段的增产作业。

The main disadvantages of the openhole completion are:

(1) For loose strata, the borehole may collapse;

(2) It is difficult to avoid tampering between layer segments;

(3) Limited options for yield-enhancing operations that do not allow for water stress.

At present, with the improvement of bare-bore packers, bare-bore multistage fracturing of horizontal wells in tight oil and gas reservoirs is becoming the main technology.

### 2.4.1.2 Perforated or Slotted Linner Completion Method

Perforated or slit-lined tubing completion is similar to barehole completion and is one of the more widely used completion methods for horizontal wells in recent years. The selection principles are:

(1) Reservoirs with no air tops, no bottom water, no water-bearing interlayers and easily collapsible interlayers;

(2) A single thick reservoir, or a multilayer reservoir with essentially the same pressure and lithology;

(3) Reservoirs that are not prepared to implement segregated segments and selectivity;

(4) Medium- and coarse-grained sand reservoirs with relatively loose lithology.

The main advantages of perforated or slit-lined completions are:

(1) Relatively low cost;

(2) The reservoir is not harmed by the cement slurry;

(3) It prevents borehole collapse.

### 2.4.1.3 Cemented Hole Completion Method

Cemented hole completion is one of the most widely and mainly used completion methods at home and abroad, including casing

裸眼完井的主要缺点是：

（1）对于疏松地层，井眼可能坍塌；

（2）难以避免层段之间的窜通；

（3）可选择的增产作业有限，不能进行水压力。

目前，随着裸眼封隔器的完善，致密油气藏水平井裸眼多级压裂正成为主体技术。

### 2.4.1.2 打孔或割缝衬管完井方式

打孔或割缝衬管完井类似于裸眼完井，是近年来水平井应用较多的完井方式之一。其选择原则为：

（1）无气顶、无底水无含水夹层及易塌夹层的储层；

（2）单一厚储层，或压力、岩性基本一致的多层储层；

（3）不准备实施分隔层段及选择性处的储层；

（4）岩性较为疏松的中、粗砂粒储层。

打孔或割缝衬管完井方式的主要优点是：

（1）成本相对较低；

（2）储层不受水泥浆伤害；

（3）可防止井眼坍塌。

### 2.4.1.3 固井射孔完井方式

固井射孔完井是国内外最为广泛和最主要使用的一种完井方式，包括套管射孔完井尾管射孔完井和尾管回接射孔完井。

perforation completion, tailpipe perforation completion and tailpipe tieback perforation completion. The selection principles are:

(1) Reservoirs with complex geological conditions such as gas tops, bottom waters, water-bearing interlayers and collapse-prone interlayers, which require the implementation of segregated layer segments;

(2) Reservoirs in which there are differences in pressure, lithology, etc., between the sublayers, requiring the implementation of sublayer testing, sublayer oil recovery, sublayer water injection, and sublayer treatment;

(3) Low-permeability reservoirs requiring large-scale hydraulic fracturing operations.

The main advantages of cemented shot hole completions are:

(1) It enables the most effective layer separation;

(2) Effective selective yield enhancement measures can be carried out.

The main disadvantages of cemented shot hole completions are:

(1) Relatively high completion cost;

(2) The reservoir is vulnerable to cement slurry;

(3) Requires a high level of skill in hole-shooting operations.

### 2.4.1.4 Openhole Sand Control Screen Tubing or Gravel-fill Completions

For reservoirs that are loosely cemented and have serious sand outcrops, bare eye sand control screen tubing or gravel-fill completion should generally be used. Gravel filling completion is to first lower the wireline screen tube into the oil formation part of the well, and then pump the pre-selected gravel on the ground to the annular space between the wireline screen tube and the borehole or

其选择原则为：

（1）有气顶、底水、含水夹层及易坍塌夹层等复杂地质条件，要求实施分隔层段的储层；

（2）各分层之间存在压力、岩性等差异，要求实施分层测试、分层采油、分层注水、分层处理的储层；

（3）要求实施大规模水力压裂作业的低渗透储层。

固井射孔完井的主要优点是：

（1）能实现最有效的层隔；

（2）可以进行有效的选择性增产措施。

固井射孔完井的主要缺点是：

（1）相对较高的完成本；

（2）储层易受水泥浆的伤害；

（3）要求水平较高的射孔操作技术。

### 2.4.1.4 裸眼防砂筛管或砾石充填完井

对于胶结疏松、出砂严重的储层，一般应采用裸眼防砂筛管或砾石充填完井方式。砾石充填完井是先将绕丝筛管下入井内油层部位，然后用充填液将在地面上预先选好的砾石泵送至绕丝筛管与井眼或绕丝筛管与套管之间的环形空间内，构成一个砾石充填层，以阻挡油层砂流入井筒，达到保护井壁、防砂入井的目的。

between the wireline screen tube and the casing with the filling liquid, constituting a gravel filling layer to block the flow of sand from the oil formation into the wellbore, so as to achieve the purpose of protecting the wall of the well, and preventing the sand from entering the well.

Due to the complexity and high cost of the gravel-fill completion process, it is rarely used unless there is a genuine need for sand control. In general, sand control is accomplished by using bare eye sand control screen tubing. The selection principles are:

(1) Reservoirs with no gas tops, no bottom water, and no water-bearing interlayers;

(2) A single thick reservoir, or a multilayer reservoir with essentially the same pressure and lithology;

(3) Reservoirs that are not prepared to implement segregated segments and selective treatment;

(4) Medium, coarse and fine-grained sand reservoirs with loose lithology and severe sand outcrops.

In addition to wireline screen tubes for bare eye sand control screen tube completions, precision microporous composite sand control screen tubes, precision perforated screen tubes, metal fiber screen tubes, star hole screen tubes, pre-filled gravel wireline screen tubes, and external guided cover sand filtering screen tubes may be selected depending on the sand out of the formation.

### 2.4.1.5 In-Tubing Gravel-fill Completion/

The principles for selecting gravel-filled completions in tubulars are:

(1) Reservoirs with gas tops, or bottom water, or complex geological conditions such as water-bearing interlayers and collapse-prone interlayers, thus requiring the

由于砾石充填完井工艺的复杂性和高成本，一般较少使用，除非确实有防砂需求。一般性的防砂均使用裸眼防砂筛管完井。其选择原则为：

（1）无气顶、无底水、无含水夹层的储层；

（2）单一厚储层，或压力、岩性基本一致的多层储层；

（3）不准备实施分隔层段及选择性处理的储层；

（4）岩性疏松且出砂严重的中、粗、细砂粒储层。

裸眼防砂筛管完井除绕丝筛管外，根据地层出砂情况，还可能选择精密微孔复合防砂筛管、精密冲缝筛管、金属纤维筛管、星孔筛管、预充填砾石绕丝筛管、外导向罩滤砂筛管等。

### 2.4.1.5 管内砾石充填完井

管内砾石充填完井选择原则为：

（1）有气顶，或有底水，或有含水夹层及易坍塌夹层等复杂地质条件，因而要求实施分隔层段的储层；

implementation of segregated segments;

(2) Reservoirs in which there are differences in pressure, lithology, etc., between the sublayers, thus requiring the implementation of selective treatment;

(3) Medium, coarse and fine-grained sand reservoirs with loose lithology and severe sand outcrops.

The sand prevention mechanism of in-tube gravel-filled completions and bare eye gravel-filled completions is exactly the same, and the main difference is that it requires selective exploitation of the reservoir and high pore density (30~40 holes/m) and large pore size ($\phi$20.0~25.4mm) injection after cementing, in order to increase the filling circulation area and facilitate filling gravel to the surrounding oil layer outside the borehole, and avoiding the mixing of gravel and formation sand to increase the resistance to seepage.

### 2.4.2 Basic Well Completion Requirements

Well completions play an important role in the oil and gas production process. The basic role of a well completion is to provide mechanical support to the borehole and a pathway for fluid flow from the reservoir to the surface, which are the minimum requirements for a well completion in the early stages of oil and gas production. In the early days completion technology evolved from barehole wells (the simplest type of completion) to casing cementing and perforating. Over time, the requirements for completions have changed to include more than just providing borehole stabilization and flow paths. Many wells produce waste fluids (e.g., gas and water) and solids (e.g., formation sands and particulates), and water and sand production

（2）各分层之间存在压力、岩性等差异，因而要求实施选择性处理的储层；

（3）岩性疏松且出砂严重的中、粗、细砂粒储层。

管内砾石充填完井与裸眼砾石充填完井的防砂机理完全相同，其主要区别是需要进行储层的选择性开采并要求固井后采用高孔密（30~40孔/m）大孔径（$\phi$20.0~25.4mm）射孔，以增大充填流通面积，并便于向孔眼外的周围油层填入砾石，避免砾石和地层砂混合增大渗流阻力。

### 2.4.2 完井基本要求

在石油和天然气生产过程中，完井起着重要作用。完井的基本作用是为井眼提供机械支撑和从储层到地面的流体流动通道，这些是油气生产初期对完井的最低要求。在早期完井技术从裸眼井（最简单的完井）发展到套管固井射孔。随着时间推移，对完井要求发生了变化，已不仅仅是提供井眼稳定和流动通道。许多井生产产出废液（如气体和水）和固体（如地层砂及微粒），出水和出砂成为石油工业中的棘手问题。控水和防砂（弱胶结地层）是完井设计中必须解决的两个常见的生产问题。为解决上述问题，开发了割缝衬管、筛管和砾石充填技术。此外，还发明了滑套和封隔器用以对多目的层进行选择性开采。完井过程包括下套管、固井、射孔、下油管和安装采油树。对于疏松地层，可能还需要

have become difficult problems in the petroleum industry. Water control and sand control (weakly cemented formations) are two common production problems that must be addressed in well completion design. To solve these problems, cuttings liners, screen pipes and gravel filling technologies have been developed. In addition, slipcovers and packers have been invented to selectively exploit multi-purpose formations. The completion process including casing, cementing, perforating, tubing and tree installation. For loose formations, gravel-fill completions may also be required.

When reservoir pressure is depleted, water injection is often used to maintain reservoir pressure and well production. Enhanced Recovery (EOR) improving fluid properties in the reservoir by injecting steam or surfactants to enhance oil and gas recovery is the primary method of enhanced recovery (EOR). The recovery rate can be maximized by controlling the injection and driving leading edge wave coefficient of the injection.

Hydraulic fracturing, matrix acidizing and acid fracturing are the main ways to increase production from wells. Well completion provide fractures in the formation for measures to increase production, distribute fracturing fluids, and generate acid pores to improve modification efficiency.

The traditional concept of well completion is that it is the last process of drilling, after casing, cementing (including perforation), etc., after drilling the target formation, and ensure the integrity of the drilled well. It is also divided into surface casing completion, technical casing completion, oil casing completion and so on. The modern concept of well completion

进行砾石充填完井。

当储层压力耗尽时，通常采用注水来维持储层压力和油井产量。提高采收率（EOR）通过注入蒸汽或表面活性剂来改善储层中流体性能，从而提高油气采收率，是提高采收率（EOR）的主要方法。可通过控制注入井的注驱前缘波及系数，尽可能地提高采收率。

水力压裂、基质酸化和酸压是油井增产的主要方式。完井为措施增产在地层中提供裂缝、分配压裂流体和产生酸蚀孔洞，以提高改造效率。

传统完井的概念认为，完井是钻井的最后一道工序，即钻完目的层后下套管、注水泥固井（包括射孔）等，是为保证钻井的完整性而进行的工程。其中又分为表层套管完井、技术套管完井、油层套管完井等。现代完井的概念认为，完井是钻井与采油的桥梁，是从制订钻采工程方案优化完井方式开始，到钻开油气层、下入套管（尾管、筛管）、注水泥固井、射孔、先期增产措施、下生产管柱、试油排液，直至投产的一项系统工程。

is that it is a bridge between drilling and oil production, a systematic project starting from the development of drilling and production engineering plan to optimize the completion method, to drilling open the oil and gas layer, lowering into the casing (tail pipe, screen pipe), cementing, perforation, advance production enhancement measures, lowering of production tubing, oil testing and drainage, and up to the commissioning of the well.

Reservoir completions include barehole completions, perforated tubing or slit-lined tubing completions, cemented perforated completions, barehole sand-proof screened tubing or gravel-filled completions, gravel-fill completion in tubing, and so on. With the development of horizontal well technology and the increase of reservoir leakage wells, a composite reservoir completion method with more field applications is called through-the-eye completion, screened tubing is used in the reservoir section, and the well section in the upper part of the reservoir is cemented (perforated or unperforated).

Reservoir completions must be based on geological characteristics, development and mining technology requirements, as well as costs and overall economic benefits to optimize the most reasonable completion method, mainly based on the following six aspects:

(1) Geological characteristics: including porosity, fracture, air top, bottom water (interlayer water, top water) etc.;

(2) Development methods: including water injection development, gas/steam injection mining, depletion mining, etc.;

(3) Well categories: including oil/gas production wells, water injection wells, gas

储层完井包括裸眼完井、打孔管或割缝衬管完井、固井射孔完井、裸眼防砂筛管或砾石充填完井、管内砾石充填完井等。随着水平井技术的发展及储层漏失井的增多，目前现场应用较多的一种复合储层完井方式叫贯眼完井，即储层段采用筛管，储层上部井段进行固井（射孔或不射孔）。

储层完井必须根据地质特点、开发开采技术要求以及成本和整体经济效益等来优选最合理的完井方法，主要依据以下6个方面来进行：

（1）地质特点：包括孔隙性、裂缝性、气顶、底水（层间水、顶水）等；

（2）开发方式：包括注水开发、注气/注汽开采、枯竭式开采等；

（3）井别：包括油/气生产井、注水井、注气井、注蒸汽井等；

injection wells, steam injection wells, etc.;

(4) Lithology: including sandstone, carbonate rock, volcanic rock, metamorphic rock, mudstone and shale;

(5) Operational measures: including fracturing, acidizing, sand control, etc.;

(6) Types of wells: including vertical wells, horizontal (sidetracked horizontal wells), and well slopes.

The basic requirements for well completion are as follows:

(1) Maximize reservoir protection and prevent damage to the reservoir;

(2) Reduce the flow resistance of oil and gas flow into the wellbore;

(3) Effectively sealing off oil, gas and water layers and preventing mutual interference between layers;

(4) Overcome well caving or sanding of producing layers, ensure long-term stable production of oil and gas wells, and prolong the life of wells;

(5) Water injection, fracturing, acidizing and other measures to increase production can be implemented;

(6) Simple process and low cost.

## 2.4.3 Completion Fluids

According to the needs of the completion process, the technical requirements for the use of completion fluids at different stages of the completion operation are different, so a series of functional completion fluids have been formed according to the different needs of the completion process, such as perforating fluids used in perforation drilling operations, cleaning fluids used in well washing, and acid fluids and fracturing fluids used in acidizing and fracturing processes.

（4）岩性：包括砂岩、碳酸盐岩、火山岩、变质岩、泥岩、页岩等；

（5）作业措施：包括压裂、酸化、防砂等；

（6）井的类型：包括垂直井、水平（侧钻水平井）、井斜等。

完井的基本要求有以下几点：

（1）最大限度地保护储层，防止对储层造成伤害；

（2）减少油气流进入井筒时的流动阻力；

（3）能有效地封隔油气水层，防止各层之间的互相干扰；

（4）克服井塌或产层出砂、保障油气井长期稳产，延长井的寿命；

（5）可以实施注水、压裂、酸化等增产措施；

（6）工艺简单、成本低。

## 2.4.3 完井液

根据完井工艺需求，在完井作业不同阶段使用完井液的技术要求也不同，所以根据完井过程的不同需求形成了系列的功能完井液，如射孔作业使用的射孔液、洗井过程中使用的清洗液和酸化压裂过程中使用的酸液和压裂液等。

Completion fluids are basically categorized into two main groups according to their composition: water-based and oil-based, and in special cases, weighted completion fluids may also be required.

Completion fluids must not only ensure the stability of oil and gas wells during the whole process of completion operations to meet the needs of completion operations, but also must protect the reservoir from injury, and its performance is directly related to the reserves and production of oil and gas fields, and its main functions including:

(1) Balance downhole pressure to ensure operational safety and casing stability;

(2) Protect the reservoir and reduce reservoir injury during well completion;

(3) Good anti-corrosion ability to slow down the corrosion of casing and downhole tools;

(4) Cleaning downhole oil, carrying and suspending solid-phase particles, keeping downhole clean;

(5) Modify the reservoir;

(6) Meet the requirements of well completion operations such as plugging, breaking rubber, unplugging and anti-sand filling;

(7) Maintain the stability of various properties in the well under long-term static conditions in the wellbore;

(8) Have no effect on the indicators of downhole geophysical logging;

(9) It can be compatible with other working fluids;

(10) Its performance should be adjusted at any time according to the geological conditions of oil and gas wells;

(11) It is safe to use and harmless to the environment and human health.

完井液体系按照组成基本可分为水基、油基两大类，特殊情况下还可能需要加重完井液。

完井液既要在完井作业全过程中保证油气井的稳定，满足完井作业需要，又必须保护储层不受伤害，其性能好坏直接关系到油气田的储量和产量，其主要功能包括：

（1）平衡井下压力，保证作业安全和套管的稳定；

（2）保护储层，减少完井期间的储层伤害；

（3）具有良好的防腐能力，可减缓对套管和井下工具的腐蚀；

（4）清洗井下油污，携带、悬浮固相颗粒物、保持井下清洁；

（5）改造储层；

（6）满足完井时堵漏、破胶、解堵和防砂充填等作业要求；

（7）在井筒中长期静置条件下保持井内各种性能的稳定；

（8）对井下地球物理测井的指标没有影响；

（9）能与其他工作液相配伍；

（10）其性能应根据油气井地质条件随时进行调整；

（11）使用安全，对环境和人体健康无害。

# Chapter 2 Completion

# Ideological and Political Case Studies

Academician Wang Demin is a renowned expert in petroleum engineering and the founder of stratified production and chemical flooding technology in Chinese oilfields. He was born in Tangshan, Hebei Province in February 1937, he graduated from the Beijing Petroleum Institute in 1960 with a major in oil production. During his internship at the Daqing Oilfield Production Command's pressure measurement group, he independently derived the "Songliao Method," which significantly improved testing efficiency and was widely used in the Daqing Oilfield. He was later transferred to the Production Technology Research Institute, where he led a special research project on layered water injection and oil production with the goal of "six divisions and four clarifications". He successively developed multi-stage packer testing methods, a series of oil-water layered testing tools and technologies, filling a domestic gap. He independently invented the eccentric layered production and injection downhole tools and technologies, elevating China's oilfield layered production and injection technology to a world-leading level, earning him the National Science Congress Award and the National Second Prize for Technical Invention. In response to the challenges of developing the superficial reservoirs in the Daqing Oilfield, he developed the "limited flow fracturing technology for thin oil layer development", enabling the utilization of over 700 million tons of difficult-to-recover geological reserves and winning the National First Prize for Science and Technology Progress. Concurrently, he organized research on "mechanical oil production energy-saving

# 思政案例

王德民院士是著名的石油工程专家，是中国油田分层开采和化学驱油技术奠基人。他1937年2月出生于河北省唐山市。1960年毕业于北京石油学院钻采系采油专业。在大庆油田采油指挥部测压组实习期间，他自行推导出"松辽法"，显著提高了测试效率，在大庆油田广泛应用；后调入采油工艺研究所，承担以"六分四清"为目标的分层注水采油专项技术研究，接连研发了多级封隔器验封方法、油水井分层测试系列工具和工艺，填补了国内空白。他自主发明了偏心式分层配产配注系列井下工具和工艺，使我国的油田分层注采技术提升至世界领先水平，获全国科学大会奖和国家技术发明二等奖；针对大庆油田开发表外储层的难点，研发"限流法压裂在薄油层开发中的应用"技术，使7亿多吨难采地质储量得以利用，荣获国家科学技术进步奖一等奖；同期组织研究"机械采油节能配套技术"，取得产量上升、耗电不增的显著效果，获国家科学技术进步奖二等奖；他领导科研团队先后完成"大庆油田高含水地层聚合物驱油技术研究""大庆油田高含水期稳油控水系统工程""大庆油田三元复合驱先导性矿场试验研究""桥式偏心分层开采配套工艺技术"等重大科研项目，分别获国家科学技术进步奖特等奖、一等奖和国家"八五"科技攻关重大科技成果奖。1994年中国工程院成立时，王德民成为首批院士。他累计取得20多件发明专利，在国际石油论坛上发表60余篇学术论文，获世界石油工程师学会（SPE）颁发的"杰出会员奖""油藏描述与开发贡献奖"，被授予终生会员荣誉。王德民院士把对祖国的热爱化作致力于科技创新的不竭动力，始终

supporting technology", achieving significant results of increased production without increased power consumption, and received the National Second Prize for Science and Technology Progress. Leading a research team, he completed major scientific research projects such as the "Daqing Oilfield High Water Cut Formation Polymer Flooding Technology Research" "Daqing Oilfield Stable Oil and Water Control System Project during High Water Cut Period" "Daqing Oilfield Ternary Composite Drive Pilot Field Test Research" and "Bridged Eccentric Layered Production Supporting Technology", winning the National Special Prize, First Prize for Science and Technology Progress, and the National "8th Five-Year Plan" Major Scientific and Technological Achievement Award. In 1994, when the Chinese Academy of Engineering was established, Wang Demin became one of the first academicians. He has obtained more than 20 invention patents, published over 60 academic papers at international petroleum forums, and received the "Outstanding Member Award" and "Reservoir Description and Development Contribution Award" from the Society of Petroleum Engineers (SPE), and was honored with lifetime membership. Academician Wang Demin has turned his love for his country into an inexhaustible drive for technological innovation, always taking the responsibility of ensuring national energy security as his mission, and continuously exploring ways to improve oil recovery. He is not only a distinguished leader in the field of chemical flooding research for tertiary oil recovery but also a pioneer in the development of quaternary oil recovery technology.

Wang Demin: To Dedicate My Life to Daqing!

以保障国家能源安全为己任，持续探索提高采收率的途径，不仅在三次采油化学驱科研领域成为功勋卓著的领军人，还是研发四次采油技术的开拓者。

王德民：愿以一生许大庆！

He is the first academician of the Chinese Academy of Engineering in the field of oil extraction in China. His work is evaluated in the academician registration form as follows: "These technologies are of great significance, have the greatest difficulty, and are the most advanced in the development of oilfields in the world." The International Minor Planet Center named asteroid 210231 "Wang Demin Star," and the naming announcement introduces his identity as the founder of stratified mining and chemical flooding technology in Chinese oilfields.

When I met the 84 years old Academician Wang Demin, he had just undergone leg surgery and was walking a bit slowly. "It doesn't affect work!" he said nonchalantly, his posture still upright. "Honors only represent the past!" he said casually, his face showing a calm and stern expression.

Only when he talked to the journalist about oil did he become eloquent, his expression gradually revealing a youthful vigor. Since he was 23 years old, he has always been engaged in "combat" with oil, and he has never forgotten the oath he made at that time: "To the front line of the great oil battle, to win the big oilfield for the country!" "At that time, we often said 'everything needs to be rebuilt,' and these were the most urgent."

In the winter of 1960, on the Songnen Plain, the water froze into ice. Wang Demin and several workers carried a 100-plus kilogram winch from one oil well to another. They used the winch to send the testing instruments down the well. However, the wellhead was frozen in the cold, and the instruments couldn't be lowered. What to do? Wang Demin and the workers wrapped the wellhead with their cotton

他是我国石油开采专业首位工程院院士，院士登记表里这样评价他的工作："这些工艺，都是世界上油田开发意义重大、难度最大、工艺最先进的技术"。国际小行星中心将210231号小行星命名为"王德民星"，命名公报中如是介绍他的身份：中国油田分层开采和化学驱油技术奠基人。

见到84岁的王德民院士时，他刚动过腿部手术不久，步履还有些迟缓。"不影响工作！"问起手术，他满不在乎，身姿依然笔挺。"荣誉只代表过去！"说起成绩，他轻描淡写，面庞清癯冷峻。

唯有和记者聊起石油，他才滔滔不绝，眉宇间渐渐透出一股俊朗少年之气。自从23岁来到大庆，他始终在和石油"过招较劲"，也不曾忘记当年立下的誓言："到大会战前线去，为国家拿下大油田！""那时常说'百废待兴'，这些都是最'待兴'的"。

1960年隆冬，松嫩平原滴水成冰。王德民和几位工人扛着100多公斤重的绞车，从一口油井挪向另一口。他们要用绞车把测试仪器送到井下。可严寒中井口冻死，仪器下不去。怎么办？王德民和工人们解开棉袄包住井口，又把冰冷的防喷管抱在怀里焐热。等原油化开，仪器下井，嘴唇早已冻紫了。此时，他刚到大庆4个多月。

coats and held the cold blowout preventer against their chests to warm it up. When the crude oil melted, the instruments could be lowered down the well, but their lips were already purple from the cold. At that time, he had only been in Daqing for four months.

Wang Demin, of Chinese-Swedish descent, was born in Tangshan, Hebei, and later moved to Beijing. He grew up in a well-off family, with his father serving as the deputy director of Tongren Hospital and his mother teaching at a university. His earliest memory is of the "July 7th Incident" when Japan invaded China, and the memory of war became his deepest pain. The awareness that "a country that is not strong will be bullied" was deeply imprinted on his young heart.

Just before the founding of the People's Republic of China, Wang Demin entered the Huwen Middle School in Beijing for his junior high education. In his second year of high school, his school established the "Wu Yunduo Class," named after the pioneer of China's new weaponry industry, and he became a member due to his excellent academic performance.

The first time he heard Wu Yunduo give a report at school, he was deeply shocked. "The title was 'Devote Everything to the Party,' and he spoke for 7 hours. Me and my classmates were in tears." Wang Demin reread the deeds of this "Chinese Paul Kurchatov" again and again, and a belief was engraved in his heart: "Go to the most difficult place, dedicate everything to the motherland!"

After graduating from high school in 1955, Wang Demin wrote several college entrance examination applications, with majors almost all related to petroleum, water conservancy and steel. The reason was simple:

中瑞混血的王德民，生于河北唐山，后举家迁至北京。自小家境优渥，父亲曾任同仁医院副院长，母亲在高校执教。出生5个月，日寇发动"七七事变"。童年时期，战乱成为他最深痛的记忆，"国家不强大就会被欺凌"的意识，也深深印在他幼小的心里。

中华人民共和国成立前夕，王德民升入汇文中学读初中。高二那年，学校组建以新中国兵器工业开拓者吴运铎命名的"吴运铎班"，品学兼优的他成为其中一员。

第一次听吴运铎来校做报告，他深受震撼。"题目是《把一切献给党》，讲了7小时。我和同学们泪流满面。"王德民把这位"中国保尔·柯察金"的事迹重温了一遍又一遍，一个信念刻在心中："到最艰苦的地方去，为祖国献出一切！"

1955年高考后，甲等生王德民挥笔写下七八个志愿，专业几乎全是石油、水利、钢铁之类。原因很简单——"那时常说'百废待兴'，而这些都是最'待兴'的"。北京石油学院录取了他。当时中国石油极缺，老百姓把汽车用的汽油、点灯用的煤油都称为"洋油"。西方专家有关"中国贫油"的论断，像山一样压在中国人头上。

"At that time, we often said 'everything needs to be rebuilt,' and these were the most urgent." Beijing Petroleum Institute accepted him. At that time, China was extremely short of oil, and the people referred to gasoline for cars and kerosene for lighting as "foreign oil." The Western experts' assertion that "China is poor in oil" was like a mountain pressing on the Chinese people.

During his five years of college, Wang Demin learned a lot of oil extraction technology knowledge. During his internship in his third year, he also participated in the Sichuan Central Oil Battle. "The results were disappointing. The country had high expectations, but it didn't produce oil for long here." He said, "I personally experienced how difficult it is to work in the oil industry, but the more difficult it is, the more it needs people to do it, and the more it relies on science."

Just before graduation, the shocking news came that China's largest oilfield, the Daqing Oilfield, had emerged. The Central Committee of the Communist Party of China and the State Council approved the launch of a major oil battle, and the days when New China would throw off the label of a "poor oil country" were not far off! The students in the petroleum engineering field were ecstatic, "Beating pots and singing songs, we made a racket all night." Following this, a batch of Wang Demin's classmates were sent to Daqing. Then, encouraging news kept coming: Daqing was a "battleground of hard work," gathering a group of "Iron Man" like Wang Jinxi who fought hard, drilling new oil wells in the shortest time…… Wang Demin was excited beyond words. Finally, when graduation arrived, he gave up the opportunity to stay

五年大学生涯，王德民学到了许多采油科技知识。大三实习，还参加了川中石油会战。"结果让人失望。国家寄予很大期望，但这里很快不出油了。"他说，"我切身体会到，搞石油太难！但越难，越需要有人做，越要靠科学。"

毕业前夕，石破天惊的喜讯不期而来：中国最大的油田——大庆油田横空出世，党中央国务院批准开展石油大会战，新中国将"贫油国"的帽子一举甩进太平洋的日子，为期不远了！石油学子们欣喜若狂，"大家敲脸盆、唱着歌，喧腾了一整夜"。紧接着，王德民的同学中有一批被派往大庆。随后，振奋人心的消息不断传来：大庆是一片"奋斗的热火"，汇聚了王进喜等一大批舍身奋战的"铁人"，他们用最短的时间打出一口口新的油井……王德民激动难抑。终于等到毕业，他放弃留校机会，义无反顾地选择了大庆："搞石油不去油田，何谈报效祖国！""大庆等不得，我也不怕那么多了。"

at the university and chose to go to Daqing without hesitation: "How can you talk about serving the motherland if you don't go to the oilfield for oil extraction?" "Daqing can't wait, and I'm not afraid of so much."

After graduation, Wang Demin's first job was as an intern in the geological room's pressure measurement group at the oilfield. During the "pressure measurement battle" in Daqing, the goal was to determine the formation pressure of hundreds of production wells. However, the international standard pressure measurement method, "Heron's Method" was not applicable, and the measurement deviation was significant. China needed its own pressure measurement method! Wang Demin initiated a research effort. He worked from morning to late at night, ate a quick meal, and then started self-studying heat conduction mathematics, hydromechanics, and Russian, until two or three in the morning; at night, he would cross vast cornfields to the library to borrow books, waking up the librarian who kindly reminded him that there were wolves in the area at night……After several months, Wang Demin systematically mastered various foreign pressure measurement methods, and through countless calculations, he began to see the answer.

On February 14, 1961, the Chinese New Year's Eve. The oilfield distributed half a jin of white flour and a bowl of minced meat to everyone to make dumplings. In the festive canteen, Wang Demin looked at the flour and minced meat in his hands and frowned. "This won't take half a day to make? No way! But I'm also hungry." He had a sudden idea, flattening the flour into two large "cakes" and making two extra-large "dumplings", boiling them for half an hour before eating them,

王德民工作后的第一个职务，是油田地质室测压组实习员。时值大庆"测压会战"，目的是弄清数百口生产井的地层压力。然而，国际通行的测压法"赫诺法"并不适用，测压偏差很大。中国得有自己的测压法！王德民发起攻关。白天工作到晚上七八点，刨两口饭就开始自学热传导数学、水力学、俄文，直到凌晨两三点；夜里穿越大片苞谷地去图书室借书，被敲门声惊醒的图书管理员好心提醒他：这片地界晚上有狼……几个月过去，王德民系统掌握了国外各种测压法，无数次推演隐隐有了答案。

1961年2月14日，除夕夜。油田发给每人半斤白面、一碗肉馅，让大家包饺子。喜气洋洋的职工食堂里，王德民看着手里的面粉和肉馅，皱起了眉头。"这不得把大半天'包'进去？不行！可我也饿啊。"灵机一动，他把面粉擀成两张"大饼"，捏了两个特大号"饺子"，煮了半小时，便不顾生熟捞起来吃下肚去，回到宿舍继续奋笔疾书。

regardless of whether they were cooked or not, and then returning to his dormitory to continue writing.

The inspiration broke through the last barrier on New Year's Eve. Wang Demin derived China's first and the world's third set of unstable well testing pressure formulas! The formula, named "Songliao Method", was applied throughout the oilfield, with an accuracy twice that of "Heron's Method". Over the next decade, Wang Demin successively developed multi-layer oil testing, layered oil and water well testing, and other processes, helping Daqing Oilfield achieve second-tier oil recovery that surpassed world-class standards. His extraordinary courage was also widely talked about – using a steel wire with a diameter of 2.2mm to replace a large pipe machine, realizing the steel wire launching of testing instruments, which was both economical and convenient. He developed eccentric production and injection devices, solving the problem of "removing all layers to get one layer" of the concentric production and injection device, increasing the water injection qualification rate by over 40 percentage points…… "In those special years, every battle involved 'expert-specialist' risks. Moreover, I have foreign blood." Wang Demin sighed, "But Daqing can't wait, and I'm not afraid of so much." "It's like finding another large oilfield."

As the spring tide surged and the reform and opening-up curtain was raised, Daqing Oilfield had maintained stable and high production for ten years. How to maintain production for another ten years or even longer? It was necessary to develop new reserves. There were new reserves, but the extraction was extremely difficult – unlike foreign oilfields that usually have only

灵感,在辞旧迎新之夜冲破最后一道屏障。王德民推导出了中国第一套、世界第三套不稳定试井测压公式!从此,被命名为"松辽法"的公式在全油田应用,精度比"赫诺法"提高两倍。其后十余年,王德民接连研制出多层试油、油水井分层测试等工艺,助力大庆油田二次采油赶超世界先进水平。而他的过人胆识,也为人们津津乐道——用直径2.2毫米的钢丝替代大型通井机,实现了测试仪器钢丝化投捞,既经济,又灵便。研发偏心配产、配水器,解决了同心配产器"取一层须拆所有层"的弊病,配水合格率提高40多个百分点……"特殊年代,每次请战都冒着'白专'风险。何况,我还有外国血统。"王德民感慨,"但大庆等不得,我也不怕那么多了。""相当于又找到了一个大油田。"

春潮澎湃,改革开放大幕拉开。此时,大庆油田已保持10年稳产高产。怎样再稳产10年甚至更久?必须开发新储量。新储量是有,但开采难度极大——与国外油田一般只有几个油层不同,大庆油田油层少则80多个,多则140多个,其中1/4是0.2~0.5米的薄油层,国际尚无开采先例。

a few oil layers, Daqing Oilfield has 80 to 140 oil layers, with a quarter being thin oil layers of 0.2 to 0.5 meters, which had no precedent for extraction internationally.

"Let the Chinese people set this precedent!" Another round of day and night work followed. Two years later, Wang Demin led the development of the "limited flow fracturing method," which could open up 20 to 30 thin oil layers or even 70 at once, turning low-permeability layers into economically recoverable oil layers. The geological reserves of the Daqing Oilfield increased by 700 million tons, "equivalent to finding another large oilfield!" Wang Demin was not satisfied. At this time, the recovery rate of the Daqing Oilfield was 40%, reaching the limit of water flooding. He shifted his focus to the global challenge of chemical flooding for tertiary oil recovery.

Many advised him to be cautious because foreign authorities had already declared that tertiary oil recovery was "future technology" and currently unattainable. "How can we know if we don't try? We must seek breakthroughs in the 'hopeless' places," Wang Demin studied numerous cases of failed chemical flooding applications in the United States, listing over 200 topics across 8 categories, and formulated a meticulous experimental plan. The proposal was eventually approved. After ten years of hard work, in 1996, polymer flooding technology was widely adopted in Daqing, creating a world record—China became the first country to implement chemical flooding for tertiary oil recovery on a large scale; the oil recovery rate in Daqing reached nearly 70%, far exceeding the level of 45% in developed countries. Wang Demin was not content to stop there. He then led the charge on quaternary oil recovery. "The

"中国人来开这个先例！"又是一番夜以继日。两年后，王德民牵头研制出"限流压裂法"，可一次压开20~30个乃至70个薄油层，使低渗透层变成了可经济开采的油层。大庆油田地质储量猛增7亿吨，"相当于又找到了一个大油田！"王德民并不满足。此时的大庆，采收率40%，已到水驱的极限值。他把目光投向世界性难题——化学驱三次采油。

很多人劝他慎重，因为国外权威早已断言：三次采油是"未来技术"，目前无法实现。"不干怎么知道呢？就要在'没希望'的地方找突破。"王德民研究了美国应用化学驱失败的大量案例，列出8大类、200多项课题，形成了周密的试验思路。方案终获批准。十年艰辛，1996年，聚合物驱油技术在大庆推广，世界纪录随之诞生——中国成为首个实现化学驱三次采油大规模应用的国家；大庆石油采收率近70%，远超发达国家45%的水平。王德民仍未止步，又率先向四次采油发起进攻。"大庆石油含水量越来越高，采上来100吨液体，98吨都是水。"王德民解释，"我们开发出同井注采工艺，在井下油水分离，只采石油不采水，成本降幅巨大。"又是十年呕心沥血。2020年底，同井注采顺利通过验收。王德民期待着它推广应用，"引发一场全球老油田复采的大变革。"

water content in Daqing's oil is increasing, and for every 100 tons of fluid pumped, 98 tons are water," Wang Demin explained. "We developed the same well injection and production technology, which separates oil and water underground, allowing for the extraction of oil without water, resulting in a significant reduction in costs." After another decade of arduous effort, the same well injection and production technology was successfully tested and approved at the end of 2020. Wang Demin anticipated its widespread application, "sparking a major global revolution in the re-exploitation of mature oilfields."

Throughout the years, a love for the field was passed down. Wang Demin's son, Wang Yan, grew up watching his father's dedication to his work. That focus instilled in him a lifelong interest in petroleum. After graduating from university, he returned to the Daqing Oilfield, eventually rising to become the chief engineer of the Second Oil Extraction Plant. Speaking of his father, Wang Yan's main impression was one word—strict. "My father was always serious and meticulous in his work and life. As a child, I often complained, but later, when I embarked on this path, I understood him." Scientific research was the best way for them to communicate. In the same well injection and production project, Wang Yan was one of the project leaders, and his elderly father, although no longer frequently visiting the well site, maintained daily calls. "Sometimes we would talk for over an hour. From the construction of a small part to the calculations of the entire system, there was no room for error."

This strictness, Wang Demin's research team member and professor at Northeast Petroleum University, Ma Wenguo, felt deeply.

耳濡目染间，一种热爱在传承。王德民的儿子王研，是看着父亲伏案工作的背影长大的。那种专注，让他从小就对石油产生了兴趣。大学毕业后，他回到大庆油田，一步步成长为采油二厂总工程师。说起父亲，王研最大的感受只有一个字——严。"父亲不苟言笑，对工作、生活极严谨。我儿时难免抱怨，后来走上这条路，才理解他。"科研是父子间最好的交流。同井注采，王研是项目负责人之一，年迈的父亲不再常到井边，但每天的电话雷打不动。"有时一打就是个把小时。小到一个零件的构造，大到整个系统的计算，都不容分毫偏差。"

这种严，王德民科研团队成员、东北石油大学教授马文国感受至深。"王院士从不听'左右''大概'。"马文国说，"一次路遇，他问：实验数据有什么变化？我说，'和上次差不多。上次数据好像是，是……'院士脱口而出：'6.51! 你再测几次，认真分析。'我又佩服，又惭愧。"

"Academician Wang never listened to 'about' or 'more or less'" Ma Wenguo said. "Once on the road, he asked about the experimental data. I said, 'It's about the same as last time. The last data was…' and the academician immediately said, '6.51! You should measure it again and analyze it carefully.' I was both impressed and ashamed."

The person Wang Demin was most strict with was himself. In his early years, despite his handsome appearance, he was well-known for being "untidy": he wore the same work clothes all the time, too busy to wash them, and had only a few to change into; his hair grew so long that it covered his eyes, but he couldn't bear to spend time on haircuts; he often stayed up late working at his desk, with thick glasses perched on his nose. As his workload increased, he wished he could "split every minute in half". To save time boiling water, he simply ate powdered milk and coffee powder to stave off hunger and stay awake; after breaking his leg and being hospitalized for three days, he returned to his post; he rarely attended social events, always bringing his own bread and fruit to eat in the office to save time going to the canteen……"He calculated that working 12 hours a day, seven days a week, was equivalent to working 8 hours a day, five days a week for 2.1 times, which was like extending his service to the country for 30 years to over 60 years," Professor Wu Wenxiang of Northeast Petroleum University recalled.

Due to his fluent English and outstanding speaking skills since childhood, Wang Demin was pushed to the forefront of China's petroleum international exchanges, winning numerous awards such as the "Pioneer Award for Tertiary Oil Recovery" from the

王德民最严厉对待的人，是自己。早年间，相貌堂堂的他却是出了名的"不修边幅"：常年一身工作服，脏了没工夫洗，也没几件可换；头发长得遮住了眼睛，也舍不得花时间理；常年熬夜伏案，鼻梁上架起厚厚的近视镜片。后来工作愈发繁重，他恨不能"一分钟掰成两半用"。为省下烧开水冲泡的时间，索性干吃奶粉、咖啡粉充饥提神；不慎摔断腿，住院三天就回岗位；应酬交际极少参加，中午总是自带面包、水果在办公室吃，省下去食堂的时间……"他算过，每天工作12小时、每周7天，就是每天8小时、每周5天的2.1倍，相当于把为国家工作30年延长到60多年。"东北石油大学教授吴文祥回忆。

由于自小英语流畅、口才出众，王德民在新中国石油国际交流中被推到前台，荣获国际石油学会"三次采油先锋奖"等诸多奖项，也多次收到出国工作的邀请，而他"根本没动过这个念头"。"我的成就都是中国的。"他说，"正因为中国还不富裕，才更需要我。"

International Petroleum Association, and he was also frequently invited to work abroad. However, he "never even considered it." "All my achievements are China's," he said, "It's precisely because China is not yet prosperous that I am needed even more."

Daqing was the place where Wang Demin vowed to spend his life. After retiring, he could have returned to Beijing, the place where he grew up, to live out his days with his brothers and sisters. But he couldn't bear to leave Daqing because, "which city has such a good oilfield that I can study at any time?" As a former sports champion in his youth, Wang Demin now turned scientific research into an endless sprint, striving to outrun the aging of the oilfield. "Quaternary oil recovery, green oil extraction, and a lot of things are waiting," he said, looking at the distant oilfield. Turning to the journalist, he asked, "You think, can I ever stop?"

(Guangming Daily / September 21, 2021 / Page 001)

大庆，是王德民许下一生的地方。退休后，他本可以回到北京，回到从小长大的地方，与哥哥姐姐等亲人相伴终老。但他舍不得离开大庆，因为，"哪个城市有这么好的大油田，让我随时去研究？"少年时代，王德民曾在北京市中学生短跑比赛中获奖。这位昔日的运动健将，如今把科研变成了一场没有终点的竞速跑，拼尽全力，也要跑赢油田的衰老速度。"四次采油、绿色采油，还有一堆事情等着啊！"望着远处的油田，王德民问记者："你说，我能停下来吗？"

（光明日报/2021年/9月/21日/第001版）

# Chapter 3  Well Cementing Design

## 3.1 Characteristics of Cementing

### 3.1.1 Characteristic

(1) Cementing operations are one-time projects that are generally difficult to remedy if the quality is poor.

(2) Cementing operation is a hidden project, the main process in the well, the construction can not be directly observed, the quality of cementing is affected by a variety of factors, depending on the accuracy of the design of cementing and quality control in the construction process.

(3) Impact on oil and gas field development and subsequent engineering. If the quality of cementing is not good, it may cause interlayer flux during the development process (especially water injection process), which will have serious impact on the normal development of oil and gas field.

(4) Cementing is a costly project.

(5) Short construction time and more processes.

### 3.1.2 Purpose of Cementing and Requirements for Cementing Quality

#### 3.1.2.1 Purpose of Cementing

(1) Seal off complex formations such as

# 第三章 固井工程设计

## 3.1 固井

### 3.1.1 特点

（1）固井作业是一次性工程，如果固井质量不好，一般情况下难以补救。

（2）固井作业是隐蔽性工程，主要流程在井下，施工时不能直接观察，固井质量受多种因素的综合影响，取决于固井设计的准确性和施工过程中的质量控制。

（3）对油气田的开发和后续工程产生影响。若固井质量不好，在开发过程中（特别是注水过程）可能造成层间窜通，对油气田的正常开发造成严重影响。

（4）固井是一项费用高的工程。

（5）施工时间短，工序多。

### 3.1.2 固井的目的及对固井质量的要求

#### 3.1.2.1 固井的目的

（1）封隔易塌、易漏等复杂地层，保证

those prone to collapse and leakage to ensure smooth drilling.

(2) Seal off the oil, gas and water layers, establish oil and gas flow out channels, and prevent interflow between production layers.

(3) Strengthening the wellbore to extend the lifespan of oil and gas wells.

(4) Installing wellhead equipment.

#### 3.1.2.2 Basic Requirements for Cementing Quality

(1) The return depth of the cement slurry and the height of the cement plug inside the casing must conform to the design requirements.

(2) All drilling fluid in the annular space of the cementing section must be displaced by the cement slurry, with no residual fluid.

(3) The cement has sufficient bond strength and a certain toughness between the casing, wellbore wall rock, and can withstand perforation, acid fracturing, and the impact of downhole pipe strings.

(4) There is no channeling after the cement has set, with no oil, gas, or water venting outside the pipe, and fluids in different pressure systems within the annulus cannot intermix.

(5) The cement can withstand long-term erosion in oil, gas, water and corrosive environments.

### 3.1.3 Cementing Construction Design

Cementing construction design cementing is the last link in each stage of drilling. Due to the inherent uncertainty in drilling, the size of the drilled wellbore is always variable, leading to uncertainties in cement slurry volumes, casing centralization, etc., which can result in discrepancies between the actual cementing operation and the design. Therefore, after each

钻井顺利进行。

（2）封隔油气水层，建立油气流出通道，防止产层间互窜。

（3）加固井筒，延长油气井寿命。

（4）安装井口装置。

#### 3.1.2.2 固井质量的基本要求

（1）水泥浆返深和套管内水泥塞高度必须符合设计要求。

（2）注水泥井段环形空间内的钻井液全部被水泥浆替走，不存在残留现象。

（3）水泥石与套管及井壁岩石之间有足够的胶结强度和一定的韧性，能经受住射孔、酸化压裂及下井管柱的冲击。

（4）水泥凝固后无窜槽现象，管外不冒油、气、水，环空内不同压力体系内的流体不能互窜。

（5）水泥石能经受油、气、水及腐蚀环境下的长期侵蚀。

### 3.1.3 固井施工设计

固井是各阶段钻井的最后一个环节。由于钻井中总是存在一定的不确定性，钻成的井眼尺寸总是不断变化，这带来水泥浆量、套管扶正等不确定性，使固井施工时与设计时产生差异。因此在各阶段钻完进尺并电测后，应进行固井施工设计。

stage of drilling is completed and electric logging is performed, cementing construction design should be conducted.

3.1.3.1 Cementing Construction Design

Cementing Engineer Before the drilling reaches the casing cementing stage, the basic information of drilling design and construction is collected. The main collected information includes: cementing design and requirements, expected depth of casing, drilling fluid density and other performance, complexities and accidents encountered in drilling, electric logging well diameter, electric logging well deviation angle and azimuth angle, electric logging formation temperature, lithology logging, comprehensive logging, etc., comprehensively analyze the risks of casing and cementing.

Well deviation angle and azimuth angle data are the basis for determining the spacing of casing centralizers. Wellbore diameter data is the basis for calculating the cement slurry volume. Drilling fluid losses encountered during drilling may be critical for the cement slurry to reach the designed return depth. Sticking conditions may affect the casing running. The density of the cement slurry is determined based on the formation pressure and actual construction conditions, and the specific formula of the cementing cement slurry is determined based on the formation temperature and construction time.

3.1.3.2 Cement Slurry Formulation and Material Design

The specific density of the cement slurry is designed based on the drilling fluid density used during drilling, the loss of drilling fluid, etc. On this basis, a specific cement slurry formula is designed according to the design

3.1.3.1 固井施工条件

固井工程师在钻达下套管固井前，收集钻井设计与施工的基本情况资料，主要包括：固井设计与要求，预计下套管井深、使用钻井液密度与其他性能、钻井中遇到的复杂与事故、电测井径、电测井斜角与方位角、电测地层温度、岩性录井、综合录井等资料，全面分析下套管与固井的风险。

井斜角、方位角数据是确定套管扶正器安放间距的依据。井径数据是计算水泥浆量的依据。钻井过程中出现的井漏可能对水泥浆能否达到按设计上返非常关键。阻卡情况则对套管下入产生影响。根据地层压力和实际施工情况确定水泥浆的密度，根据地层温度与施工时间确定固井水泥浆的具体配方。

3.1.3.2 水泥浆配方与材料设计

根据钻井过程中使用钻井液密度情况、钻井中漏失情况等设计水泥浆的具体密度，在此基础上按设计封固要求设计出具体的水泥浆配方。再取现场的水样、施工所用批次的水泥样，按地层温度（井底循环温

sealing requirements. Then, the cement slurry thickening and rheological properties tests are conducted at the site with the field water samples and the cement samples of the batches used in construction, according to the formation temperature (bottom hole circulation temperature) and pressure. The cement slurry is required to have a setting time longer than the construction time by 1 hour under the formation temperature and pressure conditions. Furthermore, the slurry should not thicken when mixed with drilling fluid, and the setting time should not significantly change even when the dosage of additives changes or the density of the cement slurry varies by no more than $0.02g/cm^3$, still meeting the construction requirements.

After the cement slurry formula is determined, the amount of cement slurry and the required amount of various additives are calculated based on the electric logging data. If dry mixing is required, it should be thoroughly and uniformly mixed with the cement in advance. Additives that are not suitable for dry mixing should be prepared as a gel liquid at the site before construction, to facilitate the cementing construction operation.

### 3.1.3.3 Casing Strength Checking

The drilling fluid density used during drilling may differ from the design provided by the client, so before cementing, it is necessary to recheck the casing strength based on the specific depth of the casing to be run and the actual drilling fluid density used. This is to avoid the casing strength not meeting the requirements due to changes in well depth and drilling fluid density.

In directional and horizontal wells, the bending stress should be considered during the casing strength check to avoid the casing

度）、压力进行水泥浆稠化与流变性试验，要求水泥浆在地层温度、压力条件下凝固时间大于施工所用时间1小时以上。并且在水泥浆与钻井液混合时不稠化，在处理剂加量变化、水泥浆密度变化不超过 $0.02g/cm^3$ 情况下，老化时间不产生显著变化，仍能满足施工要求。

在水泥浆配方确定后，再根据电测数据计算水泥浆用量和各种处理剂的需求量。如果需要干混，则提前与水泥进行充分均匀的混拌；而不宜干混的添加剂，在施工前在现场配成胶液，以便于固井施工作业。

### 3.1.3.3 套管强度校核

钻井过程中使用的钻井液密度可能与甲方钻井工程设计值有差异，因此在固井前要根据具体下套管井深和实际使用钻井液密度，对套管强度进行再次校核，以避免由于井深与使用钻井液密度变化而导致套管强度不满足要求。

在定向井与水平井中，应根据具体的井斜情况，在校核套管强度时，应考虑弯曲应力的影响，以避免由于过大的弯曲应力导致套管强度不满足要求。

strength not meeting the requirements due to excessive bending stress.

#### 3.1.3.4 Casing String Design

According to the conditions of the casing string, the casing lower structure, and other string accessories, and considering the control of casing suspension during cementing, the casing string design is carried out. When designing the casing string, the following issues should be noted:

(1) Starting from the perspective of improving cementing quality and meeting exploration and development requirements, evaluate whether to use annulus packers, and if so, design the required quantity and specific placement positions;

(2) Design float collars, float shoes (introduction shoes), and matching short joints. For possible perforation locations, sufficient standard short joints should be designed for magnetic positioning during perforation;

(3) According to the well deviation and wellbore enlargement, rationally design the types, quantities, and placement positions of casing centralizers to ensure the centralization of the casing during cementing;

(4) Other cementing tools and accessories should be designed based on specific construction conditions;

(5) Design the measures for casing running, including casing cleaning, measurement, inspection, and numbering. After the casing is on the derrick, the requirements for casing coupling and tightening operations should be designed. For high-pressure gas wells, the casing threads should also be checked for gas sealing after tightening;

(6) List the serial numbers of the casing strings into the well and the remaining casing

#### 3.1.3.4 套管串设计

根据送井套管与套管下部结构及其他管串附件情况，考虑固井时套管悬空控制，进行套管串设计。设计套管串需注意以下问题：

（1）从提高固井质量，满足勘探开发要求出发，评价是否需要采用管外封隔器，如果需要，应设计加量以及具体安放位置；

（2）设计浮箍、浮鞋（引鞋）及配合短节，在可能的射孔位置还应设计足够的标准短节，以便于射孔时的磁性定位；

（3）根据井斜情况及井径扩大情况，合理设计套管扶正器的类型、数量与安放位置保证固井时套管的居中度；

（4）其他固井工具与附件根据具体施工条件进行设计；

（5）设计套管下入措施，包括套管的清洗、丈量、检查、编号，上钻台后的对扣、上扣操作要求，高压气井套管螺纹上扣后还应进行气密封检查；

（6）列出入井套管串的编号和井场剩余套管数量等，避免下错套管。

quantities on the well site to avoid running the wrong casing.

### 3.1.3.5 Cementing Construction Measures Design

Designing specific construction measures to ensure the success of cementing operations and to enhance cementing quality. This includes:

3.1.3.5.1 Wellbore Preparation Measures

Wellbore preparation measures involves pressure sealing before casing installation, treatment of drilling fluid and wellbore cleaning tools before casing installation. Ensuring no leakage during cementing is a prerequisite for effective sealing of the sections that need to be cemented. Although leak-prevention cement slurry systems are currently available, these systems do not fundamentally solve the problem of wellbore leakage during cementing. The primary conditions for ensuring that the cement slurry returns to the specified well depth is that the formation pressure capacity meet the requirements. If regional experience indicates that the cement slurry density is higher than the drilling fluid density and leakage occurs, plugging must be performed to ensure that the formation pressure capacity meet the requirements of cementing operations. The pressure sealing plugging test must clearly specify the drilling fluid density and wellhead pressure requirements that should be achieved. To ensure that the wellbore is clear, the casing must be safely run to the predetermined location. If the electric logging takes a long time, drilling should be resumed for wellbore cleaning. During wellbore cleaning, the original drilling tool string should be used by default. However, if the stiffness of the drilling tool string is less than that of the casing, the stiffness of the bottom hole assembly should be increased. If encountering resistance

### 3.1.3.5 固井施工措施设计

提出确保固井施工成功并提高固井质量的具体施工措施。

3.1.3.5.1 井眼准备措施

井眼准备包括下套管前的承压堵漏、下套管前处理钻井液与通井工具等。固井时保证不漏是保证水泥有效封固需要封固井段的前提。虽然目前也出现了防漏水泥浆体系，但这种体系并不能从根本上解决固井时的井漏问题。保证水泥浆上返到规定的井深的首要条件就是地层承压能力能满足要求。如果地区经验表明水泥浆密度高于钻井波密度就会发生漏失，则必须进行堵漏，确保地层承压能力满足固井施工要求。承压堵漏试验必须明确试验应达到的钻井液密度与井口承压要求。确保井眼畅通才能保证套管安全下到预定位置。如果电测时间较长，应下钻进行通井。通井时原则上采用原钻具通井，但如果钻具刚度小于套管刚度时，则应提高底部钻具组合的刚度。通井时如果遇阻，应提出划眼措施。为了提高固井质量，下套管前必须处理钻井液。为提高顶替效率，应适当降低钻井液的黏度与切力。同时需要提出循环洗井要求，以尽可能冲洗掉井壁上的虚滤饼。

during wellbore cleaning, the drill string should be pulled up for reaming. To improve cementing quality, the drilling fluid must be treated before casing installation. To enhance displacement efficiency, the viscosity and shear stress of the drilling fluid should be appropriately reduced. Additionally, circulation cleaning requirements should be proposed to clean off as much of the loose filter cake from the wellbore walls as possible.

3.1.3.5.2 Casing Running Measures

Casing running measures proposing technical measures for casing running to ensure the casing is safely run to the predetermined location. This including requirements for casing lower structure connection, thread lubrication, thread connection, lowering procedures, and countermeasures for encountering resistance.

3.1.3.5.3 Cementing Grouting Measures

Cementing grouting measures proposing requirements for the on-site mixing of cement slurry, rheological design of cement injection based on cement slurry performance data, and designing the specific slurry column structure and injection parameters, including the design of spacer fluid, flush fluid, lead slurry and tail slurry. Monitoring requirements for cement slurry performance and displacement volume should be proposed during construction to ensure strict adherence to the design and prevent voids.

3.1.3.5.4 Cementing Construction Equipment Assurance Measures

Cementing equipment assurance measures designing cementing equipment, monitoring instruments, and meters to ensure successful cementing operations.

3.1.3.5.2 套管下入措施

提出下套管的技术措施，保证套管安全下到预定位置。具体包括下部结构接入要求套管入井前现场检查、螺纹脂要求、上扣要求、下放措施、遇阻后的应对措施等内容。

3.1.3.5.3 固井注水泥施工措施

提出对水泥浆现场混配要求，根据水泥浆性能数据进行流变学注水泥设计，设计具体的水泥浆的浆柱结构及注替参数，包括前置液、冲洗液、领浆与尾浆设计。施工过程中应提出水泥浆性能和替浆量监测要求，保证施工严格按设计进行，防止替空。

3.1.3.5.4 固井施工设备保证措施

设计固井施工的设备、监测仪器与仪表，以保证成功进行固井施工作业。

## 3.2 Casing String Design

Casing string design is one of the key contents of cementing engineering design. Casing string design includes casing string structure design, casing column strength design, lower casing design, casing support device placement design and so on. For tail pipe cementing, there are also feeding string design and calculation, and for inner pipe cementing, there are also inner pipe string design and so on.

### 3.2.1 Structural Design Of Casing Strings

The casing string structure includes the casing and accessories, i.e., pilot shoe (floating shoe), floating hoop, support ring, casing, outer pipe packer, grading hoop, suspension, feeder drilling tool, and inner pipe column.

The main types of casing string structures are shown in Table 3.1, and other further refinements of casing string structures are basically variations thereof.

## 3.2 套管串设计

套管串设计是固井工程设计的关键内容之一。套管串设计包括套管串结构设计、套管柱强度设计、下套管设计、套管扶正器安放设计等。对于尾管固井，还要有送入管串设计及计算，对于内管注水泥固井，还要有内管串设计等。

### 3.2.1 套管串结构设计

套管串结构包括套管及附件，即引鞋（浮鞋）、浮箍、承托环、套管、管外封隔器、分级箍、悬挂器、送入钻具、内管柱等。

主要套管串结构类型见表3.1，其他进一步的细化套管串结构基本由表3.1延伸变化而来。

Table 3.1　Type of Casing String Structure

| Symbol | Cementing method | Casing string configuration |
|---|---|---|
| 1 | conventional cementing | Introduction shoe (float shoe) + Casing (15m or more)+ Float collar (suspension ring)+Casing (+ annulus packer) (+ top joint) |
| 2 | inner string cementing | Casing string: Introduction shoe + Casing (15m or more) + Float collar(Sleeve) + Casing(+ top joint)<br>Inner string: Cementing plug + Compatible joint (+ inner casing centralizer) + Drill pipe + adaptor + Square drill pipe (or top drive) |
| 3 | double-stage cementing | Introduction shoe (float shoe)+ Casing (15m or more)+ Float collar+ suspension ring+ Casing (+ annulus packer)+ Staged cementing device+ Casing(+ top joint) |
| 4 | tail-end cementing | Introduction shoe (float shoe) + Casing (15m or more)+ Float collar+ Seated coupling + casing (+ annulus packer) + tubing hanger (+ back-up sub) + running tools (+ short drill pipe) |
| 5 | tail-string casing back-cementing | Back-up sub+ Casing (15m or more)+ Float collar (suspension ring)+Casing (+ annulus packer) (+ top joint) |
| 6 | perforated well casing cementing | Introduction shoe + screen pipe + blind flange + Casing (15m or more)+ annulus packer + staged cementing casing (+ annulus packer) (+ tieback joint) (+ tubing hanger + running tools) |

表 3.1 套管串结构类型

| 序号 | 注水泥方式 | 套管串结构类型 |
| --- | --- | --- |
| 1 | 常规注水泥 | 引鞋（浮鞋）+ 套管（15m 以上）+ 浮箍（承托环）+ 套管（+ 管外封隔器）（+ 联顶节） |
| 2 | 内管注水泥 | 套管串：引鞋 + 套管（15m 以上）+ 浮箍（插座）+ 套管（+ 联顶节）<br>内管串：注水泥插头 + 配合接头（+ 内管扶正器）+ 钻杆 + 转换接头 + 方钻杆（或顶驱） |
| 3 | 双级注水泥 | 引鞋（浮鞋）+ 套管（15m 以上）+ 浮箍 + 承托环 + 套管（+ 管外封隔器）+ 分级注水泥器 + 套管（+ 联顶节） |
| 4 | 尾管注水泥 | 引鞋（浮鞋）+ 套管（15m 以上）+ 浮箍 + 坐落接箍 + 套管（+ 管外封隔器）+ 尾管悬挂器（+ 回接筒）+ 送入钻具（+ 短钻杆） |
| 5 | 尾管回接注水泥 | 回接插管 + 套管（15m 以上）+ 浮箍（承托环）+ 套管（+ 管外封隔器）（+ 联顶节） |
| 6 | 贯眼完井注水泥 | 引鞋 + 筛管 + 盲板 + 套管（15m 以上）+ 管外封隔器 + 分级注水泥器套管（+ 管外封隔器）（+ 联顶节）（+ 尾管悬挂器 + 送入钻具） |

### 3.2.2 Casing Column Strength Design

Casing column design and selection is an important element of drilling completion preparation. The importance and significance of casing column optimization design is mainly manifested in the following two aspects:

(1) After a well is completed, the only things that ultimately remain in the well are the casing column and the cement ring, and the casing column is subjected to many years and a variety of severe working conditions;

(2) Casing costs often account for 15% ~ 35% of the total cost composition of most wells, so optimizing casing column design can greatly reduce the total cost of oil and gas wells.

Strength design of casing columns at home and abroad includes calculation methods of casing strength, analysis methods of loads acting on casing columns and optimization design methods of casing columns.

Casing strength design needs to be carried out in accordance with the specific requirements of the relevant standards and norms, the software used must be tested and certified by authoritative organizations, otherwise the designers need to take responsibility for the defects in the design.

### 3.2.2 套管柱强度设计

套管柱设计和选择是钻井完井准备的一个重要内容。套管柱优化设计的重要性和意义主要表现在以下两个方面：

（1）一口井完成后，最终留在井内的只有套管柱和水泥环，套管柱要承受多年的、各种严酷的工作条件；

（2）在大多数油井的总成本构成中，套管成本往往占到15%~35%，因此，优化套管柱设计可以大幅减少油气井的总成本。

国内外套管柱强度设计包括套管强度的计算方法、作用在套管柱上的载荷的分析方法和套管柱优化设计方法。

套管强度设计需按有关标准规范的具体要求进行，采用的软件必须经权威机构测试认证，否则设计人员需对设计出现的缺陷承担责任。而一旦软件计算出现错误，可能将产生严重的后果甚至造成严重的安全事故。

And once the software calculation error, may have serious consequences or cause serious safety accidents.

American Petroleum Institute (API) using theoretical analysis and experimental modification of the method, research a complete set of casing strength of various strength formula, including casing body and joints axial tensile strength, body and joints of the strength of the internal pressure and the body of the strength of the extrusion. For the extrusion strength of pipe body, API according to the casing extrusion damage form according to the diameter and thickness of the score is divided into four calculations. At present, API casing industry standard system has been widely adopted by China's petroleum industry standards (SY/T 5724—2008 *casing column structure and strength design*).

3.2.2.1　Calculation Method of Casing Strength

Casing strength data can usually be obtained from the following ways: one is from the relevant manual, calculation software database, product catalog and other channels to query; the other is based on steel grade, wall thickness, test data and other calculations. At present, many manuals and calculation software databases have the strength data of API standard casing, and the product catalog of casing manufacturers also provides the strength data of API standard and non-API standard casing produced. For API standard and non-API quasi-casing strength data, it can also be calculated according to its steel grade, wall thickness and other parameters. It should be noted that because the casing is subject to axial and radial combined stress, the influence of biaxial stress must be considered in strength calculation.

美国石油学会（API）采用理论分析和实验修正的方法，研究出一整套有关套管各种强度的计算公式，其中包括套管管体和接头轴向抗拉强度、管体和接头的抗内压强度和管体的抗挤强度。对于管体的抗挤强度，API 根据套管挤压破坏形式按径厚比分为四段计算。目前 API 有关套管的行业标准体系已被我国广泛采用为石油行业标准（SY/T 5724—2008《套管柱结构与强度设计》）。

3.2.2.1　套管强度计算方法

套管强度数据通常可以从以下途径获得：一是从有关手册、计算软件数据库、产品目录等渠道进行查询；二是根据钢级、壁厚、试验数据等进行计算得到。目前很多手册、计算软件数据库均为 API 标准套管的强度数据，套管生产厂家的产品目录也提供所生产的 API 标准及非 API 标准套管的强度数据。对于 API 标准及非 API 准套管强度数据，也可以根据其钢级、壁厚等参数计算得出。需要注意的是，由于套管受轴向与径向复合应力作用，强度计算必须考虑双轴应力的影响。

For special casing, the strength data published by the manufacturer can be used, but the strength standard should be explained in the design and where the strength data come from, so as to prompt drilling personnel to pay attention. Because each casing performance data obey the random distribution law, casing manufacturers must set their own internal control indicators to ensure that their casing 100% reaches the specified data, usually the internal control indicators are higher than the specified data 5%~10%, so as to ensure that each casing performance below the specified value is less than one in 10000 or even lower.

### 3.2.2.2 Calculation Method of Casing Load

The loads on the casing mainly include internal pressure, external compression, and axial force, and the specific calculation methods are as follows:

#### 3.2.2.2.1 Effective Internal Pressure

For the surface casing and technical casing/tail-string of gas wells, the maximum internal pressure at the casing shoe is calculated based on the maximum drilling fluid density used for the next operation,

$$p_{bs} = 0.00981 \rho_{max} H_s \quad (3-1)$$

$p_{bs}$—The maximum internal pressure at the casing shoe, MPa;

$\rho_{max}$—Maximum drilling fluid density for the next drilling operation, g/cm³;

$H_s$—Casing depth, m.

$$p_{bh} = \frac{p_{bs}}{e^{1.1155 \times 10^{-4}(H_s - h)\gamma_g}} \quad (3-2)$$

$p_{bh}$—Maximum internal pressure at any depth in the wellbore, MPa;

$\gamma_g$—Any depth in the wellbore, m;

对于特殊套管可以使用厂家公布的强度数据，但需在设计中说明其强度应达到的标准及强度数据来自何处，以便提示钻井施工人员注意。由于套管性能数据服从随机分布规律，因此套管生产厂商为保证自己的套管100%达到规定的数据，必须设定自己的内控指标，通常内控指标要高于规定数据5%~10%以上，这样才能保证每一根套管性能低于规定值的可能性小于万分之一甚至更低。

### 3.2.2.2 套管柱载荷计算方法

套管所受载荷主要包括内压、外挤与轴向力，具体载荷计算方法如下：

#### 3.2.2.2.1 有效内压力

对于气井的表层套管和技术套管与技术尾管按下一次使用的最大钻井密度计算套管鞋处的最大内压力，即：

$$p_{bs} = 0.00981 \rho_{max} H_s \quad （3-1）$$

式中 $p_{bs}$——套管鞋处的最大内压力，MPa；

$\rho_{max}$——下次开钻最大钻井液密度，g/cm³；

$H_s$——套管下入深度，m。

$$p_{bh} = \frac{p_{bs}}{e^{1.1155 \times 10^{-4}(H_s - h)\gamma_g}} \quad （3-2）$$

式中 $p_{bh}$——任意井深处套管最大内压力，MPa；

$\gamma_g$——任意井深，m；

$h$—Natural Gas Relative Density, 0.50~0.55.

The effective internal pressure is calculated using the following formula:

$$p_{be} = p_{bh} - 0.00981\rho_c h \quad (3-3)$$

or

$$p_{be} = 0.00981\left[\frac{\rho_{max}H_s}{e^{1.1155\times10^{-4}(H_s-h)\gamma_g}} - \rho_c h\right] \quad (3-4)$$

$p_{be}$—Effective internal pressure at any depth in the wellbore, MPa;

$\rho_c$—Formation Water Density, 1.03 ~ 1.06g/cm³.

For the production casing and production tail-string of a gas well, considering the casing fully filled with natural gas, the maximum internal pressure at the bottom of the well is:

$$p_{bs} = 0.00981\rho_m H_s \quad (3-5)$$

$p_{bs}$—Maximum internal pressure at the bottom of the well, MPa;

$\rho_m$—Drilling fluid density during cementing, g/cm³.

The maximum internal pressure at any depth in the wellbore is calculated by Equation (3-2), but according to Equation (3-5).

$$p_{bh} = \frac{p_{bs}}{e^{1.1155\times10^{-4}(H_s-h)\gamma_g}}$$

The effective internal pressure is calculated by Equation (3-3), but according to Equation (3-5).

$$p_{be} = p_{bh} - 0.00981\rho_c h$$

For the surface casing and technical casing/tail-string of an oil well, the maximum internal pressure at any depth in the wellbore is:

$$p_{be} = p_{bh} - 0.00981\rho_{max} h \quad (3-6)$$

$h$——天然气相对密度，0.50~0.55。

有效内压力用下式进行计算：

$$p_{be} = p_{bh} - 0.00981\rho_c h \quad （3-3）$$

或

$$p_{be} = 0.00981\left[\frac{\rho_{max}H_s}{e^{1.1155\times10^{-4}(H_s-h)\gamma_g}} - \rho_c h\right] \quad （3-4）$$

式中　$p_{be}$——任意井深处套管有效内压力，MPa；

$\rho_c$——地层水密度，1.03 ~ 1.06g/cm³。

对于气井的生产套管和生产尾管按套管内全充满天然气考虑，井底最大内压力为：

$$p_{bs} = 0.00981\rho_m H_s \quad （3-5）$$

式中　$p_{bs}$——井底最大内压力，MPa；

$\rho_m$——固井时钻井液密度，g/cm³。

任一井深处的最大内压力按式（3-2）计算，只是$p_{bs}$按式（3-5）计算，即：

$$p_{bh} = \frac{p_{bs}}{e^{1.1155\times10^{-4}(H_s-h)\gamma_g}}$$

有效内压力按式（3-3）计算，只是$p_{bs}$按式（3-5）计算，即：

$$p_{be} = p_{bh} - 0.00981\rho_c h$$

对于油井表层套管和技术套管与技术尾管，任一井深处的套管最大内压力：

$$p_{be} = p_{bh} - 0.00981\rho_{max} h \quad （3-6）$$

The effective internal pressure is calculated using the following formula:

$$p_{be} = p_{bh} - 0.00981\rho_c h \quad (3-7)$$

$$p_{be} = 0.00981(\rho_{max} - \rho_c)h \quad (3-8)$$

For the production casing and production tail-string of an oil well, if the well is not produced with tubing, the maximum internal pressure is calculated using the following formula:

$$p_{bs} = G_p H_s \quad (3-9)$$

$G_p$—Oil or formation pressure gradient, MPa/m.

The maximum internal pressure at any depth in the wellbore is calculated using the following formula:

$$p_{bh} = \frac{p_{bs}}{e^{1.1155\times10^{-4}(H_s-h)\gamma_g}} = \frac{G_p H_s}{e^{1.1155\times10^{-4}(H_s-h)\gamma_g}}$$

For the production casing and production tail-string of an oil well that is produced with tubing, the maximum internal pressure is calculated using the following formula:

$$p_{bh} = G_p H_s + 0.00981\rho_w h \quad (3-10)$$

$\rho_w$—Cementing fluid density during well completion, g/cm³.

Effective internal pressure:

$$p_{be} = p_{bh} - 0.00981\rho_c h \quad (3-11)$$

The effective internal pressure in a directional well should be calculated after converting the measured depths of the curved and inclined sections to vertical depths.

3.2.2.2.2 Effective External Compression Pressure

For the surface casing, technical casing, and technical tail-string in a non-plastic creep

有效内压力用下式计算：

$$p_{be} = p_{bh} - 0.00981\rho_c h \quad （3-7）$$

$$p_{be} = 0.00981(\rho_{max} - \rho_c)h \quad （3-8）$$

对于油井的生产套管和生产尾管不用油管生产的用下式计算最大内压力：

$$p_{bs} = G_p H_s \quad （3-9）$$

式中 $G_p$——油层或地层压力梯度，MPa/m。

任一井深处的最大内压力，用下式计算：

$$p_{bh} = \frac{p_{bs}}{e^{1.1155\times10^{-4}(H_s-h)\gamma_g}} = \frac{G_p H_s}{e^{1.1155\times10^{-4}(H_s-h)\gamma_g}}$$

对于油井的生产套管和生产尾管用油管生产的用下式计算最大内压力：

$$p_{bh} = G_p H_s + 0.00981\rho_w h \quad （3-10）$$

式中 $\rho_w$——固井时完井液密度，g/cm³。

有效内压力：

$$p_{be} = p_{bh} - 0.00981\rho_c h \quad （3-11）$$

定向井有效内压力应将弯曲段和斜直段的测量井深换算为垂直井深后再进行计算。

3.2.2.2.2 有效外挤压力

对非塑性蠕变地层表层套管和技术套管与技术尾管，有效外挤压力用下式计算：

formation, the effective external compression pressure is calculated using the following formula:

$$p_{ce} = 0.00981[\rho_m - (1-k_m)\rho_{min}]h \quad (3\text{-}12)$$

$p_{ce}$—Effective external compression pressure at any depth in the wellbore, MPa;

$\rho_{min}$—Minimum drilling fluid density for the next drilling operation, g/cm³;

$k_m$—Pumpout factor (ranging from 0 to 1, with 1 indicating full pumpout).

For the surface casing, technical casing, and technical tail-string in a plastic creep formation, the effective external compression pressure is calculated using the following formula:

$$p_{ce} = \left[\frac{v}{1-v}G_v - 0.00981(1-k_m)\rho_{min}\right]h \quad (3\text{-}13)$$

$v$—Formation rock Poisson's ratio;

$G_v$—Overburden pressure gradient.

For the production casing and production tail-string in a non-plastic creep formation, the effective external compression pressure is calculated using the following formula:

$$p_{ce} = 0.00981[\rho_m - (1-k_m)\rho_w]h \quad (3\text{-}14)$$

For the production casing and production tail-string in a plastic creep formation, the effective external compression pressure is calculated using the following formula:

$$p_{ce} = \left[\frac{v}{1-v}G_v - 0.00981(1-k_m)\rho_w\right]h \quad (3\text{-}15)$$

The effective external pressure in a directional well should be calculated after converting the measured depths of the curved and inclined sections to vertical depths.

$$p_{ce} = 0.00981[\rho_m - (1-k_m)\rho_{min}]h \quad (3\text{-}12)$$

式中 $p_{ce}$——任意井深处套管有效外挤压力，MPa；

$\rho_{min}$——下次钻井最小钻井液密度，g/cm³；

$k_m$——掏空系数（取0~1，取1时为全掏空）。

对塑性蠕变地层表层套管和技术套管与技术尾管，有效外挤压力用下式计算：

$$p_{ce} = \left[\frac{v}{1-v}G_v - 0.00981(1-k_m)\rho_{min}\right]h \quad (3\text{-}13)$$

式中 $v$——地层岩石泊松比（0.3~0.5）；

$G_v$——上覆岩层压力梯度，MPa/m。

对非塑性蠕变地层生产套管和生产尾管，有效外挤压力用下式计算：

$$p_{ce} = 0.00981[\rho_m - (1-k_m)\rho_w]h \quad (3\text{-}14)$$

对塑性蠕变地层生产套管和生产尾管，有效外挤压力用下式计算：

$$p_{ce} = \left[\frac{v}{1-v}G_v - 0.00981(1-k_m)\rho_w\right]h \quad (3\text{-}15)$$

定向井有效外压力应将弯曲段和斜直段的测量井深换算为垂直井深后再进行计算。

3.2.2.2.3 Effective Axial Force

The effective axial force in the well is calculated using the following formula:

$$T_e = \sum_{i=1}^{n} \Delta L_i q_i \left(1 - \frac{\rho_m}{\rho_s}\right) \quad (3\text{-}16)$$

$T_e$—Effective axial force of a casing string segment, kN;

$L_i$—Length of the $i$th casing string segment, m;

$q_i$—Weight of the $i$th casing string segment, kN/m;

$\rho_s$—Steel density, g/cm³.

The effective axial force in a directional well needs to be obtained through friction calculations for different working conditions and well segments.

3.2.2.2.4 Calculation of Pipe Bending Force

Bending force is an additional axial force. In directional and horizontal wells, the axial force in the upper well section should be further increased by the bending force, while the lower well section should consider the additional compressive force caused by the bending force. Calculation formula:

$$F_b = 2.32 \times 10^{-3} D_c q_c \theta \quad (3\text{-}17)$$

$F_b$—Casing bending force, kN;

$q_c$—Casing string mass, kg/m;

$\theta$—Wellbore full-angle change rate.

3.2.2.3 Casing Design Calculation Method

The purpose of casing design is to design a casing string that meets the operational requirements of the well throughout its life cycle (surface casing, technical casing only for the next layer before installation) under the most economic conditions. In the design of cementing engineering, casing strength design

3.2.2.2.3 有效轴向力

井的有效轴向力用下式进行计算：

$$T_e = \sum_{i=1}^{n} \Delta L_i q_i \left(1 - \frac{\rho_m}{\rho_s}\right) \quad （3\text{-}16）$$

式中 $T_e$——段套管有效轴向力，kN；
$L_i$——第 $i$ 段套管长度，m；
$q_i$——第 $i$ 段套管线重量，kN/m；
$\rho_s$——钢材密度，g/cm³。

定向井的有效轴向力需要通过摩阻计算来获得不同工况不同井段的套管轴向力。

3.2.2.2.4 管曲力计算

弯曲力是附加的轴向力，在定向井与水平井中，上部井段轴向力应再加上弯曲力，而下部井段应考虑弯曲力引起的附加压缩力。计算公式：

$$F_b = 2.32 \times 10^{-3} D_c q_c \theta \quad （3\text{-}17）$$

式中 $F_b$——套管弯曲力，kN；
$q_c$——套管线质量，kg/m；
$\theta$——井眼全角变化率。

3.2.2.3 套管设计计算方法

套管设计的目的是在最经济的条件下设计出满足油井生命期（表层套管、技术套管只针对下入下一层套管前）使用要求的套管柱。在固井工程设计中，套管强度设计计算分套管强度校核及套管优化设计两种。

calculation is divided into two types: casing strength checking and casing optimization design.

Casing strength checking involves checking the strength of the selected casing string based on formation pressure data, and providing its safety factor for comparison with the required safety factor. This is currently the most commonly used method by cementing designers or engineers after the casing string has been basically selected by the engineering construction unit for development wells.

Three-axis stress intensity safety factor of the casing string:

(1) Eccentricity coefficient: $S=1.00\sim1.125$;

(2) Anti-internal pressure coefficient: $S=1.05\sim1.15$;

(3) Anti-tension coefficien: $S=1.60\sim2.00$.

Casing optimization design involves selecting the casing with the lowest steel grade and thinnest wall thickness that meets the design safety factor based on the given casing dimensions and formation pressure conditions, starting from the bottom up against the strength of external compression. At the same time, the strength against tensile and internal pressure is checked. If the tensile strength or internal pressure strength does not meet the requirements, a casing with a higher steel grade or thicker wall thickness is selected, and the design is switched to the tensile strength or internal pressure strength, followed by a check against external compression strength. This continues until the design requirements are met. This method is commonly used for exploration wells, new block wells, and is carried out by drilling engineering designers or by the engineering construction unit or a designated unit by the engineering construction unit.

套管强度校核就是根据地层压力数据，对已选择的套管柱进行强度校核，给出其安全系数并与要求的安全系数进行对比。这是目前针对开发井在工程建设单位基本选定套管后固井设计人员或固井工程师最常用的方法。

套管柱三轴应力强度设计安全系数：

（1）抗挤系数：$S=1.00\sim1.125$；

（2）抗内压系数：$S=1.05\sim1.15$；

（3）抗拉系数：$S=1.60\sim2.00$。

套管优化设计就是根据给出的套管尺寸及地层压力等条件数据，先按抗挤强度自下而上先选择满足设计安全系数的最低钢级、最薄壁厚的套管，同时进行抗拉强度和抗内压强度校核，若抗拉强度或抗内压强度不能满足要求时，则选择比前一段高一钢级级或壁厚的套管，改为抗拉强度或抗内压强度设计，并进行抗挤强度校核，一直到满足设计要求为止。这是目前探井、新区块井等通常所使用的方法，由工程建设单位或工程建设单位委托的钻井工程设计人员进行。

After years of development, the current steps for casing optimization design are detailed in most textbooks, manuals, and other references. There are also many software tools available for casing optimization design and casing strength checking. Designers can generally complete this part of the work without needing to consider the detailed design steps. It is important to note that casing optimization design not only addresses the lowest cost issue but also takes into account inventory, management, and transportation concerns. It is not advisable to design too many different steel grades and wall thicknesses of casing. Generally, unless there are special requirements, each layer of casing should not have more than 3 sections, and the shortest length of each section should not be less than 500 meters.

### 3.2.3 Casing Running Design Calculation

With the increase in horizontal wells, large-offset wells, and complex wells, many wells require casing running design. This includes, but is not limited to, calculations for casing running friction, casing bending force, floating casing running, casing running feasibility analysis, casing deformation in curved sections, casing running speed calculations.

#### 3.2.3.1 Casing Running Friction Calculation

Currently, there are two types of models used for casing running friction calculations: soft rod model and hard rod model. The soft rod model can be used for most directional and horizontal wells. Friction calculations should ensure that the casing has sufficient force to be run down. If the friction exceeds the force of the casing's descent, further measures to reduce resistance should be considered.

经过多年的发展，目前套管优化设计计算步骤在大部分教材、手册等资料中均有详细的介绍，套管优化设计及套管强度校核方面的软件也很多，设计人员基本上不需要考虑详细的设计步骤即可以完成此部分工作。需要注意的是，套管优化设计不仅要考虑成本最低问题，还应考虑库存、管理、运输等方面的问题。不能设计过多的不同钢级、壁厚的套管，一般来说，除非有特殊需要，每层套管不能多于3段，每段最短长度不能小于500m。

### 3.2.3 下套管设计计算

随着水平井、大位移井、复杂井的增多，很多井需要进行下套管设计。这包括但不限于套管下入摩阻计算、套管弯曲力计算、漂浮下套管计算、套管下入可行性分析计算、弯曲井段套管变形计算和套管下入速度计算等。

#### 3.2.3.1 下套管摩阻计算

目前下套管摩阻计算所使用的模型有软杆计算模型和硬杆计算模型之分。大多数定向井、水平井可以使用软杆计算模型。摩阻计算应保证套管有足够的下行力，如果摩阻大于套管下行力，则应考虑采取进一步的减阻措施。

#### 3.2.3.2 Floating Casing Running Calculation

To overcome the difficulty of running casing in long horizontal sections of horizontal wells, a floating coupling is often added to the casing string in the horizontal section to reduce the friction between the casing and the wellbore wall.

When running casing in horizontal well, the load on the hook is equal to the difference between the casing's buoyancy and the friction of the casing being lowered. By using friction calculation software, first calculate the casing running friction and the load on the hook. Through friction calculations, if the load on the hook is very small or even less than 0, it indicates that the friction is too high and floating couplings need to be installed to reduce the friction and ensure the smooth running of the casing.

Assuming a floating coupling is installed 200m above the bottom of the well, the equivalent linear weight of this section of casing needs to be calculated and substituted for the original casing line weight in the friction calculation software to calculate the load on the hook at the wellhead. According to the size of the load on the hook at the wellhead, it can be determined whether the casing can be successfully run down; if not, the installation height of the floating coupling is increased by 100m in sequence until the calculated load on the hook at the wellhead can ensure the casing is successfully run down.

#### 3.2.3.3 Casing Running Feasibility Analysis Calculation

In directional and horizontal wells, the feasibility analysis calculation for casing running must consider two aspects: the minimum curvature radius that the casing can pass through and

#### 3.2.3.2 漂浮下套管计算

为克服长水平段水平井下套管较困难的问题，常在水平段套管串中加漂浮接箍，以降低水平段套管与井壁的摩阻。

水平井下套管时，大钩载荷等于套管浮重与下放套管摩阻之差。利用摩阻计算软件，先计算下套管摩阻及大钩载荷。通过摩阻计算，如果大钩载荷很小甚至小于0，说明摩阻太大，需要安装漂浮接箍来减小摩阻，以保证套管的顺利下入。

假设一漂浮接箍安装位置距井底200m，这段套管需要计算当量线重，并代替原摩阻计算软件中的套管线重，计算井口大钩载荷。根据井口大钩载荷的大小判断套管是否能顺利下入；如果不能，则依次向上增加100m漂浮接箍的安装高度，直到计算出的井口大钩载荷能保证套管顺利下入为止。

#### 3.2.3.3 套管下入可行性分析计算

在定向井及水平井中，套管下入可行性分析计算要从套管能通过的最小曲率半径以及套管柱与定向井相容性等两个方面来进行分析。

the compatibility of the casing string with the directional well.

Calculate the minimum curvature radius that the casing can pass through based on the casing size, steel grade, and bending strength:

$$R = \frac{ED_c}{2000Y_P}K_1K_2 \quad (3\text{-}18)$$

$R$—Allowable bending radius of the casing, m;

$E$—Elastic modulus of steel, $206 \times 10^6$ kPa;

$D_c$—Outside diameter of the casing, mm;

$Y_P$—Yield limit of steel, kPa (Only related to the steel grade);

$K_1$—Bending safety factor, dimensionless (1.2~1.25);

$K_2$—Safety factor at the thread connection, dimensionless (2.0~2.5).

Calculate the minimum curvature radius of the wellbore:

$$R_0 = \frac{180 \times 30}{\pi K} \quad (3\text{-}19)$$

$R_0$—Minimum curvature radius of the wellbore, m;

$K$—Maximum dogleg severity of the wellbore, (°)/30m。

If $R_0 \geqslant R$, Then the casing can be run down.

## 3.3 Cementing Mud and Injection Design

Cement slurry and injection replacement design includes cement slurry, antecedent fluid (flushing fluid and isolation fluid) design and injection replacement design, i.e. slurry structure design, usage design, formulation design and pressure stabilization design.

根据套管尺寸、钢级和抗弯强度计算套管能通过的最小曲率半径:

$$R = \frac{ED_c}{2000Y_P}K_1K_2 \quad (3\text{-}18)$$

式中 $R$——套管允许的弯曲半径，m;

$E$——钢材的弹性模量，$206 \times 10^6$ kPa;

$D_c$——套管外径，mm;

$Y_P$——钢材的屈服极限（只与钢级有关）;

$K_1$——抗弯安全系数，无量纲（1.2~1.25）;

$K_2$——螺纹连接处的安全系数，无量纲（2.0~2.5）。

计算井眼的最小曲率半径:

$$R_0 = \frac{180 \times 30}{\pi K} \quad (3\text{-}19)$$

式中 $R_0$——井眼最小曲率半径，m;

$K$——井眼最大狗腿度，(°)/30m。

校核：如果 $R_0 \geqslant R$，则套管可下入。

## 3.3 固井水泥浆及注替设计

水泥浆及注替设计包括水泥浆、前置液（冲洗液及隔离液）设计及注替设计，即浆体结构设计、使用量设计、配方设计和压稳设计等。

## 3.3.1 Design of Rinsing and Isolation Solutions

Flushing fluid and isolation fluid are generally referred to as pre-fluid.

### 3.3.1.1 Rinse Solution Performance Indicators

The performance specifications of the rinse solution are required as follows:

(1) Density of base liquid: $1.00\sim1.08\text{g/cm}^3$;

(2) Flow requirements: in the injection replacement displacement should be able to make the flushing liquid to turbulent flow;

(3) Thermal stability: in the cycle temperature, after 10h aging test, the performance change should not exceed 10%;

(4) Flushing rate: the flushing rate of drilling fluid is more than 95%.

### 3.3.1.2 Performance Index of Isolation Fluid

The isolation fluid performance specifications are required as follows:

(1) The density can be adjusted in the range of $1.08\sim2.0\text{g/cm}^3$, and the design density should be between the density of cement slurry and the density of drilling fluid;

(2) At the circulating temperature, the dynamic plastic ratio is 0.05~0.12;

(3) Loss of water standard: in the temperature range of 27~103°C, under the condition of pressure difference of 7MPa, 30min loss of water less than 250mL; in the temperature higher than 103°C, under the condition of pressure difference of 7MPa, 30min loss of water should be less than 150mL;

(4) The compatibility test thickening time is not less than the cement paste thickening time;

(5) Good thermal stability at high temperature.

## 3.3.1 冲洗液及隔离液设计

冲洗液及隔离液一般统称为前置液。

### 3.3.1.1 冲洗液性能指标

冲洗液性能指标要求如下：

（1）基液密度：$1.00\sim1.08\text{g/cm}^3$；

（2）流态要求：在注替排量下应能使冲洗液达到紊流；

（3）热稳定性：在循环温度下，经过10h老化试验，性能变化不应超过10%；

（4）冲洗率：对钻井液的冲洗率为95%以上。

### 3.3.1.2 隔离液性能指标

隔离液性能指标要求如下：

（1）密度可调节范围$1.08\sim2.0\text{g/cm}^3$，设计密度应在水泥浆密度与钻井液密度之间；

（2）在循环温度下，动塑比为0.05~0.12；

（3）失水量标准：在27~103°C温度范围内、压差7MPa条件下，30min失水量小于250mL；在温度高于103°C、压差7MPa条件下，30min失水量应小于150mL；

（4）相容性试验稠化时间不少于水泥浆稠化时间；

（5）高温热稳定性好。

### 3.3.2 Cement Slurry Design

Cement slurry design is the foundation of cementing technology, and successful cementing operation is closely related to the reasonable design of cement slurry performance In order to ensure the construction safety and improve the quality of cementing, it is necessary to reasonably design the dosage of cement slurry, conventional performance, slurry column structure, slurry system, slurry rheology, slurry anti-scrambling performance, etc. before cementing operation. The cement slurry design is mainly based on:

(1) Formation pore pressure, formation fracture pressure and formation leakage pressure;

(2) Stratigraphy and stratigraphic fluid properties;

(3) Wellbore and circulation temperatures;

(4) Length of cement sealing section;

(5) Well type and cementing construction process.

#### 3.3.2.1 Cement Paste Dosage and Conventional Performance Design

3.3.2.1.1 Cement Slurry Dosage Design

Calculate the cement slurry dosage based on the height of the annular cement seal required by geology and production, including the volume of cement slurry in the annular seal section, the volume of cement plugs, and the volume of suspended cement plugs (the volume of upper cement plugs will also be considered for tailpipe cementing).

3.3.2.1.2 Cement Slurry Density Design

The density of regular density (1.75~2.10g/cm$^3$) cement paste can be adjusted by changing the water cement ratio, and the density of low density (1.30~1.75g/cm$^3$) cement

### 3.3.2 水泥浆设计

水泥浆设计是注水泥技术的基础，成功的固井作业与合理的水泥浆性能设计密切相关为了保证施工安全，提高固井质量，固井作业前需要合理设计水泥浆用量、常规性能、浆柱结构、水泥浆体系、水泥浆流变性、水泥浆防窜性能等。水泥浆设计依据主要为：

（1）地层孔隙压力、地层破裂压力和地层漏失压力；

（2）地层及地层流体性质；

（3）井底温度和循环温度；

（4）水泥封固段长度；

（5）井型和固井施工工艺。

#### 3.3.2.1 水泥浆用量及常规性能设计

3.3.2.1.1 水泥浆用量设计

根据地质及生产需要的环空水泥封隔高度进行水泥浆用量计算，包括环空封固段的水泥浆体积、水泥塞体积、悬空水泥塞体积（尾管固井还要考虑上水泥塞体积）。

3.3.2.1.2 水泥浆密度设计

常规密度（1.75~2.10g/cm$^3$）水泥浆的密度可以通过改变水灰比来调节，低密度（1.30~1.75g/cm$^3$）水泥的密度要加入减轻料或高活性的超粉料等外料来调节高密度

has to be adjusted by adding extraneous materials such as mitigating materials or high-activity super-powdered materials. The density of high density (>2.10g/cm³) cement has to be adjusted by adding extraneous admixtures such as weighted materials or high-activity ultra-fine powdered materials.

The density of cement slurry is determined comprehensively according to a number of factors such as formation pressure, borehole condition, annular sealing height, cementing process and so on during design. If the borehole is stable, there is no leakage, and the formation pressure coefficient is lower than 1.90g/cm³, the design will try to use conventional density cement paste. In other cases, according to the geological conditions and other comprehensive consideration of the choice of low density or high density cement paste. It should be fully considered when designing the density of cement slurry:

(1) Density difference meets the requirement of topdressing, only the average density of topdressing fluid is greater than the density of drilling fluid for effective topdressing;

(2) To meet the limitations of downhole pressure conditions, the sum of static liquid column pressure and flow resistance should be less than the formation rupture pressure (or leakage pressure), so as not to pressure leakage formation. At the same time, it must also be greater than the formation pore pressure to avoid oil and gas water intrusion;

(3) The water cement ratio should be appropriate, usually if the density of cement slurry is too low, although the fluidity is good, but the compressive strength of cement stone is often difficult to meet the requirements, while the cement slurry solidification may also form

（>2.10g/cm³）水泥的密度要加入加重材料或高活性的超细粉体材料等外掺料来调节。

设计时根据地层压力、井眼状况、环空封固高度、固井工艺等多项因素综合确定水泥浆的密度。如果井眼稳定、无漏失、地层压力系数低于1.90g/cm³，设计尽量使用常规密度水泥浆。其他情况下根据地质条件等综合考虑选用低密度或高密度水泥浆。在设计水泥浆密度时要充分考虑到：

（1）密度差满足顶替要求，只有顶替液平均密度大于钻井液密度才能有效顶替；

（2）满足井下压力条件的限制，静液柱压力及流动阻力之和应小于地层破裂压力（或漏失压力），才不至于压漏地层。同时，还必须大于地层孔隙压力，以避免油气水侵窜；

（3）水灰比要合适，通常如果水泥浆密度过低，虽然流动性能好，但水泥石抗压强度往往难以满足要求，同时水泥浆凝固时还可能形成析水带，造成固井质量差。如果水泥浆密度过高，往往流动性能差，混拌和泵送困难，流动阻力较高，将给固井施工带来隐患严重时还会产生憋泵事故。

precipitation zone, resulting in poor cementing quality. If the density of cement slurry is too high, often led to poor fluidity, mixing and pumping difficulties, high flow resistance, will bring hidden dangers to the cementing construction of serious cases will also produce holding pump accident.

The effective jacking of annular drilling fluid by cement slurry is not only affected by the degree of casing centering (geometry of annular surface), jacking displacement, cement strength (static shear) of mud and cement slurry, and density difference, etc., but also the jacking flow state is an important influencing factor. When the displacement is certain, the flow profile of the cement slurry fluid depends on the flow state, which in turn depends on the rheological parameters. There are three flow patterns of fluid, which are plug flow, laminar flow and turbulent flow in order of their flow rate from small to large. The study shows that when the density difference between cement slurry and drilling fluid is less than $0.2g/cm^3$, pro-flow topping is used; and when the density difference is between $0.2\sim0.5g/cm^3$ both chilled and turbulent flow have better topping efficiency.

3.3.2.1.3  Cement Slurry Stability Design

Stability is one of the important performance indicators of cement paste. The degree of densification of the cement column formed by a less stable cement paste is very uneven from top to bottom. In large inclined wells and horizontal wells, this unevenness is more prominent, from the high side to the low side of the borehole, the degree of densification and cementation of the cement stone is weakening, which has an adverse effect on the sealing quality of the cement ring. At the same time, poorly stabilized cement slurry,

水泥浆对环空钻井液的有效顶替，除了受套管居中度（环形面的几何形状）、顶替排量、泥浆与水泥浆的胶凝强度（静切力）、密度差值等影响外，顶替流态也是一个重要的影响因素。当排量一定时，水泥浆流体的流动剖面取决于流动状态，而流态又取决于流变参数。液体的流态有三种流型，按其流速由小到大依次为塞流、层流和紊流。研究表明：当水泥浆与钻井液密度差小于 $0.2g/cm^3$ 时，采用层流顶替；当密度差在 $0.2\sim0.5g/cm^3$ 时，塞流和紊流都有较好的顶替效率。

3.3.2.1.3　水泥浆稳定性设计

稳定性是水泥浆的重要性能指标之一。稳定性较差的水泥浆所形成的水泥柱其致密程度从上到下非常不均匀。在大斜度井及水平井中，这种不均匀性更加突出，从井眼高边到低边，水泥石的致密程度及胶结程度在不断减弱，对水泥环的封固质量有着不良的影响。同时，稳定性差的水泥浆，一般情况下游离液也多，同样会在水泥柱中形成油窜、气窜、水窜的通道，影响水泥环的封固质量。

in general, also has more free liquid, which will likewise form channels for oil, gas and water to escape in the cement column, affecting the sealing quality of the cement ring.

The stability test of cement slurry can be done by BP settlement test method. The difference in density between the top and bottom of the cement slurry column is required to be no greater than $0.05g/cm^3$, and cement slurry characterization is required to be higher for large inclined wells and horizontal wells.

Another assessment of cement paste stability is free liquid. For conventional cementing operations, the free liquid content is required to be less than 1%, and for wells with large gradients and horizontal wells, the water precipitation or free liquid is required to be zero.

3.3.2.1.4  Design of Cement Paste Water Loss

Controlling water loss is the key to successful implementation of cementing. From the point of view of protecting the oil layer, it is also necessary to control the amount of water loss, and the water loss of cement slurry is also an important design parameter for the gas layer cementing to prevent gas flushing and the control of cement slurry "weightlessness", from the point of view of construction safety, the amount of water loss is too large and will pose a threat to the construction. Cement paste in the permeability of good sandstone water loss will cause the local cement paste water cement ratio becomes small, this time the cement paste not only increases the consistency, but also shorten the time of thickening, and in serious cases, it will cause can not be topped. Therefore, controlling the water loss of cement paste has become a problem that cannot be ignored.

水泥浆的稳定性测试可用 BP 沉降试验方法。要求水泥浆柱顶部和底部的密度差不大于 $0.05g/cm^3$，大斜度井、水平井对水泥浆定性的要求更高。

水泥浆稳定性的另一个考核指标是游离液。对于常规注水泥作业要求游离液含量小于 1%，对于大斜度井和水平井，要求析水或游离液为 0。

3.3.2.1.4  水泥浆失水设计

控制失水是成功实施注水泥的关键。从保护油层角度出发，也需要控制失水量，对气层固井的防气窜及水泥浆"失重"控制来说，水泥浆失水也是一个重要的设计参数，从施工安全角度出发，失水量过大会对施工造成威胁。水泥浆在渗透性好的砂岩失水过大会造成局部水泥浆的水灰比变小，此时水泥浆不仅稠度增加，而且稠化时间缩短，严重时造成无法顶替。因此，控制水泥浆失水已成为一个不可忽视的问题。

There are two stages of water loss in cement slurry: the dynamic water loss during the cement top up process and the static water loss during the waiting stage. Due to the scouring effect of the liquid on the well wall during cement slurry injection, the thickness of the cement cake formed by dynamic water loss does not increase indefinitely (it will reach equilibrium relatively quickly within a certain range). The same is not true for the cement cake formed by static water loss, which will probably increase in cake thickness as water is lost into the formation.

水泥浆失水分两个阶段：一是注水泥顶替过程的动态失水，二是候凝阶段的静失水。由于水泥浆注替过程中液体对井壁的冲刷作用，动失水所形成的水泥饼厚度并不是无限制增加（会在某一范围内较快地达到平衡）。静失水所形成的水泥饼却不一样，它可能会随着失水进入地层，其滤饼厚度也将不断增加。

3.3.2.1.5  Design of Cement Paste Thickening Time

3.3.2.1.5  水泥浆稠化时间设计

As the cement paste continues to hydrate, it continues to thicken until it loses its flow. The thickening time of a cement paste is the time required for the paste to lose flow during the flow process. The flow properties of cement slurries are generally expressed in Consistency (in Bc) according to API standards. When a cement paste reaches a consistency of 100 Bc in a simulated mixing test, the cement has lost its flow properties and the measured time is defined as the thickening time.

随着水泥浆的不断水化，水泥浆不断变稠，直至丧失流动性。水泥浆的稠化时间是指水泥浆在流动过程中丧失流动所需的时间。按 API 标准，水泥浆的流动性能一般用稠度表示（单位 Bc）。当水泥浆在模拟搅拌试验中，其稠度达到 100Bc 时，水泥已丧失流动性能，所测得的时间定义为稠化时间。

In order to ensure the safety of cement injection construction, the cement slurry must have a certain pumpable time. In the design of cement slurry thickening time need to pay attention to the following points:

为了保证注水泥施工安全，水泥浆必须具备一定的可泵送时间。在设计水泥浆稠化时间时需要注意以下几点：

(1) In order to ensure construction safety, the thickening time should be greater than the construction time of more than 60min, in order to prevent accidents. However, if the thickening time is too long, it is easy to occur in the process of oil, gas and water runoff;

（1）为保证施工安全，稠化时间应该大于施工时间 60min 以上，以防止出现意外情况。但如稠化时间过长，在候凝过程中易发生油窜、气窜、水窜；

(2) Reasonable design of cement slurry system, preferred cement slurry formula,

（2）合理设计水泥浆体系，优选水泥浆配方，缩短稠化过渡时间，有效防止环空气窜提高固井质量。当水泥浆顶替到位后，随着水化的不断进行，水泥浆内部形成空间网架结构，水泥浆部分重量悬挂于井壁，随着水泥浆凝固过程中的体积收缩，会出现"胶凝失重"，无法压稳地层，这时如果水泥浆内部阻力较小，气窜很可能发生。因此需要缩短水泥浆过渡时间，亦即缩短

shorten the transition time of thickening, and effectively prevent the annulus from air movement to improve the quality of cementing. When the cement slurry top replacement in place, with the continuous hydration, the cement slurry internal space network structure is formed, part of the weight of the cement slurry suspended in the well wall, with the volume contraction of the cement slurry solidification process, there will be a "gel loss", can not be stabilized stratum, if the internal resistance of the cement slurry is smaller, the gas flurry is likely to occur. Therefore, it is necessary to shorten the transition time of cement paste, that is to say, to shorten the time from the loss of weight of cement paste to the inability to stabilize the stratum to the formation of a stronger structure inside the cement paste (the cementing strength reaches 240Pa);

(3) For high-temperature and high-pressure well cementing, Portland cement is very sensitive to micro-chemical reactions and micro-physical reactions in the formula, so the test should all use on-site slurry water and on-site cement ash samples, and make a good thickening test under the high point of the temperature (circulating temperature increased by 5℃).

3.3.2.1.6　Compressive Strengths

Compressive strength is the pressure acting on a unit area of a damaged cement specimen. Higher compressive strength cements are generally of better quality. Cement systems with high compressive strengths have a tendency to be brittle under repeated high stresses. Generally, medium strength (13.7~20.6MPa) cement stones have better sealing performance. According to experience, when the compressive strength of cementite

水泥浆失重到无法压稳地层至水泥浆内部形成较强的结构（胶凝强度达到240Pa）的时间；

（3）对于高温高压井固井，波特兰水泥对配方中微化学反应和微物理反应十分敏感，因此试验要全部采用现场配浆水和现场水泥灰大样，并且做好温度高点（循环温度增加5℃）下的稠化试验。

3.3.2.1.6　抗压强度

抗压强度是指破坏水泥试样单位面积所作用的压力。抗压强度较高的水泥石固井质量般较好。在高应力反复作用下，抗压强度高的水泥体系存在易脆裂的问题。一般中等强度（13.7~20.6MPa）的水泥石具有较好的密封性能。根据经验，当水泥石的抗压强度达到3.45MPa时已能支撑套管所形成的轴向载荷。为适应油层开发和射孔的要求，水泥石的抗压强度值要求在7.0~14.0MPa之间。当强度低于7.0MPa时，水泥胶结和密封性

reaches 3.45MPa it can already support the axial load formed by casing. In order to meet the requirements of oil formation development and perforation shooting, the compressive strength value of cementite is required to be between 7.0~14.0 MPa. When the strength is lower than 7.0MPa, the cement cementing and sealing performance is poor. When the strength is higher than 14.0MPa, the cement stone appears brittle and breaks easily when shooting holes. If the strength of the cement stone is lower than the rupture strength of the formation to be fractured, not only will it fail to achieve the expected effect of fracturing, but it will damage the cement ring around the casing. Therefore, the strength of cement stone should be determined according to the need of sealing the purpose layer.

### 3.3.2.2 Pressure Stabilized and Tamper Proof Design

The design of cementing pressure stabilization and tamperproof needs to consider the following conditions: first, the annular pressure of water injection slurry and slurry replacement process can not be higher than the leakage pressure; second, the annular pressure of water injection slurry and slurry replacement process can not be lower than the formation pressure (especially when the low-density pre-fluid enters into the annular phase); and third, to shorten the process of cementing slurry as much as possible, so as soon as possible to form the strength, and reduce the tampering of the oil, gas and water runoff in the formation.

To solve the above problems, one is to work on the slurry structure, such as the use of segmented cement slurry, the use of different densities or different thickening time of the cement slurry; the second is a low-

能差。当强度高于14.0MPa时，水泥石出现脆性，射孔时容易破裂。如果水泥石的强度低于所需压裂地层的破裂强度，不仅达不到压裂的预期效果，反而会破坏套管周围的水泥环。因此水泥石强度应根据封固目的层的需要来确定。

### 3.3.2.2 压稳及防窜设计

固井压稳及防窜设计需要综合考虑以下条件：一是注水泥浆、替浆过程环空压力不能高于漏失压力；二是注水泥浆、替浆过程环空压力不能低于地层压力（特别是当低密度前置液进入环空阶段）；三是尽量缩短水泥浆固化过程，尽快形成强度，减少地层油窜、气窜、水窜。

要解决好上述问题，一是在浆体结构上下功夫，如采用分段水泥浆、使用不同密度或不同稠化时间的水泥浆；二是低密度冲洗液要严格控制密度和使用量；三是要在水泥浆稠化时间上严格控制，这包括前面要求的化时间大于施工时间的60min，但尽量不能大于90min，并尽量缩短过渡时间；四是要用精确度较高的"U形管效应"模拟器来模拟注替过程中指定位置不同时间环空当量循环密度ECD来进行压稳校核；五是进行水泥浆防窜设计和评价。

density flushing fluid to strictly control the density and the amount of use; the third is to be in the cement slurry thickening time under strict control, which includes the previous requirements of the time is greater than the construction of time 60min, but try not to be greater than 90min and try to shorten the transition time; four is to use the high accuracy of the "U-tube effect" simulator to simulate the process of injection alternation specified location of different time ring space equivalent cyclic density ECD to carry out the pressure stability calibration; and five is to carry out the cement slurry tamperproof design and evaluation.

Regarding the design and evaluation of cement slurry anti-fouling there are currently many methods, and the following methods are more related to the design of cement slurry performance.

(1) Cement paste performance factor method (SPN).

The cement paste performance factor method takes into account the effects of the rate of static cementitious strength development, the true rate of water loss from the cement paste downhole, and the borehole geometry on the ability to prevent gas flow. The occurrence of air fouling in the annulus is a result of the combined effect of cement paste static cementitious strength development and cement paste volume loss, and does not depend on the value of cement paste permeability.

(2) Modified cement paste performance factor method (SPNx).

The cement paste performance factor method (SPN) only considers the static gel strength development rate of the cement paste and the real water loss downhole without

关于水泥浆防窜设计和评价目前方法很多，与水泥浆性能设计关系比较大的有以下方法。

（1）水泥浆性能系数法（SPN）。

水泥浆性能系数法考虑了静胶凝强度发展速率、井下水泥浆真实失水速率及井眼几何形状对防气窜能力的影响。环空发生气窜是水泥浆静胶凝强度发展和水泥浆体积损失共同作用的结果，并不取决于水泥浆渗透率值的高低。

（2）修正的水泥浆性能系数法（SPNx）。

水泥浆性能系数法（SPN）只考虑了水泥浆的静胶凝强度发展速率和井下真实失水未考虑水泥浆密度、井眼几何形状、水泥浆柱长度等因素，不能全面地预测环空气窜的发生，所以引入了一个考虑现场实际因素的系数 $F_b$，形成了修正 SPNx 系数。

considering the density of the cement paste, the borehole geometry, the length of the cement paste column and other factors, which can not comprehensively predict the occurrence of the annular air runoff, so a coefficient Fb considering the actual factors in the field was introduced to form the modified SPNx coefficient.

In practice, if the SPNx value of cement paste is large, the pressure stabilization factor Fb can be increased. on the other hand, if the Fb value of a certain well design is relatively large, the SPN value can be appropriately relaxed to reduce the difficulty of cement paste formulation optimization and experimentation.

### 3.3.2.3 Cement Slurry Leak Prevention and Plugging Design

Leaking wells can be divided into three categories according to the nature of the leakage: fracture and cavernous leakage, permeability leakage and formation fracture leakage. Fracture and cavern leakage generally occurs in submerged caves or faults, permeability leakage generally occurs in highly permeable sandstone formations, and formation rupture leakage generally occurs in formations with low formation rupture pressure coefficients, and when the sum of the annular static fluid column pressure and flow resistance exceeds the formation rupture pressure during cementing, it will lead to cement slurry leakage. In order to prevent well leakage during cementing operation, it is necessary to carry out pressurized leakage plugging before cementing operation to improve the formation pressure-bearing capacity as much as possible, and carry out cementing design and construction based on the formation pressure-bearing capacity after leakage plugging. In the cementing process, multi-stage injection of cement and the use of multi-coagulation cement

实际应用时，如果水泥浆 SPNx 值偏大，则可增加压稳系数 $F_b$。从另一方面说，如果某口井设计的 $F_b$ 值比较大，则可适当放宽 SPN 值，减少水泥浆配方优选与实验难度。

### 3.3.2.3 水泥浆防漏堵漏设计

漏失井按照漏失性质可分为三类：裂缝及溶洞性漏失、渗透性漏失和地层破裂性漏失。裂缝及溶洞性漏失一般发生在潜山溶洞或断层，渗透性漏失一般发生在高渗透砂岩地层，地层破裂性漏失一般发生在地层破裂压力系数较低的岩层，当固井过程中环空静液柱压力与流动阻力之和超过地层破裂压力时就会引起水泥浆漏失。为了防止固井作业过程中发生井漏，需要在固井作业前先进行承压堵漏，尽可能提高地层承压能力，并依据堵漏后的地层承压能力进行固井设计施工。在固井工艺上可以采取多级注水泥、使用多凝水泥浆体系等方法防止漏失。设计时为了防止因漏失造成水泥浆低返而影响环空封固质量，根据漏失类型有以下几种典型方法。

slurry system can be adopted to prevent leakage. In order to prevent the quality of annular sealing from being affected by the low return of cement slurry due to leakage, there are the following typical methods according to the type of leakage.

3.3.2.3.1 Design of Pressurized Leakage Plugging Tests

Before drilling and casing, increase the density of drilling fluid or pressurize the wellhead annulus to confirm that the formation can withstand the highest ECD during cementing, and if it cannot meet the requirement of no leakage during the cementing process, the drilling fluid can be used directly to conduct a pressurized plugging test, which can be used to increase the leakage pressure of the formation by squeezing the plugging agent into the formation, so as to ensure the success of the cementing workover.

3.3.2.3.2 Isolating Fluid Design

The use of potassium silicate, sodium silicate plus phosphate, or a mixture of potassium silicate and sodium silicate as a prelude enables the formation to withstand a high static liquid column pressure. This is because this type of antecedent fluid, after entering a highly permeable formation, can contact calcium ions in the formation water to form a calcium silicate gel. If there is insufficient calcium ions in the formation, calcium chloride solution can be pumped in properly before pumping the prelude, and the high concentration of calcium ions in the cement slurry can quickly seal the formation. If it belongs to the fracture stratum, some short fiber materials can be added or hollow drift beads can be added to reduce the density in order to plug the gap and reduce the pressure of static liquid column in the annulus.

3.3.2.3.1 承压堵漏试验设计

钻完进尺准备下套管前，通过提高钻井液密度或井口环空加压方式，确认地层能承受固井时最高 ECD。如果不能满足固井过程中不发生漏失，可以直接以堵漏钻井液进行承压堵漏试验，通过向地层挤入堵漏剂来提高地层的漏失压力，确保固井施工成功。

3.3.2.3.2 隔离液设计

使用硅酸钾、硅酸钠加磷酸盐或硅酸钾和硅酸钠的混合物作前置液，能使地层承受较大的静液柱压力。因为该类型的前置液进入高渗透地层之后，能与地层水中的钙离子接触生成硅酸钙凝胶。如果地层内的钙离子不足，可在泵入该前置液之前适当泵入氯化钙溶液，水泥浆中高浓度的钙离子可迅速封堵地层。若属裂缝性地层，可加入一些短纤维材料或加空心漂珠降低密度，以便堵塞缝隙和降低环空静液柱压力。

3.3.2.3.3　Low-density Cement Paste Design

Low-density cement paste with suitable strength configured with bentonite, diatomite, expanded perlite, water glass, siliceous filler, etc., with a minimum density limit of 1.31g/cm³, is now widely used. However, low density cement paste is often applied with certain limitations due to larger water cement ratio and external admixtures, lower compressive strength and higher permeability. In recent years at home and abroad has been studied two kinds of ultra-low density cement paste, one is a high strength hollow beads cement paste, the other is a foam cement paste. The high strength hollow beads cement paste density can reach as low as 0.96g/cm³. If the density of 0.36g/cm³ with high strength compression resistant hollow glass beads can be used to reduce the density of cement paste to 0.85g/cm³, and its compressive strength can be comparable with the normal density of cement paste. can be comparable with the normal density cement paste, and the permeability is only 1/10 of the cement stone formed by the normal density cement paste.If the two systems of high-strength hollow beads cement paste and foam cement paste are used in conjunction, the density can be up to 0.72g/cm³.

3.3.2.3.4　Thixotropic Cement Paste

Thixotropic cement slurry has good fluidity in the process of injecting top replacement, and forms a rigid, self-supporting gel structure immediately after the pumping stops, so that the leakage problem can be solved effectively. Therefore, thixotropic cement paste is an important technical means to solve the malignant well leakage problem. The application of thixotropic cement paste in leakage prevention and plugging mainly including the following aspects:

3.3.2.3.3　低密度水泥浆设计

以膨润土、硅藻土、膨胀珍珠岩、水玻璃、硅质充填物等材料配置的具有合适强度的低密度水泥浆，最低密度极限是1.31g/cm³，目前已广泛使用。但是低密度水泥浆由于水灰比和外掺料较大，抗压强度较低，渗透性较高，其应用受到一定的限制。近年来国内外已研究出了两种超低密度水泥浆，一种是高强度空心微珠水泥浆，另一种是泡沫水泥浆；高强度空心微珠水泥浆密度最低可以达到0.96g/cm³。如果采用密度为0.36g/cm³的具有高强度抗压缩的空心玻璃微珠，可将水泥浆密度降低至0.85g/cm³，其抗压强度可与正常密度的水泥浆相当，渗透率仅为正常密度水泥浆形成的水泥石的1/10。如果将高强度空心微珠水泥浆和泡沫水泥浆这两种体系配合使用，密度可达到0.72g/cm³。

3.3.2.3.4　触变性水泥浆

触变性水泥浆在注入顶替过程中流动性能良好，泵送停止后则立即形成具有刚性、能自身支持的胶凝结构，从而可以有效解决漏失问题。因此，触变性水泥浆是解决恶性井漏问题的一项重要技术手段。触变性水泥浆在防漏堵漏上的应用主要包括以下方面：

(1) Cementing operations in leaky formations and dealing with well leaks during drilling;

(2) When remedial cement squeezing is carried out in permeable strata, thixotropic cement slurry can be used as the pilot slurry to achieve the purpose of increasing the squeezing pressure and improving the success rate of squeezing;

(3) Suitable for cementing operations in weak formations.

3.3.2.3.5　Fiber Cement Plugging

Fiber cement slurry is mixed with a certain proportion and length of fiber material in the cement slurry base formula, which is uniformly dispersed in the cement slurry under the action of high speed mixing. When the fiber cement slurry system is pumped into the leakage layer, it is easy to form a net-like bridging structure in the leakage channel through bridging effect, and the bridging structure is more stable under a certain pressure difference. Fiber cement slurry can achieve better plugging effect and has been successfully applied in oil well plugging operation.

3.3.2.4　Cement Slurry System Design

The design of cement paste system requires comprehensive consideration of factors such as formation pressure, oil and gas water content, hole penetration distribution, complexity and accidents during drilling, and cementing requirements. When designing, it is required to evaluate the influence of these factors on cementing construction, and on this basis, the cement slurry system is preferred to ensure the success of cementing construction and improve the quality of cementing.

Common cement paste systems are listed below:

(1) Conventional cement paste system:

（1）漏失层的注水泥作业和处理钻井过程中的井漏；

（2）在渗透性地层进行补救挤水泥时，可以采用触变性水泥浆作为先导浆，以达到增加挤注压力和提高挤水泥成功率的目的；

（3）适用于薄弱地层的固井作业。

3.3.2.3.5　纤维水泥堵漏

纤维水泥浆就是在水泥浆基础配方中混入一定比例和长度的纤维材料，在高速搅拌作用下均匀分散在水泥浆里。当纤维水泥浆体系泵注到漏层时，容易在漏失通道中通过桥接作用形成网状架桥结构，架桥结构在一定压差下较为稳定。纤维水泥浆能达到较好的封堵效果，已经成功应用于油井封堵作业。

3.3.2.4　水泥浆体系设计

水泥浆体系设计需综合考虑地层压力、含油气水情况、孔渗分布、钻井期间复杂与事故、封固要求等因素。在进行设计时，要求评价这些因素对固井施工的影响，在此基础上优选水泥浆体系，以保证固井施工成功，提高固井质量。

常见水泥浆体系如下：

（1）常规水泥浆体系：水泥＋降失水剂＋分散剂＋缓凝剂；

cement + water loss reduction agent + dispersant + retarder;

(2) High-strength low-density cement paste system: cement + reducing agent + reinforcing material + water loss reduction agent + dispersant + retarder;

(3) Leak-proof cement paste system: cement + leak-proof plugging agent + water loss reduction agent + dispersant + retarder;

(4) Cementitious slurry system: cement + fiber material + water loss reduction agent + dispersant + retarder;

(5) Leak-proof and toughened cement paste system: cement + fiber material + leak-proof and plugging agent + water loss reduction agent + dispersant + retarder;

(6) Latex anti-fouling cement paste system: cement + latex + dispersant + retarder;

(7) High-density cement paste system: cement + weighting agent + reinforcing material + water loss reduction agent + dispersant + retarder;

(8) Anti-salt water slurry system: cement + anti-salt water loss reduction agent/salt intrusion inhibitor + dispersant + retarder;

(9) High-temperature cement paste system: cement + anti-temperature water loss reduction agent + anti-strength decay agent + dispersant + high-temperature retarder;

(10) Low-temperature early-strength cement paste system: cement + early-strength agent + water loss reduction agent + dispersant + retarder;

(11) Ultra-low-temperature early-strength cement paste system: low-temperature early-strength cement + water loss reduction agent + dispersant + retarder;

(12) Early-strength low-density cement paste system: cement + early-strength agent +

（2）高强低密度水泥浆体系：水泥+减轻剂+增强材料+降失水剂+分散剂+缓凝剂；

（3）防漏水泥浆体系：水泥+防漏堵漏剂+降失水剂+分散剂+缓凝剂；

（4）增水泥浆体系：水泥+纤维材料+降失水剂+分散剂+缓剂；

（5）防漏增韧水泥浆体系：水泥+纤维材料+防漏堵漏剂+降失水剂+分散剂+缓凝剂；

（6）胶乳防窜水泥浆体系：水泥+胶乳+分散剂+缓凝剂；

（7）高密度水泥浆体系：水泥+加重剂+增强材料+降失水剂+分散剂+缓凝剂；

（8）抗盐水泥浆体系：水泥+抗盐降失水剂/盐侵抑制剂+分散剂+缓凝剂；

（9）高温水泥浆体系：水泥+抗高温降失水剂+抗强度衰减剂+分散剂+高温缓凝剂；

（10）低温早强水泥浆体系：水泥+早强剂+降失水剂+分散剂+缓凝剂；

（11）超低温早强水泥浆体系：低温早强水泥+降失水剂+分散剂+缓凝剂；

（12）早强低密度水泥浆体系：水泥+早强剂+减轻剂/（起泡剂+泡剂）+降失水剂+分散剂+缓凝剂；

reducing agent / (foaming agent + bubbling agent) + water loss reduction agent + dispersant + retarder;

(13) Anti-fouling cement paste system: cement + anti-fouling material + water loss reduction agent + dispersant + retarder;

(14) Micro-expansion cement paste system: cement + expansion agent + water loss reduction agent + dispersant + retarder;

(15) Ultrafine cement paste system: ultrafine cement + water loss reduction agent + dispersant + retarder;

(16) Slag cement slurry system: slag cement + water loss reducing agent + dispersant + retarder, etc.

# Ideological and Political Case Studies

Academician Su Yinao is a renowned expert in oil and gas drilling engineering. Over the years, he has undertaken and presided over many national key scientific and technological projects, and has profound expertise in drilling technology research and application in directional wells, cluster wells, and horizontal wells. His innovative achievements in drilling mechanics, wellbore trajectory control, and downhole tool research are at the international advanced level, and have formed a system used for significant production benefits. He creatively introduced engineering cybernetics and aerospace guidance technology into drilling engineering, pioneering new fields, and proposed the new concept of "underground control engineering," which he conducted pioneering basic research on. The "underground control engineering" he created has now become a new branch under the first-level discipline of petroleum and natural gas

（13）抗窜水泥浆体系：水泥+防窜材料+降失水剂+分散剂+缓凝剂；

（14）微膨胀水泥浆体系：水泥+膨胀剂+降失水剂+分散剂+缓凝剂；

（15）超细水泥浆体系：超细水泥+降失水剂+分散剂+缓凝剂；

（16）矿渣水泥浆体系：矿渣水泥+降失水剂+分散剂+缓凝剂等。

# 思政案例

苏义脑院士是著名的油气钻井工程专家，多年来承担和主持了多项国家重点科技项目攻关，在定向井、丛式井、水平井等钻井技术研究与应用方面有深厚造诣，在钻井力学、井眼轨道控制和井下工具研究中多项创新成果居国际先进水平，形成体系且生产效益显著。他创造性地把工程控制论和航天制导技术引入钻井工程，开拓新领域，提出"井下控制工程"这一新概念并做开拓性基础研究，创建的"井下控制工程"现已成为我国学位教育石油天然气工程这个一级学科下的新分支；主持研制P5LZ四大系列导向钻县和K7LZ系列空气螺杆钻具，主持导向钻井工艺技术、高陡构造防斜打快技术研究，均取得较好的经济效益；他主持设计全国第一口薄油层中曲率水平井轨道控制方案并实施成功；他主持研发成功拥有我国自主知识产权的近钻头地质导向钻井系统，"给钻头装上眼睛""使钻头闻着油味儿走"的梦想变为现实，使钻头能

engineering in China's degree education. He led the development of the P5LZ series of guide drilling tools and the K7LZ series of air screw drilling tools, and presided over the research on guide drilling technology and high-steep structure anti-deviation and rapid drilling technology, all of which have achieved good economic benefits. He led the design of the first thin oil layer medium-curvature horizontal well trajectory control plan in China and successfully implemented it. He led the development of the near-bit geological steering drilling system with China's independent intellectual property rights, realizing the dream of "giving the drill bit eyes" and "letting the drill bit smell its way to the oil," enabling the drill bit to freely penetrate thousands of meters of rock underground and accurately enter the oil and gas layer, making a remarkable contribution to enhancing China's core competitiveness in drilling technology. His project results have won two first prizes and one second prize in the National Science and Technology Progress Award, one second prize in the National Technical Invention Award, five first prizes or above at the ministerial and provincial level, one National Patent Excellence Award, and he has been awarded the title of "Chinese Ph.D. Recipient with Outstanding Contributions" by the state, the first National Postdoctoral Award, and the title of "National Outstanding Postdoctoral Scholar". He has also been honored with the title of "Model Worker of the Central Enterprises", the Special Model Worker of the China National Petroleum Corporation, and has received the "First Model Worker of the Iron Man Award", the Ho Leung Ho Lee Science and Technology Award, and the Guanghua Engineering Science and Technology Award.

自如地钻穿底下几千米的岩石并准确地进入油气层，为提高我国钻井技术的核心竞争力做出了卓越贡献。其项目成果获国家科学技术进步奖一等奖2项，等奖1项，国家技术发明奖二等奖1项；省部级一等奖以上5项，国家专利优秀奖1项；他被国家授予"做出突出贡献的中国博士学位获得者"称号，并荣获全国首届博士后奖和"全国优秀博士后"称号；还获得中央企业劳动模范、中国石油天然气集团有限公司特等劳动模范、"首届铁人奖章获得者"称号、何梁何利科学技术奖和光华工程科技奖等。

Journalist: Professor Su, we all know that you have established a second-level discipline, underground control engineering. How did you come up with the idea of combining underground drilling technology with aerospace guidance technology?

Su Yinao: During 1984-1987, while I was pursuing my Ph.D. and participating in the national "7th Five-Year Plan" key scientific and technological project "Research on Directional and Cluster Wells Drilling Technology", inspired by the successful application of engineering cybernetics in missile guidance and aerospace telemetry, I had the idea to introduce engineering cybernetics into petroleum drilling engineering to achieve the combination of aerospace guidance and drilling trajectory control. The three angles of wellbore trajectory control in oil and gas wells (i.e., inclination angle, azimuth angle, tool face angle) correspond to the three angles of attitude control in aircraft (i.e., pitch angle, yaw angle, roll angle), and their control is scientifically consistent. Therefore, after completing the defense of my Ph.D. thesis, "Analysis of Several Mechanical Problems in Drilling with Downhole Power Tools and Preliminary Study on Predictive Control of Directional Well Trajectory", in June 1988, I entered the postdoctoral research station at Beijing University of Aeronautics and Astronautics, under the guidance of Professor Huang Kelie, a renowned expert in aerospace mechanics in China. During my two years at the station, in addition to completing the follow-up experiments and finalizing the national "7th Five-Year Plan" key scientific and technological project topic "Theory and Technology of Directional Well Trajectory

记者：苏院士您好，我们都知道您曾经创立过一门二级学科—井下控制工程学。您当时怎么会想到让地下钻井技术与天上制导技术相结合？

苏义脑：1984—1987年，我在攻读博士学位和参加国家"七五"重点科技项目"定向井丛式井钻井技术研究"攻关期间，为了进一步提高井眼轨道的控制水平和质量，受工程控制论在导弹制导和航天测控中成功应用的启示，产生了把工程控制论引入石油钻井工程、实现航天制导与钻井轨道控制相结合的想法。油气井井眼轨道控制的3个角（即井斜角、方位角和工具面角）和航空器姿态控制的3个角（即俯仰角、偏航角和滚转角）一一对应，对它们的控制在科学的本质上是一致的。所以，我在完成博士学位论文《用井下动力钻具钻井时若干力学问题的分析和定向井轨道预测控制的初步研究》答辩后，1988年6月，即带着要尝试探索新途径的想法进入了北京航空航天大学博士后流动站，师从我国著名航空力学专家黄克累教授。在站两年中，我除了完成国家"七五"重点科技项目专题"定向井井眼轨道控制理论与技术研究"的后续实验和结题任务及撰写专著《井斜控制理论与实践》之外，就集中精力学习航空航天控制知识，探索提出并致力开拓"井眼轨道制导控制理论与技术"这一新领域，寄希望于实现"井下闭环控制"和"用手段解决问题"。

Control" and writing the monograph "Theory and Practice of Inclination Control", I focused on learning aerospace control knowledge and exploring the proposal and development of the new field of "Wellbore Trajectory Guidance and Control Theory and Technology", hoping to achieve "downhole closed-loop control" and "solve problems with means."

Pioneering a new research direction, especially when breaking through traditional concepts to explore a new field, the pressure encountered is imaginable. The initial path is difficult and cautious, from the judgment of the nature of the problem to the introduction or establishment of new concepts, from the contemplation and definition of the connotation of the new field to the decomposition of a series of research topics, from the derivation and determination of system models, equations, and boundary conditions to the conception and design of a patent scheme, all are accompanied by repeated wavering, reflection, self-questioning and self-verification, and are basically carried out and completed in an "amateur" manner.

Despite the fact that, when I left Beihang University in June 1990, I had already established the framework of this new field and completed the first patent in this area, the new direction had not yet been widely recognized. At the end of 1991, when I discovered from technical news reports that some foreign colleagues were also beginning to engage in research in this direction and had adopted the idea and terminology of "Closed Loop Control", I further solidified my confidence in continuing forward. Around 1993, based on the exploratory work of the previous years, considering the commonalities of various

开辟一个新的研究方向，特别是突破传统观念去开拓一个新领域时，所遭遇的压力是可想而知的。开始的路走得艰难而又谨慎，从问题性质的判断到新概念的引入或建立，从对新领域内涵的思索和界定到一系列研究课题的分解，从系统模型、方程、边界条件的推演和确定到某项专利方案的构思和设计，无不伴随着反复徘徊、反思、自我诘问和自我验证，并且基本上是以"业余"方式进行和完成的。

尽管在1990年6月我从北航出站时已经建立了这一新领域的框架并完成了在这方面的第一个发明专利，但是这一新方向还没有得到广泛认同。1991年末，当从技术消息报道中发现国外一些同行也在或开始致力于这一方向的研究并也采用了"Closed Loop Control（闭环控制）"的思路和提法，我进一步坚定了继续前进的信心。直到1993年前后，在前几年探索工作的基础上，考虑到油气井各种工艺环境的共性及都存在控制问题的普遍性，我和研究团队又把研究对象从钻井轨道控制进一步扩展到各种井下工艺过程，把认识从对具体问题的研究提高到对理论与技术体系乃至于学科分支的考虑，于是产生了"井下控制工程学"这一提法。随着时间推移，这一新领域逐步获得较多同行专家和有关领导的理解和支持。1995年3月，我应邀在《中国科学报》上发表了《正在兴起的"井下控制工程学"》一文；同年，"井眼轨道遥控技术研究"被立为中国石油天然气总公司"九五"前沿技术攻关项目；1997年，中国石油工程学会钻井基础理论学组在讨论钻井专业学科方向时，第一次把"钻井控制工程"列为油气钻井工程新分支；1999年，在多方努力下，"CGDS地质导向钻井系统研制"被列为中国石油天然气集团公司的科研项目。

technological environments in oil and gas wells and the universality of the presence of control issues, I and my research team further expanded the research scope from drilling trajectory control to various downhole process technologies, elevating our understanding from the study of specific problems to the consideration of theoretical and technological systems, and even to the discipline branches, which led to the concept of "Underground Control Engineering". Over time, this new field gradually gained understanding and support from many colleagues and related leaders. In March 1995, I was invited to publish an article titled "The Emerging 'Underground Control Engineering'" in the "China Science Daily"; that year, "Remote Control Technology for Wellbore Trajectory" was established as a cutting-edge technology research project for the "9th Five-Year Plan" of the China National Petroleum Corporation; in 1997, the Drilling Basic Theory Group of the China Petroleum Engineering Society, when discussing the direction of the drilling professional discipline, listed "Drilling Control Engineering" for the first time as a new branch of oil and gas drilling engineering; in 1999, with the efforts of many parties, "Research and Development of CGDS Geological Guidance Drilling System" was listed as a scientific research project of the China National Petroleum Corporation.

After 10 years of hard work and research, we successfully developed the "CGDS-1 Geological Guidance Drilling System" with China's independent intellectual property rights and achieved industrialization. This achievement won the Second Prize of the National Technical Invention Award in 2009.

历经10年艰苦攻关，研发成功具有我国自主知识产权的"CGDS-1地质导向钻井系统"并实现了产业化，该成果荣获2009年度国家技术发明二等奖。2008年2月，"井下控制工程"被教育部公示为学位教育中"石油天然气工程"这个一级学科下新增的二级学科，获准独立招收硕士、博士生。我带领的这支集理论研究、技术开发和工程服务于一体的专业团队，目前也成为中国石油钻井工程技术研究院所属的专业研究所；几个相关高等院校也在开设井下控制方面的选修课程，有的院校也在组建相关的研究团队。进入21世纪以来，油气井下控制技术已成为国际石油工程中最具发展活力的热点之一；在我国，从油气钻井工程轨道控制为起点的这一新领域，也在向采油采气等相关专业扩展。

In February 2008, "Underground Control Engineering" was announced by the Ministry of Education as a new second-level discipline under the first-level discipline of "Petroleum and Natural Gas Engineering" in degree education, and was approved to independently recruit master's and doctoral students. The professional team I lead, which integrates theoretical research, technology development, and engineering services, has also become a professional research institute under the China Petroleum Drilling Engineering Technology Research Institute. Several related universities have also begun to offer elective courses in underground control, and some institutions are forming related research teams. Since the beginning of the 21st century, the technology of underground control in oil and gas wells has become one of the most dynamic areas of development in international petroleum engineering. In China, this new field, starting from the trajectory control of oil and gas drilling engineering, is also expanding into related fields such as oil production and gas extraction.

Journalist: The guided screw drilling tools that you have designed and developed have now become an indispensable main tool in the national horizontal well breakthrough efforts; the geological guidance drilling technology you led in developing has broken through foreign technological blockades and monopolies, and is hailed as the "Two Bombs and One Satellite" of drilling technology, marking a significant milestone in the history of China's drilling technology development. What do you think contributed to these achievements?

记者：您主持设计和研制的导向螺杆钻具，现在已成为全国水平井攻关中不可或缺的主要工具；您领衔研发的地质导向钻井技术打破了国外的技术封锁与垄断，被誉为钻井技术中的"两弹一星"，在我国钻井技术发展史中具有里程碑意义。这些成绩的取得，您认为得益于什么？

Su Yinao: After being elected a member of the Chinese Academy of Engineering in 2003, I was once interviewed by a journalist. As soon as we met, he asked me a question: "As a successful person, please talk about the secret of success". I said firstly that I am not a successful person, but if you let me talk about the experience of my work, I can share some insights. I remember I mentioned three insights at that time: the starting point of progress is pursuit; the key to development is innovation; and the secret to success is perseverance. After the successful research of the geological guidance drilling technology, when the research group held a summary meeting, I summarized it with two words: dedication and perseverance. It means that whatever you do, you should be dedicated, dedicated to research, dedicated to leading the team, and dedicated to strategic planning. Innovation is a difficult thing; it requires boldness and perseverance. Innovation is about negating or transforming the old, and in the initial stages of innovation, it is bound to be "the few who understand are rare", do not expect "many who respond", this is a test of the perseverance of innovators. Only by breaking the routine, boldly imagining, rigorously proving, persevering, overcoming difficulties, and persevering can we possibly achieve success. Innovation must be tested by practice; only when it is proven correct in practice can it be considered innovative. Theoretical and technological innovations must ultimately be transformed into productivity to be considered complete. This is also the consensus among my team members and something we constantly remind ourselves of. Chairman Mao once said in his article "*Strategic Issues of Guerrilla Warfare in the Anti-Japanese

苏义脑：我2003年当选中国工程院院士以后，曾有一个记者来采访我，他一见面就给我提了一个问题："作为一名成功人士，请谈谈成功的秘诀是什么？"我说首先声明我不是成功人士，您如果让我谈工作的体会那我可以谈一些。记得当时我谈了三点体会：进步的起点在于追求；发展的关键在于创新；成功的秘诀在于坚持。当年地质导向钻井技术研究成功以后，课题组开总结会时我总结了四个字：用心、坚持。就是无论做什么事都要用心，用心搞研究，用心带队伍，用心做布局。而创新是一件很难的事情，它要求大胆和坚持。创新是对旧物的否定或改造，在创新初期必然是"和者盖寡"，不要奢望"应者云集"，这是对创新者定力的考验。只有打破常规，大胆设想，严密求证，锲而不舍，攻坚克难，坚持前行，才有可能取得成功。创新要经受实践的检验，只有被实践证明正确才是创新。理论创新和技术创新最终要转化为生产力，才算完成创新。这也是我和团队成员的共识，并时刻以此自勉。毛主席在《抗日游击战争的战略问题》中曾说过这样一句话："有利的情况和主动的恢复，产生于'再坚持一下'的努力之中。"我对这句话的印象和体会都太深了。从我父亲去世时考大学开始，到历时10年研究地质导向钻井技术时遇到的种种阻碍，这几十年一路走来，每当遇到非常艰难的情况，我就想到再咬咬牙，再坚持一下。很多时候我都是自己默默书写蒲松龄的那副对联勉励自己："有志者，事竟成，破釜沉舟，百二秦关终属楚；苦心人，天不负，卧薪尝胆，三千越甲可吞吴"。而地质导向钻井技术等研究实践也证明，任何成绩的取得都来自锲而不舍的坚持。

*War*": "Favorable situations and the recovery of initiative come from the effort of 'holding on a little longer'". I have a very deep impression and understanding of this sentence. Since my father passed away and I took the college entrance examination, to the various obstacles encountered during the 10-year research on geological guidance drilling technology, over the past few decades, whenever I encounter extremely difficult situations, I think of holding on a little longer, persevering. Many times, I silently write Pu Songling's couplet to encourage myself: "For those with ambition, success will be achieved; breaking the cauldron, burning the boats, the two hundred and twenty-one passes of Qin will finally belong to Chu; for those with perseverance, the heavens will not fail them; enduring hardship, biting the bullet, the three thousand Yue troops can conquer Wu". And the research and practical experience of geological guidance drilling technology has also proven that any achievement comes from persevering dedication.

Excerpted from "Writing Scientific Research Papers in the Lab and at the Well Site" by Tang Daling, published in the 4th issue of "China Petroleum Enterprise" in 2022.

节选自唐大麟发表于《中国石油企业》2022年第4期的《"把科研论文写在实验室和井场上"——访中国工程院院士、油气钻井工程专家苏义脑》。

# Chapter 4 HSE Knowledge Related to the Oil and Gas Industry

# 第四章 油气行业 HSE 相关知识

## 4.1 Introduction to HSE-Related Knowledge

HSE stands for Health, Safety and Environment. The Health, Safety and Environment Management System abbreviated as HSE Management System or simply HSE MS, is a commonly used management system in the international oil and gas industry. It refers to the management system established by enterprises based on occupational health and safety and environmental management system standards.

The oil industry is a high-risk industry. The exploration, transportation, and refining and processing of oil are inseparable from flammable and explosive substances, toxic and harmful substances, high-temperature and high-pressure containers, also have the characteristics of continuous operation and numerous points and long lines. Therefore, the HSE Management System is particularly important in oil companies.

The standard "Health, Safety and Environmental Management Systems for the Petroleum and Natural Gas Industry"

## 4.1 HSE 相关知识简介

HSE 是健康（Health）、安全（Safety）和环境（Environment）的缩写。健康、安全与环境管理体系简称为 HSE 管理体系，或简单地用 HSE MS（Health Safety and Environment Management System）表示，是国际石油天然气工业通行的管理体系，是指企业依据职业健康安全（Occupational Health and Safety）、环境（Environment）管理体系标准建立的管理体系。

石油行业是一个高危行业，石油的开采、运输和炼化加工离不开易燃易爆物、有毒有害物、高温高压容器，同时具有连续作业、点多线长的特点，因此健康、安全与环境（HSE）管理体系在石油企业中尤为重要。

《石油天然气工业健康、安全与环境管理体系》标准（SY/T 6276—2014）由 7 个一级要素、26 个二级要素构成。2004 年，根据中国石油天然气集团公司（简称中国石油）HSE 管理实际，考虑到 HSE 管理标准与国家标准 GB/T 24001《环境管理体系要求及使用指南》、GB/T 28001《职业健康安全管理体系要求》兼容的需要，并参照 GB/T

(SY/T 6276-2014) consists of 7 primary elements and 26 secondary elements. In 2004, based on the actual HSE management of China National Petroleum Corporation (CNPC), considering the compatibility of HSE management standards with national standards GB/T 24001 "Environmental Management Systems – Requirements and Guidelines for Use" and GB/T 28001 "Occupational Health and Safety Management System Requirements", and referring to the relevant requirements of GB/T 19001 "Quality Management System Requirements" standard, CNPC revised the SY/T 6276 standard, The Q/CNPC 104.1-2004 standard "Health, Safety and Environmental Management Systems Part 1: Specification" (issued on July 29, 2004 and implemented on October 1, 2004) was promulgated. This standard adopts the structure of the GB/T 24001 standard, integrates relevant requirements of environmental management systems and occupational health and safety management systems, and has stronger universality.

19001《质量管理体系要求》标准有关要求，中国石油对SY/T 6276标准进行了修订，颁布了Q/CNPC 104.1—2004标准《健康、安全与环境管理体系 第1部分：规范》(2004年7月29日颁布，2004年10月1日实施)该标准采用了GB/T 24001标准的结构，整合了环境管理体系、职业健康安全管理体系有关要求，具有更强的通用性。

## 4.2 Master HSE Risk Identification for Drilling Operations

## 4.2 掌握钻井作业HSE风险识别

### 4.2.1 Related Knowledge

### 4.2.1 相关知识

#### 4.2.1.1 Basic Concepts

(1) Risk.

Risk refers to the combination of the likelihood and consequences of a specific hazardous event occurring.

#### 4.2.1.1 基本概念

（1）风险。

风险是指某一特定危害事件发生的可能性与后果的组合。

Broadly speaking, risk is an environment or state that refers to a potential environmental condition beyond human control, or the possibility of damage or failure.

(2) Hazards.

Harm refers to the potential damage that may occur, including causing illness and trauma, causing damage to property, factories, products, or the environment, resulting in production losses or increased burden (SY/T 6276-2014).

In HSE management, hazards refer to the hazards of health, safety and environment.

(3) Hazard sources.

Hazard sources refer to the root causes that may cause casualties, property damage or environmental damage. They can be a hazardous piece of equipment, facility, system or a hazardous part of a piece of equipment, facility, or system.

(4) Hazardous factors.

Harmful factors refer to the elements in an organization's activities, products, or services that may cause injury or illness to personnel, property damage, damage to the work environment, harmful environmental impacts, or a combination of these factors, including root causes and conditions. Harmful factors are the factors that lead to events and accidents, including unsafe human behavior and unsafe conditions of objects.

(5) Identification of Hazardous Factors.

Hazard identification refers to the process of identifying the existence of health, safety, and environmental hazards and determining their characteristics. The "characteristics" here refer to the essence of the hazardous factors, whether they are human or material factors, and the regularity of the occurrence of hazardous factors.

广义的风险是一种环境或状态，它是指超出人的控制之外的某种潜在的环境条件，即指有遭到损害或失败的可能性。

（2）危害。

危害是指可能引起的损害，包括引起疾病和外伤，造成财产、工厂、产品或环境破坏，导致生产损失或增加负担（SY/T 6276—2014）。

在HSE管理中，危害是指健康、安全与环境三方面的危害。

（3）危险源。

危险源是指可能造成人员伤亡、财产损失或环境破坏的根源，可以是存在危险的一件设备、一处设施、一个系统，也可能是一件设备、一处设施、一个系统中存在危险的一部分。

（4）危害因素。

危害因素是指组织的活动、产品或服务中可能导致人员伤害或疾病、财产损失、工作环境破坏、有害的环境影响或这些情况组合的要素，包括根源和状态。危害因素是导致事件、事故的因素，包括人的不安全行为和物的不安全状态。

（5）危害因素辨识。

危害因素辨识是指识别健康、安全与环境危害因素的存在并确定其特性的过程。这里的"特性"是指危害因素的本质，如是人的因素还是物的因素，危害因素出现的规律性等。

(6) Accident hazards.

Accident hazard is a commonly used term in safety management in China.

Accident hazards generally refer to unsafe behaviors of people, unsafe states of objects, and management deficiencies in the production system that can lead to accidents. In the production process, based on the understanding of accident occurrence and prevention laws, in order to prevent accidents from occurring, standards, regulations, rules, etc. can be formulated for the state of objects, human behavior, and environmental conditions in the production process. If the state of objects, human behavior, and environmental conditions during the production process cannot meet these standards, regulations, rules, etc., accidents may occur. Therefore, accident hazards are a broader term, and hazard factors are synonymous, including people (personnel), machinery (machinery, equipment), materials (materials), environment (working environment), management (safety management), and other aspects. Identifying hidden dangers means identifying accident hazards from the above aspects.

(7) Risk management.

Risk management is the qualitative and quantitative analysis of the hazards present in a system, and the evaluation of the likelihood and severity of the consequences of the system's hazards. Based on the evaluation results, risk reduction measures are formulated for hazards, especially major hazard factors, and emergency response plans are formulated to achieve the management of risks and their impacts.

（6）事故隐患。

事故隐患是我国安全管理中常用到的一个术语。

事故隐患泛指生产系统中可导致事故发生的人的不安全行为、物的不安全状态和管理上的缺陷。在生产过程中，凭着对事故发生与预防规律的认识，为了预防事故的发生，可制订生产过程中物的状态、人的行为和环境条件的标准、规章、规定、规程等。如果生产过程中物的状态、人的行为和环境条件不能满足这些标准、规章、规定、规程等，就可能发生事故。因此，事故隐患是一个更为宽泛的名词，和危害因素是近义词，包括"人"（人员）、"机"（机械、设备）、物（物料）、"环"（作业环境）、"管"（安全管理）等方面。查隐患就是要从以上方面找出事故隐患。

（7）风险管理。

风险管理是对系统存在的危险性进行定性和定量分析，得出系统发生危险的可能性及其后果严重程度的评价，根据评价结果，对危害尤其是重大危害因素制定风险削减措施，编制应急反应计划，以实现对风险及其影响的管理。

## 4.2.1.2 The Main Risks of Drilling and Related Operations

During drilling operations, there may be technical service operations from relevant contractors, which can result in HSE risks that can affect the overall situation. Therefore, when conducting risk identification, it is not only necessary to identify common operational risks, but also to identify related operational risks.

### 4.2.1.2.1 Joint Operational Risks

(1) Blowouts and uncontrolled blowouts may cause the escape of hydrocarbons from the formation.

(2) Fire and explosion. The escape of hydrocarbon compounds, especially light oil and combustible gases, as well as the leakage of gasoline, diesel, lubricating oil, engine oil, etc., can cause fire and explosion hazards. In addition, fires also include barracks fires, electrical fires, and fires of flammable fibers or other items on site.

(3) High altitude workers fall.

(4) High altitude objects (such as hooks, traveling blocks, overhead cranes, derricks and their accessories, second floor platform accessories) or lifting heavy objects fall.

(5) Mechanical injury, such as object impact caused by personnel during construction operations (such as operating pliers).

(6) Electric shock injury.

(7) Poisoning, such as food poisoning, chemical poisoning, hydrogen sulfide poisoning, etc.

(8) Noise damage.

(9) Traffic accidents.

(10) Hazards caused by adverse weather or natural disasters, such as mountain floods, earthquakes, lightning strikes, etc.

## 4.2.1.2 钻井及相关作业的主要风险

钻井作业过程中，存在相关承包方的技术服务作业，产生的HSE风险会影响整个全局。因此，在进行风险识别时，不但要识别出共同作业风险，也要识别出相关作业风险。

### 4.2.1.2.1 共同作业风险

（1）井喷及井喷失控可能造成地层碳氢化合物的逸出。

（2）火灾及爆炸。地层碳氢化合物特别是轻质油、可燃性气体逸出，汽油及柴油、润滑油、机油等泄漏，造成火灾爆炸危险事故。此外，火灾还包括营房火灾、电气火灾、现场易燃纤维或其他物品着火等。

（3）高空作业人员坠落。

（4）高空物品（如大钩、游动滑车、天车、井架及井架附件、二层台附件）或起吊重物坠落。

（5）机械伤害，如人员施工操作（如操作大钳）过程中造成物体打击。

（6）触电伤害。

（7）中毒，如食物中毒、化学品中毒、硫化氢中毒等。

（8）噪声伤害。

（9）交通事故。

（10）恶劣天气或大自然灾害造成的危险，如山洪、地震、雷击等。

(11) Environmental pollution, including the destruction of vegetation caused by road construction and well sites, as well as the pollution of atmospheric pollutants from industrial wastewater, domestic wastewater, and harmful gases.

(12) The risks of offshore drilling, such as the hazards of adverse weather such as waves and typhoons, platform tilting, collapse, ship collision and loss of navigation.

(13) The risks brought by the social environment, such as the invasion of criminals, war, riots, etc.

4.2.1.2.2 Related Job Risks

(1) Risk of logging operations: radioactive injury, risk of accidental firing of perforating bullets, risk of logging instruments falling into the well, etc.

(2) Risk of logging operation: leakage of standard natural gas cylinders used, rough handling may cause fire and explosion, use of toxic materials such as trichloromethane may pose a risk of poisoning, and use of strong acidic substances may cause skin corrosion or burn hazards.

(3) Risk of directional well operation: injury to personnel by inclinometer winch, danger of directional well tools falling into the well, etc.

(4) Risk of cementing operation: Leakage of high-pressure manifold may cause casualties, serious channeling, and the risk of blowout due to unsealed high-pressure oil, gas, and water layers.

(5) Risk of oil testing operation: pipeline rupture, joint leakage, wellhead Christmas tree puncture, pressure explosion, etc.

(6) The pollution of wastewater, waste residue and exhaust gas generated from related homework to the environment.

（11）环境污染，包括修建道路和井场对植被的破坏，作业污水、生活污水及有害气体对大气的污染。

（12）海上钻井的风险，如海浪、台风等恶劣天气的危害，平台倾斜、倒塌、撞船及迷航等。

（13）社会环境带来的风险，如不法分子侵袭、战争、骚乱等。

4.2.1.2.2 相关作业风险

（1）测井作业风险：放射性伤害、射孔弹误发伤人危险、测井仪器落井危险等。

（2）录井作业风险：使用的天然气样标准瓶泄漏、野蛮装卸可能造成火灾爆炸、使用三氯甲烷等有毒物料可能造成中毒危险、使用强酸性物质可能造成人员皮肤腐蚀或烧伤危险等。

（3）定向井作业风险：测斜绞车伤人、定向井工具落井危险等。

（4）固井作业风险：高压管汇泄漏可能造成人员伤亡、严重窜槽、未封住高压油气水层发生井喷危险。

（5）试油作业风险：管线爆裂、接头泄漏、井口采油树刺漏、压爆等。

（6）相关作业产生的废水、废渣、废气对环境的污染。

The risks of drilling shown in Table 4.1.　　　钻井作业主要风险见表 4.1。

Table 4.1　The Main Risks of Drilling Operations

| Drilling operations | Primary hazards | Primary impacts |
| --- | --- | --- |
| Construction of wellsite | Destruction of vegetation | Destruction of ecological environment |
| Construction of offshore drilling platform | Causing local environmental damage in the marine environment | Impact on coral reefs and marine life |
| Drilling | Drilling equipment generates noise | Impact on the normal lives of humans and animals |
| Tripping | Blowout | Threat to personal and property safety |
| Lowering | Leakage | Contamination of groundwater sources |
| Wellhead operations | Dropped objects and accidents | Endangering personal safety |
| Loss of well control | Fire | Threat to personal and equipment safety, pollution of the environment |
| Hydrogen sulfide and gas/oil leakage | Toxicity and explosions | Threat to personal and equipment safety, pollution of the environment |
| Drilling fluid additives and raw materials | Corrosion and irritation of the skin, dust, toxicity | Endangering human health |
| Drilling fluid and operation wastewater | Environmental damage | Impact on crops and plant growth around the well and contamination of groundwater |
| Cementing operations | Lost circulation triggering blowout | Threat to personal and property safety |
| Well logging operations | Radioactive source leakage | Endangering human health, polluting the environment |
| Well testing operations | Oil, hydrocarbon gas spillage, fire and explosion, | Threat to personal and property safety |
| Disposed drill cuttings and waste slurry | Environmental damage | Destruction of ecological environment |
| Equipment maintenance and servicing | Generation of waste and oil spills | Pollution of the environment |
| Campsite | Generation of fousehold waste | Pollution of the environment |
| Drought around the wellsite leading to dry plant fires | Fire | Endangering human and property safety, impacting habitat animals |

表 4.1　钻井作业主要风险

| 项目 | 主要危害 | 主要影响 |
| --- | --- | --- |
| 修建井场 | 破坏植被 | 破坏生态环境 |
| 修建海上钻井平台 | 造成海洋环境局部破坏 | 影响珊瑚礁和海洋生物 |
| 钻进 | 钻井设备产生噪声 | 影响人和动物的正常生活 |
| 起钻 | 井喷（潜在） | 威胁人身及财产安全 |
| 下钻 | 井漏（潜在） | 污染地下水源 |
| 井口操作 | 落物及意外事故 | 危害人体健康 |
| 井喷失控 | 着火（潜在） | 威胁人和设备安全，污染环境 |
| 硫化氢及油气逸出 | 毒性、着火、爆炸 | 威胁人和设备安全，污染环境 |

续表

| 项目 | 主要危害 | 主要影响 |
|---|---|---|
| 钻井液处理剂及原材料 | 腐蚀刺激皮肤、粉尘、毒性 | 危害人体健康 |
| 钻井液及作业污水 | 破坏环境 | 影响井场周围农作物、植物生长，污染地下水 |
| 固井作业 | 水泥失重诱发井喷 | 威胁人身及财产安全 |
| 测井作业 | 放射源泄漏（潜在） | 危害人体健康，污染环境 |
| 试油作业 | 原油、烃类气体逸出、火灾爆炸 | 污染环境，威胁人身及财产安全 |
| 排出的钻屑及废浆 | 破坏环境 | 破坏生态环境 |
| 设备维护、保养 | 产生废弃物、油污 | 污染环境 |
| 营地 | 产生生活垃圾 | 污染环境 |
| 井场周围干燥植物着火 | 火灾 | 危害人财安全，影响栖息的动物 |

## 4.2.2 HSE Risk Control Objectives for Drilling Operations

When setting management goals, the principles of rationality, objectivity, verifiability, and feasibility should be followed. The HSE management objectives for drilling operations include two parts: overall objectives and specific objectives. The former is a major principle objective, while the latter is a specific or even quantifiable objective.

### 4.2.2.1 Overall Objective

(1) Regularly promote, educate, and train employees on HSE continuously improving their awareness and level of health, safety, and environmental protection.

(2) Integrate HSE management throughout the entire drilling construction process to minimize various risks.

(3) Create a safe and healthy work environment, ensure the health and safety of every employee, and improve work quality.

(4) Eliminate or minimize environmental pollution, protect the ecological environment, and minimize the impact of drilling operations on the environment.

## 4.2.2 钻井作业 HSE 风险控制目标

在制订管理目标时，应遵循合理性、客观性、可验证性和可实现性的原则。钻井作业 HSE 管理目标包括总体目标和具体目标两部分，前者为大的原则性目标，后者为具体甚至是可量化的目标。

### 4.2.2.1 总体目标

（1）经常对员工进行健康、安全与环境保护方面的宣传、教育与培训，不断提高员工的健康、安全与环境保护的意识和水平。

（2）将健康、安全与环境保护管理工作贯穿于钻井施工的全过程，使各种风险降低至最低程度。

（3）创造安全和健康的工作环境，确保每位员工的健康与安全，提高工作质量。

（4）杜绝或尽可能减少环境污染，保护生态环境，把钻井作业对环境的影响降低到最低程度。

(5) Striving towards the goal of no accidents, no pollution, and establishing a first-class corporate image.

#### 4.2.2.2 Specific Objective

Based on the overall objective and combined with the actual situation of the well, specific, achievable, or should be achieved HSE management objective are formulated, such as:

(1) Eliminate major personal injury accidents;

(2) Eliminate blowout and uncontrolled accidents;

(3) Eliminate major environmental pollution accidents;

(4) Eliminate fire and explosion accidents;

(5) Other accident rates;

(6) Sewage discharge volume;

(7) Pollution control rate;

(8) The standard discharge rate of sewage;

(9) Employee medical examination pass rate;

(10) Employee HSE training qualification rate, etc.

### 4.2.3 HSE Emergency Classification for Drilling Operations

Through risk analysis, propose emergency response plans for preventing and dealing with various unexpected accidents and potential accidents during drilling operations, and conduct strict training and simulation exercises in accordance with the requirements of the emergency response plans to improve the emergency response capabilities of employees. When various emergency situations occur during drilling operations, it can ensure the safety of employees and national property, and minimize various losses and impacts.

（5）向无事故、无污染、树立一流企业形象的目标迈进。

#### 4.2.2.2 具体目标

根据总体目标，结合本井实际，制订出具体的、可达到或应该达到的健康、安全与环境管理目标，如：

（1）杜绝重大人身伤亡事故；

（2）杜绝井喷及井喷失控事故；

（3）杜绝重大环境污染事故；

（4）杜绝火灾、爆炸事故；

（5）其他事故率；

（6）污水排放量；

（7）污染治理率；

（8）污水达标排放率；

（9）员工体检合格率；

（10）员工 HSE 培训合格率等。

### 4.2.3 钻井作业 HSE 应急分类

通过风险分析，提出预防、处置钻井作业中各类突发事故和可能发生事故险情的应急反应计划，并且按照应急反应计划的要求，进行严格的训练和模拟演习，提高员工的应急处理能力。当钻井作业中发生各种紧急情况时，能确保员工和国家财产的安全，最大限度地降低各种损失和影响。

According to the process characteristics and working environment characteristics of drilling operations, emergency response can be divided into 5 categories, as shown in Table 4.2.

根据钻井作业的工艺特点和作业环境特点，应急反应可分为5类，见表4.2。

Table 4.2  Emergency Response Types

| Sequence number | Emergency types | Emergency scope |
| --- | --- | --- |
| 1 | Drilling operation emergencies | Blowout, loss of well control, fire, explosion, etc. |
| 2 | Personal injury | Burns, mechanical injuries, object impact, fall from heights, electrical shock, traffic accidents, etc. |
| 3 | Acute poisoning | $H_2S$, CO and food poisoning |
| 4 | Hazardous substance leakage | Leakage of petroleum, fuels, and other toxic substances |
| 5 | Natural disasters | Flooding, severe typhoons, storms, sandstorms, lightning, landslides, earthquakes, etc. |

表 4.2  应急反应类型

| 序号 | 应急类型 | 应急范围 |
| --- | --- | --- |
| 1 | 钻井作业突发事件 | 井喷、井喷失控、火灾、爆炸等 |
| 2 | 人身安全 | 烧伤、机械伤害、物体撞击、高处坠落、触电、交通事故 |
| 3 | 急性中毒 | $H_2S$，CO及饮食中毒 |
| 4 | 有害物质泄露 | 油料、燃料及其他有毒物质泄露 |
| 5 | 自然灾害 | 山洪、强台风、暴风雨、沙尘暴、雷击、山体滑坡、地震等 |

## 4.3  On Site Medical Emergency

## 4.3  现场医疗急救

### 4.3.1  On Site Medical Emergency Procedures

### 4.3.1  现场医疗急救程序

(1) If a person is injured, immediately stop the injury-causing operation, observe the condition of the injured person, and immediately report to the health worker (if poisoned, immediately transfer the injured person to a safe area for first aid).

（1）发现人员受伤，立即停止致伤作业，观察受伤者情况，并立即报告卫生员（中毒则立即将伤者转移至安全地带再急救）。

(2) After the health worker arrives, provide emergency treatment according to the situation.

(3) The health worker decides to send the patient to the hospital and the procedure based on the severity of the injury.

(4) The platform manager implements vehicles, routes, hospitals, nursing staff, etc. based on the decision of the health worker.

(5) The platform manager reports to the company, and if necessary, the company contacts the emergency center and takes rescue measures.

(6) Nursing staff must report the changes and treatment status of patients in the hospital to the higher-level medical and health department.

(7) The drilling team shall immediately investigate the cause after sending away the injured person, implement rectification or take preventive measures, and resume operations.

(8) Submit accident situation and handling report.

### 4.3.2　On Site Medical First Aid Measures

#### 4.3.2.1　Emergency Treatment Measures for $H_2S$ Poisoning

(1) Quickly transfer the poisoned person to a place with fresh air and leave the contaminated area.

(2) Immediately inhale oxygen, administer glucose water, and perform cardiopulmonary resuscitation in patients with cardiac arrest or respiratory arrest.

(3) Intravenous injection of 50% glucose water with vitamin C.

(4) For eye symptoms, 2% sodium bicarbonate solution can be used to wash the eyes or chloramphenicol eye drops.

（2）卫生员到场后根据情况进行急救处理。

（3）卫生员根据伤情决定送往医院及程序。

（4）平台经理根据卫生员的决定，落实车辆、线路、医院、护理人员等。

（5）平台经理向公司汇报，必要时公司应与急救中心联系并采取救援措施。

（6）护理人员必须将病员在医院的变化、治疗情况向上级医疗卫生部门汇报。

（7）井队送走伤员后立即查找原因，落实整改或采取防范措施后恢复作业。

（8）提交事故情况及处理报告。

### 4.3.2　现场医疗急救处理措施

#### 4.3.2.1　$H_2S$ 中毒急救处理措施

（1）迅速将中毒者转移到空气新鲜处，脱离污染区。

（2）立即吸氧、推注葡萄糖水，有心跳呼吸停止者立即做心肺复苏术。

（3）静脉推注50%葡萄糖水加维生素C。

（4）对眼部症状，可用2%碳酸氢钠液洗眼或氯霉素眼药水。

(5) After the patient's vital signs stabilized, they were transferred to the hospital for hospitalization and treatment.

(6) On site first aid for patients with hydrogen sulfide poisoning is very important. Blindly transferring or moving patients too much should be avoided to prevent delaying the rescue time, increasing death or worsening the condition.

### 4.3.2.2 Emergency Treatment Measures for CO Poisoning

(1) Get out of the environment.

(2) Oxygen inhalation.

(3) Send to the hospital for treatment depending on the situation.

### 4.3.2.3 Emergency Treatment Measures for Electric Shock Injuries

(1) Cut off the power.

(2) If the heartbeat or breathing stops, mouth to mouth breathing and cardiac compressions can be performed

(3) Drink sugar water.

(4) Use drugs that regulate nerves.

(5) Use sedatives.

### 4.3.2.4 Emergency Treatment Measures for Coma

(1) Targeting treatment for the underlying cause.

(2) Using supportive therapy, which involves maintaining airway patency, establishing infusion channels, correcting acid-base imbalances, and maintaining normal blood systems.

(3) Apply awakening agents.

(4) According to the condition, if necessary, transfer to the nearest hospital for treatment.

（5）病人生命体征平稳后，转送至医院住院治疗。

（6）硫化氢中毒病人的现场急救十分重要，切忌盲目转送或过多地搬动病人，以防贻误抢救时机，增加死亡或恶化病情。

### 4.3.2.2　CO中毒急救处理措施

（1）脱离环境。

（2）吸氧。

（3）视情况送往医院治疗。

### 4.3.2.3　电击伤急救处理措施

（1）切断电源。

（2）如心跳、呼吸停止，可进行口对口人工呼吸和心脏按压术。

（3）饮糖水。

（4）用调节神经的药物。

（5）用镇静药。

### 4.3.2.4　昏迷急救处理措施

（1）针对病因治疗。

（2）运用支持疗法，即维持呼吸道通畅，建立输液通道，纠正酸碱失衡，维持血液系统正常。

（3）应用苏醒剂。

（4）根据病情，必要时转就近医院治疗。

### 4.3.2.5 Emergency Treatment Measures for Shock

(1) Lie flat and slightly raise the lower limbs, providing oxygen and warmth.

(2) Targeting treatment for the underlying cause。

(3) Apply supportive therapy.

(4) Transfer to another hospital in a timely manner based on the condition.

### 4.3.2.6 Emergency Treatment Measures for Fractures

(1) Fix the injured limb (using triangular bandages, bandages, first aid kits, wooden clips, stretchers).

(2) If there is bleeding in an open fracture, stop the bleeding first, then cover the wound with sterile gauze or clean cloth, and wrap it before fixation.

(3) Apply painkillers.

(4) Quickly transfer to the nearest hospital for treatment.

### 4.3.2.7 First Aid Measures for Burns, Acid, Alkali and Chemical Burns

(1) Remove from the injury site (extinguish the fire on the injured person). If acid, alkali or other chemicals cause injury, rinse with clean water for a long time.

(2) If necessary, relieve pain and calm down.

(3) Correct shock.

(4) Correct dehydration.

(5) Transfer to another hospital for treatment.

### 4.3.2.8 Drowning Emergency Treatment Measures

(1) Quickly rescue the drowning person from the water, open their mouth, remove foreign objects from their respiratory tract.

### 4.3.2.5 休克急救处理措施

（1）平卧稍抬高下肢，给予吸氧、保暖。
（2）针对病因治疗。
（3）运用支持疗法。
（4）根据病情及时转院。

### 4.3.2.6 骨折急救处理措施

（1）固定伤肢（用三角巾、绷带、急救包、木板夹、担架）。
（2）开放性骨折如有出血情况，先止血后应用无菌纱布或干净布类覆盖伤口，并加以包扎后再固定。
（3）应用止痛剂。
（4）迅速转往就近医院治疗。

### 4.3.2.7 烧烫伤及酸、碱和化学品灼伤急救处理措施

（1）脱离致伤场所（灭掉伤员身上的火），若是酸、碱等化学品致伤，应用清水长时间冲洗。
（2）必要时止痛、镇静。
（3）纠正休克。
（4）纠正脱水。
（5）转院治疗。

### 4.3.2.8 溺水急救处理措施

（1）迅速将溺水者营救出水，打开其口腔，清除其呼吸道异物。

(2) Pour water according to the situation.

(3) Artificial respiration and chest compressions are performed simultaneously.

(4) Apply relevant emergency medication.

(5) Send to the hospital for treatment as appropriate.

4.3.2.9 Emergency Measures for Food Poisoning

(1) Inducing vomiting, gastric lavage and catharsis.

(2) Targeting treatment for the underlying cause.

(3) Correct electrolyte and acid-base balance disorder.

(4) Transfer to another hospital as appropriate.

4.3.2.10 Emergency Treatment Measures for Drug Allergies

(1) Subcutaneous injection of 1mL adrenaline

(2) Subcutaneous injection of 10mg dexamethasone.

(3) Infusion.

(4) Oxygen inhalation.

4.3.2.11 Measures for Handling Acute Infectious Diseases

(1) Immediately isolate the patient.

(2) Symptomatic treatment.

(3) Disinfect the site.

(4) Transfer to hospital for treatment.

4.3.2.12 Emergency Treatment Measures for Heat Stroke Caused by High Temperature

(1) Leave the hot environment and rest in a well ventilated and cool place.

(2) Cooling down.

(3) Correct water, electrolyte, and acid-base disorder.

(4) According to the condition, seek medical treatment if necessary.

（2）根据情况进行倒水处理。

（3）人工呼吸与胸外心脏按压同时进行。

（4）应用有关急救药物。

（5）酌情送医院治疗。

4.3.2.9 食物中毒急救处理措施

（1）催吐、洗胃、导泻。

（2）针对病因治疗。

（3）纠正电解质与酸碱平衡紊乱。

（4）酌情转院。

4.3.2.10 药物过敏急救处理措施

（1）皮下注射肾上腺素 1mL。

（2）皮下注射地塞米松 10mg。

（3）输液。

（4）吸氧。

4.3.2.11 急性传染病处理措施

（1）立即对病人进行隔离。

（2）对症处理。

（3）对现场进行消毒。

（4）转医院治疗。

4.3.2.12 高温中暑急救处理措施

（1）离开高热环境，到通风良好和阴凉的地方休息。

（2）降温。

（3）纠正水、电解质与酸碱紊乱。

（4）根据病情，必要时送医院治疗。

## Post-class exercises

The oil and gas industry can be classified into three major segments: upstream, midstream and downstream.

Upstream companies are involved in the exploration, development, and production of crude or natural gas, also called exploration and production or E&P.

Midstream operations create linkage between producing areas, which are typically remote, and the consumer base. Midstream activities include the processing, storage, marketing, and bulk transportation of petroleum commodities.

Downstream industry participants include oil refineries, petrochemical plants, natural gas distribution companies, distributors of petroleum products, and retail outlets.

## 课后练习

参考译文：

油气行业可以分为三大主要领域：上游、中游和下游。

上游公司涉及原油或天然气的勘探、开发和生产，也称为勘探和生产（Exploration and Production, E&P）。

中游业务在生产区域（通常是偏远地区）和消费群体之间建立联系。中游活动包括石油商品的处理、储存、营销和大规模运输。

下游行业的参与者包括炼油厂、石化厂、天然气分配公司、石油产品分销商和零售店。

# Chapter 5 Typical Accidents and Handling Measures

# 第五章 典型事故及处理措施

## 5.1 Drill Bit Accidents

Learning objectives: Master the types, causes, and accident signs of drill bit damage.

### 5.1.1 Tooth-wheel Drill Bit Accident

#### 5.1.1.1 The Types of Damage of Tooth-wheel Drill Bits

The failure types of tungsten carbide teeth in roller cone drill bits mainly include three forms: fracture, shedding and wear.

5.1.1.1.1 Fracture of Alloy Teeth

The fracture phenomena of alloy teeth include tooth breakage, falling blocks of different sizes, local cracking, etc. The failure modes are as follows:

(1) Impact peeling;
(2) Impact thermal fatigue;
(3) Impact fatigue.

5.1.1.1.2 The Detachment of Alloy Teeth

(1) Wear on the outer surface of the teeth.

The wear on the outer surface of the tooth wheel gradually exposes the alloy tooth root, and the bonding surface between the alloy tooth and the alloy tooth hole on the outer surface of the tooth wheel decreases. The

## 5.1 钻头事故

本章学习目标为掌握钻头损坏的类型、原因及钻头发生损害时的事故征兆。

### 5.1.1 牙轮钻头事故

#### 5.1.1.1 牙轮钻头的损伤类型

牙轮钻头硬质合金齿失效类型主要有断裂、脱落和磨损三种形式。

5.1.1.1.1 合金齿的断裂

合金齿的断裂现象包括齿折断、大小不同的掉块、局部崩裂等，其失效形式有以下几种：

（1）冲击剥落；
（2）冲击热疲劳；
（3）冲击疲劳。

5.1.1.1.2 合金齿的脱落

（1）牙轮外表面的磨损。

牙轮外表面的磨损使合金齿根逐渐显露，合金齿与牙轮外表面合金齿孔的结合面减少。牙轮表面的偏磨可能是钻柱振动使钻头横向振动或偏心旋转所造成的（使某一边局部瞬时受力过大造成偏磨），也可能

eccentric wear on the surface of the cone may be caused by the lateral vibration or eccentric rotation of the drill bit caused by the vibration of the drill string (resulting in excessive instantaneous force on one side), or it may be caused by wear at the joint between the teeth and the hole.

(2) Loose alloy teeth.

The material of the dental wheel is relatively soft, and under the repeated action of large impact and shear loads, it is easy to undergo plastic deformation due to the hard alloy teeth, causing gaps between the alloy teeth and the dental wheel alloy holes or reducing the weakening and loosening of the contact surface between the alloy teeth and the holes. Especially for the edge teeth, due to the lateral deviation of the drill bit and the much greater rock shear effect on it compared to the center teeth, this loosening is more severe.

(3) The erosion effect on the drill bit.

After the high-pressure liquid flow reaches the bottom of the well, it forms a solid-liquid two-phase fluid carrying many small rock cuttings. The small rock cuttings are discharged into the outer edge of the tooth as abrasives, continuously eroding the tooth and alloy teeth. When the fluid encounters the obstruction of the alloy teeth, it forms a bypass flow, causing the boundary layer to detach on the back of the alloy teeth, forming eddy currents. This kind of flow and eddy currents can cause wear on the interface between alloy teeth and tooth wheel, ultimately leading to tooth loss.

(4) The tightening force is low and the interference is not enough, which is mainly related to the fit clearance during processing.

是齿与孔结合处受磨损造成的。

（2）合金齿的松动。

牙轮材料相对较软，在较大冲击和剪切载荷的反复作用下受较硬的合金齿作用容易发生塑性变形，使合金齿与牙轮合金孔产生间隙或使合金齿与孔接触面积减小弱化而松动。尤其是边缘齿，由于钻头横向偏摆和所受岩石剪切作用较中心齿大得多，这种松动更严重。

（3）对钻头的冲蚀作用。

高压液流到达井底后，形成携带许多小岩屑的固液两相流体，向牙轮外缘漫流排出小岩屑则作为磨料不断磨蚀牙轮和合金齿，当流体遇到合金齿阻挡时形成绕流，在合金齿的背面产生附面层脱离，形成涡流。这种绕流和涡流会导致合金齿与牙轮界面磨损最终导致掉齿。

（4）固紧力较低，过盈量不够，这主要与加工时的配合间隙有关。

5.1.1.1.3 Wear of Alloy Teeth

Alloy teeth experience two body (alloy teeth, rock) abrasive wear and three body (alloy teeth, rock debris, rock) abrasive wear during the working process. Due to different rock types, there are both hard abrasive wear and soft abrasive wear. The abrasive particles on the surface of alloy teeth have both short sliding and rolling. There are several forms of wear for alloy teeth:

(1) High stress crushing abrasive wear;

(2) The wear caused by the sliding and rolling of abrasive particles on the surface of alloy teeth, including pushing, rolling, and cutting;

(3) Scour wear;

(4) Scraping effect.

5.1.1.1.4 Bearing Accidents

The inner cavity of the drill bit cone and the neck of the cone form a semi enclosed area with a row of rollers. This structure is a pair of rolling bearings, which play a role in controlling the free rotation of the cone around the axis of the cone. The main reason for the failure here is that the drill bit is subjected to complex forces during operation, and the side is subjected to stress caused by shear reactions due to rock changes during rotation. At the same time, high-frequency impacts bring about certain vibrations and friction, making it easy for stress to concentrate and produce plastic deformation, leading to bearing failure. The main forms of bearing accidents include fracture, plastic deformation, and wear. The probability of bearing wear is high, but there is a process of bearing wear, from tight fitting to loose fitting, to ball drop, tooth jamming, tooth eccentric wear, and finally tooth detachment, which is a relatively long process. There are two main mechanisms of its occurrence.

5.1.1.1.3 合金齿的磨损

合金齿在工作过程中存在两体(合金齿、岩石)磨料磨损和三体(合金齿、岩屑、岩石)磨料磨损。由于岩性不同,同时存在硬磨粒磨损和软磨粒磨损。磨粒在合金齿表面既有短程滑动又有滚动。合金齿的磨损形式有以下几种:

(1)高应力碾碎性磨粒磨损;

(2)磨粒在合金齿表面滑滚产生的推挤、碾压、切削作用磨损;

(3)冲刷磨损;

(4)刮削作用。

5.1.1.1.4 轴承事故

钻头牙轮内腔与牙轮轴颈形成一个半封闭区域,区域内有一排滚柱,这一结构就是一副滚动轴承,起到支配牙轮绕牙轮轴线自由转动的作用。此处失效主要是钻头工作时受力较复杂,回转时侧面受到由于岩石变化的剪切反力作用造成的应力,同时高频冲击又带来一定的振动和摩擦,使应力很容易集中而产生塑性变形,导致轴承失效。轴承事故的主要形式有断裂、塑性变形和磨损。轴承磨损发生的概率较大,但轴承的磨损有个过程,由紧配合到松配合到滚珠掉落、牙轮卡死、牙轮偏磨,直至牙轮脱落是一个较长的过程。其发生机理主要有下述两种。

(1)Mismatch in hardness of tooth wheel, tooth wheel shaft and rollers.

The material of the tooth wheel, tooth wheel shaft and roller is different, and each part plays a different role, requiring different hardness. As for the teeth and teeth of the bearing inner and outer shell, the hardness should not be too high, and the hardness of the rollers should not be too low. If the hardness of the roller is too high, the tooth wheel and tooth wheel shaft will wear out prematurely. If the hardness of the roller is too low, the roller will wear out prematurely, leading to excessive bearing clearance and failure. From this, it can be seen that the hardness of the tooth wheel, tooth wheel shaft and roller should be matched reasonably.

(2) Unreasonable fit gap.

The fitting space of standard bearings is very small and the requirements are also strict, but as a bearing structure for roller drills, it cannot be treated completely according to standard bearings. The clearance between the teeth, teeth shaft, and roller should not be too small, otherwise the teeth will get stuck and cannot rotate freely; If the clearance is too large, mud and sand will enter the bearing, and the cooling and lubrication cannot keep up, it will accelerate the abnormal wear of the bearing, leading to bearing failure.

5.1.1.1.5　Other Accidents

(1) Drill bit "mud package". Shale or mudstone becomes sticky after absorbing water from drilling fluid, which easily adheres to the drill bit and causes "mud package".

(2) Outer diameter wear. Due to poor cleaning effect at the bottom of the well and severe repeated crushing, the outer row teeth and back cone teeth are severely worn, and

（1）牙轮、牙轮轴、滚柱的硬度不匹配。

牙轮、牙轮轴与滚柱的材质不同，每个零件所起的作用也不同，要求其硬度不同。作为轴承内、外壳的牙轮和牙轮轴而言，硬度不宜过高，而滚柱的硬度不宜过低。若滚柱的硬度太高，牙轮、牙轮轴会过早磨损；若滚柱的硬度过低，滚柱过早磨损而导致轴承间隙太大而失效。由此可知，牙轮、牙轮轴与滚柱的硬度并不是越硬越好，而是应该匹配合理。

（2）配合间隙不合理。

标准轴承的配合间晾很小，要求也严格，但作为牙轮钻头的轴承结构，不能完全按照标准轴承来对待。牙轮、牙轮轴与滚柱之间的配合间隙不宜过小，否则牙轮被卡住，不能自如转动；若配合间隙过大，泥砂进入轴承内，加之冷却、润滑跟不上会加速轴承的不正常磨损，从而导致轴承失效。

5.1.1.1.5　其他事故

（1）钻头"泥包"。页岩或泥灰岩从钻井液中吸水后变黏，极易黏附在钻头上造成"泥包"。

（2）外径磨损。由于井底清洁效果差，重复破碎严重，造成外排齿和背锥齿严重磨损、非常规的井眼或井斜原因引起钻柱异常振动，造成外排齿和背锥齿严重磨损。

abnormal vibration of the drill string caused by unconventional wellbore or wellbore inclination reasons, resulting in severe wear of the outer row teeth and back cone teeth.

(3) Grind the heart. At the bottom of the well, there are peeling rock blocks from the upper rock layer, or the rock in the formation has strong abrasiveness and hardness, and the broken drill bit at the bottom of the well is irregular, which can easily lead to grinding. The main feature of grinding is the absence of the core of the drill bit, which in severe cases can cause the locking device of the cone to fail and the cone to fall into the well.

(4) The drill bit blocks the water hole and drops the nozzle.

(5) Accidents such as falling tooth wheel and getting stuck teeth also occur from time to time.

### 5.1.1.2 The Causes of Tooth-wheel Drill Bit Accidents

There are various reasons that can cause accidents with tooth-wheel drills, such as poor drill quality, illegal operation, inaccurate judgment, and excessive use. Some of these factors are independent, while others influence each other. Based on on-site experience, the main causes of drill bit accidents are analyzed as follows.

#### 5.1.1.2.1 Quality Issues with the Product

When the tooth-wheel drill bit operates within the manufacturer's specified parameters, it still experiences roller cracking, weld seam cracking, palm fracture, connection thread fracture, seal failure, early bearing wear, tooth loss and water hole loss. Apart from certain external factors, these phenomena, are mainly caused by certain defects in product quality.

（3）磨心。井底有上部岩层剥落岩块或者地层岩石研磨性强、硬度大，钻头破碎井底不规则容易导致磨心。磨心的主要特征是起出钻头心部缺失，严重者造成牙轮锁紧装置失效，牙轮落井。

（4）钻头堵水眼，掉喷嘴。

（5）掉牙轮、牙轮卡死等事故也时有发生。

### 5.1.1.2 牙轮钻头事故的发生原因

引发牙轮钻头事故的原因多样，如钻头质量不好、违章操作、判断不准、超时使用等。这些因素有些是独立的，有些又相互影响，根据现场经验，钻头事故主要原因分析如下。

#### 5.1.1.2.1 产品质量问题

牙轮钻头在厂家规定的参数内运行时，仍然发生牙轮裂开、焊缝裂开、巴掌断裂、连接螺纹断裂、密封失效、轴承早期磨损、掉齿、掉水眼等。除某些外界因素影响外，这些现象主要是产品质量上存在某种缺陷而引起的。

#### 5.1.1.2.2 The Selected Drill Bit Loader is Unreasonable

Bruising the tooth wheel and teeth during fastening, or failing to fasten according to the specified torque, causing looseness or tightness. In addition, if the tooth wheel is not properly moved before drilling, it may cause the tooth wheel to get stuck, and the tooth wheel may fall out of the well after eccentric wear during drilling.

#### 5.1.1.2.3 Improper Use of Parameters

Excessive drilling pressure and rotation speed can cause premature fatigue of the drill bit; uneven drilling results in impact loads on the drill bit; slipping and stuttering drills can cause the drill bit to withstand a huge load exceeding its design capacity in a short period of time, which is enough to cause drill bit accidents. It is impossible for all the tooth wheels of tooth-wheel drill bit to fall into the well at the same time. One or half of the cones always fall into the well first, and if the operator does not check, the surface pump pressure changes, and there is already a complete drilling, skip drilling, or a sharp decrease in drilling depth underground, they are still unwilling to pull out and need to repeatedly test, which will break the second and third tooth wheels into the well and worsen the accident.

#### 5.1.1.2.4 Improper Operation

When encountering obstacles or jamming during the drilling process, improper operation and sudden lifting and pressing can not only easily cause jamming, but also damage the drill bit. In the late 1950s, accidents such as pulling out the cone during drilling and encountering sunlight frequently occurred in the Yumen Oilfield, and in the

#### 5.1.1.2.2 选用的钻头装卸器不合理

上扣时碰伤牙轮及牙齿或者没有按规定上扣扭矩上扣，过松或者过紧。此外，若开始钻进前未活动好牙轮，造成卡牙轮，钻进中偏磨后也会发生牙轮掉井。

#### 5.1.1.2.3 使用参数不当

过高的钻压或过高的转速，使钻头过早地产生疲劳现象；送钻不均，使钻头产生冲击负荷；溜钻、顿钻又使钻头在短期内承受超过其设计能力的巨额载荷，这些都足以造成钻头事故。牙轮钻头的牙轮不可能在同时间内全部落井，总是先有一个或半个落井，而操作者不检查，地面泵压发生变化，井下已有整钻、跳钻或进尺锐减的情况下，仍不甘心起钻，而要反复试探，这样就把第二个、第三个牙轮蹩入井下，使事故更加恶化。

#### 5.1.1.2.4 操作不当

起下钻过程遇阻、遇卡时，操作不当，猛提猛压，除容易造成卡钻外，也容易使钻头受损。如20世纪50年代末在玉门油田经常发生起钻遇阳时拔掉牙轮的事故；70年代初在江汉油田经常发生下钻时顿掉牙轮的事故，除钻头质量问题之外，操作不当也是发生事故的主要原因。

early 1970s, accidents such as pulling off the cone during drilling were common in the Jianghan Oilfield. In addition to quality issues with the drill bit, improper operation was also the main cause of accidents.

5.1.1.2.5　Overtime Use

Any drill bit has a reference lifespan under a certain drilling method. The quality of drill bits produced by the same manufacturer is not uniform, and their service life also varies. So the usage time of the drill bit introduced by the manufacturer cannot be the only basis, but should be analyzed and judged based on the actual working situation of the drill bit underground, and the drilling time should be determined.

5.1.1.2.6　Improper Control of Drill Bit Usage Time in Screw Drills

The tooth wheel accidents that occur when the screw drill tool is oriented or twisted account for 40% to 60% of the total number of drill bit accidents. Because the screw drill tool does not rotate, it is relatively difficult for the ground to judge the downhole drill bit, and it is even more necessary to pay close attention to its changes.

5.1.1.2.7　Lack of Operator Experience

Experienced operators can easily detect abnormal phenomena in the drill bit, such as slow footage, increased torque, uneven rotation speed, broken drilling and jumping drilling.

5.1.1.2.8　Improper Connection

The specification of the connecting thread between the drill bit and the joint is inconsistent, or the internal thread is enlarged under high torque, or uncontrolled reverse driving causes the entire drill bit to fall into the well.

5.1.1.2.5　超时使用

任何钻头在一定的钻井方式下都有一个可供参考的使用寿命。同一个厂家生产的钻头，其质量也非一致，工作寿命也是有区别的。所以不能以厂家介绍的钻头使用时间作为唯一的依据，而应根据钻头在井下的实际工作情况分析判断，以决定起钻时间。

5.1.1.2.6　螺杆钻中钻头使用时间控制不当

螺杆钻具在定向、扭方位时所发生的牙轮事故占钻头事故总数的40%~60%，因为螺杆钻钻具不转，地面对井下钻头的判断相对困难，更需要密切注意其变化。

5.1.1.2.7　操作者经验欠缺

有经验的操作者，不难发现钻头异常现象，如进尺减慢、扭矩增大、转速不匀、蹩钻、跳钻等。

5.1.1.2.8　连接不规范

钻头与接头连接螺纹的规范不一致，或者是在大扭矩下将内螺纹胀大，或者不加控制的打倒车造成整个钻头落井。

5.1.1.2.9  Stuck Drilling Causes Drill Bit Accidents

Due to mechanical failures, tool failures, operational errors, or drilling tools getting stuck during milling, it is possible for the drilling tools to hit the bottom of the well, exerting significant impact force on the drill bit, resulting in some or all of the drill bit falling into the well.

5.1.1.2.10  Falling Objects from the Wellhead

Except for normal drilling, the wellhead is often in an unclosed state. There are more opportunities for wellhead tools and their parts to fall into the well during drilling, connecting, and other operations, with larger ones such as slips and elevators, and smaller ones such as hammers and pry bars. Some of these objects fall into the bottom of the well, some adhere to the wellbore or sit on the steps, some are horizontal in the middle of the wellbore, blocking the passage of the drill string, causing accidents such as drill bit scratches and tooth collapse.

5.1.1.3  Symptoms of Tooth-wheel Drill Bit Accidents

The complexity of tooth-wheel drill bit accidents determines that the signs of tooth-wheel drill bit accidents are also complex and varied, with different manifestations. There are mainly several typical accident signs as follows.

5.1.1.3.1  Tooth Polishing

The signs of tooth polishing are a reduction in the load on the rotary table, no bouncing of the square drill rod, a significant decrease in drilling speed or no footage, a stable weight indicator without swinging, and normal pump pressure:

5.1.1.3.2  Bearing Damage

The signs of bearing damage are periodic alarm jumps on the rotary table, with light

5.1.1.2.9  顿钻造成钻头事故

由于机械故障，或工具失效，或操作失误，或套铣中途遇卡的钻具，都有可能使钻具顿入井底，给钻头施以极大的冲击力，造成部分或整个钻头的落井事故。

5.1.1.2.10  从井口落入物件

除正常钻进外，井口往往处于不封闭状态。在起下钻、接单根及其他作业过程中，井口工具及其零件落入井中的机会较多，大者如卡瓦、吊卡，小者如手锤、撬杠等。这些物件有的落入井底，有的黏附在井壁或坐落在壁阶上，有的横在井筒中间，阻断了钻柱运行的通道，导致钻头划伤、崩落牙齿等事故的发生。

5.1.1.3  牙轮钻头事故的征兆

牙轮钻头事故的复杂性决定牙轮钻头事故的征兆也复杂多变，表现不一，主要有下述几种比较典型的事故征兆。

5.1.1.3.1  牙齿磨光

牙齿磨光的征兆是转盘负荷减轻，方钻杆无蹩跳，钻速明显下降或无进尺，指重表指示平稳无摆动、泵压正常。

5.1.1.3.2  轴承损坏

轴承损坏的征兆是转盘出现周期性蹩跳，钻压小蹩跳轻，钻压大则蹩跳重，钻

jumps when drilling pressure is low, and heavy jumps when drilling pressure is high. The drilling speed decreases, the pump pressure is normal, and the weight indicator pointer swings.

#### 5.1.1.3.3　Tooth Wheel Jamming

The signs of tooth wheel jamming are an increase in the load on the turntable, a jump in the turntable chain, poor performance of the square drill rod, stopping the turntable and reversing it, a decrease in drilling speed, and severe swinging of the weight indicator pointer.

#### 5.1.1.3.4　Tooth Loss

The signs of tooth loss are an increase in the load on the turntable, severe jumping of the turntable chain, stopping the turntable and reversing it, severe drilling failure, swinging the pointer of the weight indicator back and forth, a significant decrease in drilling speed or no footage, changing the direction of lifting the drill string, and a change in the direction of downward exploration, with an elevation difference of about one tooth height.

#### 5.1.1.3.5　Drill Bit "Mud Package"

The signs of drill bit "mud package" are an increase in the load on the rotary table, a jumping phenomenon, a decrease in drilling speed, varying degrees of sticking when lifting the drill string, an increase in pump pressure, and in severe cases, pump failure.

### 5.1.2　PDC Drill Bit Accident

#### 5.1.2.1　The Failure Modes of PDC Drill Bit

Polycrystalline diamond composite drill bit (PDC drill bit for short) is an integrated drill bit that does not have moving parts like a roller drill bit. Its failure mode mainly manifests as the failure of the diamond cutting edge, and

速下降，泵压正常，指重表指针有摆动。

#### 5.1.1.3.3　牙轮卡死

牙轮卡死的征兆是转盘负荷增大，转盘链条跳动，方钻杆有蹩劲，停转盘打倒车，钻速下降，指重表指针摆动严重。

#### 5.1.1.3.4　掉牙轮

掉牙轮的征兆是转盘负荷增大，转盘链条严重跳动，停转盘打倒车，蹩钻严重，指重表指针来回摆动，钻速明显下降或无进尺，上提钻柱变换方向，下探方入有变化，高差约为一牙轮高度。

#### 5.1.1.3.5　钻头"泥包"

钻头"泥包"的征兆是转盘负荷增大，有跳钻现象，钻速下降，上提钻柱有不同程度的挂卡，泵压上升，严重时蹩泵。

### 5.1.2　钻头事故

#### 5.1.2.1　钻头的失效形式

聚晶金刚石复合片钻头（简称 PDC 钻头）是一体性钻头，没有牙轮钻头那样的活动部件，它的失效形式主要表现为金刚石切削刃的失效，也有部分由于胎体磨损或者脱焊导致钻头报废。结合现场经验，归

some drill bits are scrapped due to matrix wear or detachment. Based on on-site experience, the main failure modes of PDC drill bit cutting edges are summarized as follows.

5.1.2.1.1　Smooth Wear and Tear

The process of smooth wear: during the cutting process, because the hardness of the hard alloy matrix is lower than that of diamond, the hard alloy matrix is the first to be worn. As a result, the diamond near the hard alloy matrix loses its effective support, forming a diamond "lip" edge; under the action of cutting force, the lip bears tensile stress, which leads to the initiation and propagation of tensile stress cracks, ultimately resulting in the fracture of the lip and the rupture of the unbroken diamond layer diamond sheet. Once the entire contact surface of the diamond sheet is damaged, the hard alloy matrix effectively contacts the rock again, and then the hard alloy matrix is preferentially worn off, forming a cycle of smooth wear process. Due to the formation of the "lip" region, the contact area between the remaining diamond and the rock decreases, resulting in an increase in cutting force per unit area and the formation of the self sharpening effect, which is beneficial for maintaining the effective cutting ability of the drill bit. Compared to other forms of failure, the smooth wear process is slow and belongs to the normal form of failure.

5.1.2.1.2　Macroscopic Fracture

Macroscopic fracture is manifested as the rupture of a large-sized diamond layer, and its crack originates from the cylindrical surface of the diamond sheet. It is the most severe form of failure for PDC cutting teeth. When the drill bit encounters hard rocks or rock layers with significant changes in lithology during

纳 PDC 钻头切削刃主要失效形式如下所述。

5.1.2.1.1　平滑磨损

平滑磨损的过程：切削过程中，因为硬质合金胎体硬度低于金刚石，所以率先遭磨损的是硬质合金胎体，这样临近硬质合金胎体的金刚石就失去了硬质合金胎体的有效支撑，形成金刚石"唇"边；在切削力的作用下，唇边承受着拉应力，并导致拉应力裂纹萌生、扩展，最终唇边断裂，导致未破裂的金刚石层金刚石片的破裂，一旦金刚石片整个接触面均遭到破坏，则硬质合金胎体又重新有效地接触岩石，接着又发生硬质合金胎体优先被磨损掉，形成平滑磨损过程的循环。正由于"唇"区域的形成，使余下的金刚石与岩石的接触面积减小，使单位面积的切削力增大，而形成自锐效应，自锐效应有利于保持钻头的有效切削能力。比较其他失效形式，平滑磨损过程是缓慢的，属于正常的失效形式。

5.1.2.1.2　宏观断裂

宏观破裂表现为大尺寸的金刚石层的破断，其裂纹起源于金刚石片的圆柱面。它是 PDC 切削齿破坏最为严重的一种失效形式，由于钻头在钻进过程中遇到硬质岩石或岩性变化较大的岩层时，钻头受到较大的冲击载荷，尤其是 PDC 钻头切削齿与岩石接触面较小时，致使切削齿在短时间

the drilling process, the drill bit is subjected to significant impact loads, especially when the contact surface between the PDC cutting teeth and the rock is small, causing the cutting teeth to bear overload in a short period of time, resulting in large-scale macroscopic damage and the drill bit being scrapped.

On site usage experience has shown that when there are damaged drill bit fragments or rigid objects at the bottom of the well that have not been salvaged and cleaned in a timely manner, it can also cause abnormal impacts on the working drill bit, resulting in macroscopic fracture, tooth breakage, tooth loss, tooth detachment, and other specific manifestations of macroscopic fracture.

5.1.2.1.3  Stripping

Stripping is commonly known as peeling off. During the cutting process, the cutting teeth rise to a high temperature due to frictional heat, and when the drill bit briefly detaches from contact with the rock layer due to vibration, it is rapidly cooled by the drilling fluid. Due to the difference in thermal expansion coefficients between the layers of PDC drill bits, there is a significant difference in thermal expansion and contraction between the layers, resulting in significant internal stress. When it exceeds the bonding strength of the bonding layer, it causes peeling, resulting in the disappearance of the cutting edge and loss of cutting ability. This can be manifested as the detachment of the bonding layer and the detachment of the polycrystalline layer.

Stripping failure of drill bits is often accompanied by macroscopic fracture failure, and it can be considered that the short-term impact overload borne by the drill bit is also one of the factors that promote peeling failure.

内承受超负荷而导致发生大尺度的宏观破裂，从而使钻头报废。

现场使用经验表明，当井底存在有破损的钻头碎块或刚性物而未被及时打捞清理时。也会导致工作钻头遭受非正常的冲击，使钻头发生宏观断裂、断齿、齿缺损、齿脱落等都是宏观断裂的具体表现形式。

5.1.2.1.3  剥离

剥离俗称掉片。在切削过程中，切削齿因摩擦热而升到高温，而当钻头因振动等短时脱离与岩层接触时，又被钻井液急冷。由于PDC钻头各层间热膨胀系数差异，导致PDC钻头各层热胀冷缩的差异，造成极大的内应力，当其超过黏结层结合强度时，就造成剥离，致使刃口不复存在而失去切削能力，具体可以表现为黏结层脱落和聚晶层脱落。

剥离失效的钻头，常伴随有宏观破裂的失效，可认为钻头承受的短时冲击超负荷也是促使剥离失效的因素之一。

5.1.2.1.4 Thermal Cracking

Thermal cracking is manifested by the formation of a certain depth of network turtle cracks on the surface of PDC materials. Thermal cracking occurs in both the hard alloy matrix layer and the diamond layer. It is the result of alternating cold and hot stress, especially on the surface of materials, where the highest cold and hot stress leads to the initiation of hot cracks on the material surface. The generation of thermal cracking first requires the formation of a large area of wear surface to generate sufficient heat, and then it will occur on the surface of the material due to sufficient thermal stress in multiple cold and hot cycles. Thermal cracking is also an inevitable result of wear failure.

In addition, PDC drill bits have other types of drill bit accidents, such as nozzle blockage and nozzle drop caused by drill bit "mud package"; drill bit body detachment, resulting in drill bit scrapping, etc.

5.1.2.2 The Causes of PDC Drill Bit Accidents

From the damage characteristics of PDC drill bits, it can be seen that when working underground, PDC drill bits are not only affected by the reaction forces and rock wear caused by normal cutting of the formation, but also by additional forces. This additional force is the main cause of abnormal damage to PDC drill bits. After statistical analysis of the usage of PDC drill bits, it is believed that the failure of PDC drill bits is due to the following reasons.

(1) There is an issue with the product quality. Poor drilling process or immature diamond coating technology.

(2) Choosing the wrong drill bit loader or damaging the diamond cutting edge. When

5.1.2.1.4 热龟裂

热龟裂表现为PDC材料表面形成一定深度的网状龟裂纹。在硬质合金胎体层及金刚石层均会发生热龟裂。它是冷热应力交变作用的结果，尤其是材料表面，冷热应力最高故热裂纹萌生于材料表面。热龟裂的产生首先要形成大面积的磨损平面而产生足够的热量，接着才会在多次冷热循环中，因热应力足够大，导致材料表面热龟裂的发生。热龟裂也是磨损失效的必然结果。

此外，PDC钻头还有其他方面的钻头事故，如钻头"泥包"引起堵喷嘴、掉喷嘴等；钻头胎体脱落导致钻头报废等。

5.1.2.2 钻头事故的发生原因

从PDC钻头的损坏特征来看，PDC钻头在井下工作时，除了正常切削地层而受到的反作用力和地层岩石磨损外，还受到了额外作用力的影响，这个额外作用力是造成PDC钻头非正常损坏的主要原因。在统计和分析PDC钻头的使用情况后，认为PDC钻头的失效有以下原因。

（1）产品质量有问题。钻头接工艺差或金刚石涂层工艺欠成熟。

（2）选用错误钻头装卸器或者碰伤了金刚石切削刃。接PDC钻头要使用专用的钻头装卸器，且不能使钻头工作面上的金刚石颗粒与金属碰撞。

connecting PDC drill bits, a dedicated drill bit loader should be used, and the diamond particles on the working surface of the drill bit should not collide with the metal.

(3) Stratigraphic reasons. There are many gravel or hard interlayers in the formation.

① The lower part of the formation contains gravel layers of varying thicknesses. When PDC bits drill into this well section, the PDC teeth often produce phenomena such as fragmentation and fragmentation when cutting gravel, which is the main reason for the early damage of PDC bit cutting teeth in upper formation.

② The presence of numerous hard interlayers in the formation is the main cause of PDC drill bit composite fragmentation, chipping, or tooth column cutting fractures. Because during the drilling process of PDC drill bits, when drilling from soft formations to hard interlayers, the crown contour of the PDC drill bit causes the cutting teeth on the surface of the drill bit to have different hardness in contact with the formation, resulting in uneven force on the cutting teeth when cutting the formation. This causes the drill bit to experience complete drilling and skip drilling, and most of the load acting on the drill bit is concentrated on a few cutting teeth cutting the hard interlayers The instantaneous load generated by jumping drilling causes these cutting teeth to shatter or break due to high force (especially when cutting certain hard points in hard formations, the instantaneous load generated is enough to cause the cutting teeth to shatter or break). If the PDC composite sheet is not firmly welded to the drill bit matrix, it will cause the composite sheet to fall off when subjected to a large instantaneous

（3）地层原因。地层中存在砾石或者硬夹层较多。

① 地层的下部含有不等厚度的砾石层，PDC 钻头钻进该井段，PDC 齿切削砾石时常常产生崩片、碎裂等现象是造成 PDC 钻头切削齿在上部地层先期损坏的主要原因。

② 地层中硬夹层较多是 PDC 钻头复合片碎裂、掉片或齿柱式切削断裂的主要原因。因为 PDC 钻头在钻进过程中，从软地层钻至硬夹层时，由于 PDC 钻头的冠部轮廓使得钻头表面的切削齿接触地层的硬度不一样，造成切削齿削地层时受力不均，使钻头出现整钻、跳钻现象，作用在钻头上的载荷大部分集中在切削硬夹层的几个切削齿上，蹩钻、跳钻产生的瞬时载荷导致这几个切削齿因受力较大而碎裂或折断（特别是在切削硬地层中某些硬质点时，瞬时产生的载荷足以造成切削齿的碎裂或切削齿折断）。如果 PDC 复合片和钻头基体焊接得不牢固，当受到很大的瞬时载荷作用时，就造成了复合片掉片现象。如果地层存在倾角并且倾角较大，钻头钻遇硬夹层时蹩钻、跳钻现象更加严重，对切削齿造成更严重的损坏。钻头钻进硬地层所受的反作用力集中在复合片边缘，钻进软地层的反作用力分布在切削齿面上，因此，即使钻井参数相同，钻进硬地层比钻进软地层容易损坏切削齿。

load. If there is a dip angle in the formation and the dip angle is large, the phenomenon of broken drilling and jumping when the drill bit encounters a hard interlayer is more serious, causing more serious damage to the cutting teeth. The reaction force received by the drill bit when drilling into hard formations is concentrated at the edge of the composite sheet, while the reaction force when drilling into soft formations is distributed on the cutting tooth surface. Therefore, even with the same drilling parameters, drilling into hard formations is more likely to damage the cutting teeth than drilling into soft formations.

(4) When encountering obstacles or jamming during the drilling process, improper operation, forceful lifting and pressing, may damage the diamond. Although PDC drill bit have high hardness and wear resistance, PDC has poor thermal stability, strong brittleness, and cannot withstand impact loads.

(5) Operational errors caused serious slipping and stuttering of the PDC drill bit, leading to accidents such as chip loss, tooth breakage, and accelerated tooth wear. Due to the enormous impact force exerted by the drill bit, some or even the entire drill bit may fall into the well.

(6) Excessive use or significant decrease in drilling speed due to changes in lithology. The increase in pump pressure, heavy drilling, and serious jumping of the drill bit were not taken seriously, leading to PDC drill bit cutting tooth damage, severe tire wear and other drill bit accidents.

(7) Unclean or falling objects at the bottom of the well. Improper bottomhole shaping or falling objects such as pliers and small tools can cause PDC drill bit accidents such as chipping, tooth collapse, and tooth breakage.

（4）下钻过程遇阻、遇卡时，操作不当，猛提猛压，碰坏金刚石。虽然PDC钻头具有高硬度和高抗磨性，但是PDC热稳定性差，脆性强，不能承受冲击载荷。

（5）操作失误造成严重溜钻、顿钻，从而引发PDC钻头掉片、断齿、齿磨损加速等事故。由于顿钻施与钻头巨大的冲击力，甚至会导致部分乃至整个钻头落井的事故。

（6）超时使用或由于岩性变化等，出现钻速明显降低。泵压升高，繁钻、跳钻严重等现象没有引起重视，引发PDC钻头切削齿损坏、胎体磨损严重等钻头事故。

（7）井底不清洁或者有落物。井底造型不当或者井下有钳牙、小工具等落物，引发PDC钻头掉片、崩落齿、断齿等钻头事故。

(8) PDC drill bit parameters are incorrect. If PDC drill bits are selected with high drilling pressure and high rotation speed, it can cause accidents such as tooth breakage, excessive grinding of teeth, detachment of polycrystalline layers. When using PDC drill bits, if the specific water power is selected too high, it will cause severe erosion of the drill bit matrix, leading to the detachment of the cutting teeth or nozzle. If the specific water power is selected too low, it is not conducive to the cleaning and cooling of the composite piece, especially in soft formations, it is easy to cause drill bit mud pockets, and in hard formations, it is easy to accelerate the thermal wear of the composite piece. If the nozzle arrangement is not appropriate, the sprayed fluid will reflect and directly erode the cutting teeth after hitting the bottom of the well.

### 5.1.2.3 Symptoms of PDC Drill Bit Accidents

Based on the above reasons, it can be understood that the causes of damage to PDC cutting teeth are diverse and not singular, and their manifestations are also complex and diverse. Combined with on-site events, it can be concluded that the summary is as follows:

(1) The lithology of the formation has not changed, but the mechanical drilling speed has significantly decreased, indicating severe damage to the cutting teeth;

(2) After pressurization, the pump pressure increases, indicating a decrease in the cross-section of the water tank due to wear of the drill bit "mud package" or cutting teeth and matrix.

(3) Abnormalities in underground conditions, such as alarm drilling, jumping drilling, increased torque, decreased or no footage, indicate that there are falling objects hitting teeth or encountering grinding rock layers or hard interlayers during drilling,

（8）PDC钻头参数不当。PDC钻头如果选择高钻压、大转速，会引发断齿、齿过度磨碎、聚晶层脱落等事故。使用PDC钻头时如果比水功率选择过大，就会造成钻头基体严重冲蚀，致使切削齿柱或喷嘴脱落。如果比水功率选择过小，对复合片清洗冷却不利，特别是在软地层中易造成钻头泥包，在硬地层中易造成复合片的热磨损加速。如果喷嘴布置位置不合适，喷射的流体冲击井底后会反射直接冲蚀切削齿。

### 5.1.2.3 钻头事故的征兆

综合上述原因，可以了解到导致PDC切削齿的破坏原因多变且并不单一，其表现形式也复杂多样，结合现场事件，简单归纳如下：

（1）地层岩性没有变化，但机械钻速明显下降，表明切削齿损坏严重；

（2）加压后泵压升高，表明钻头"泥包"或者切削齿和胎体磨损而使水槽截面减小。

（3）井下情况异常，如蹩钻、跳钻、扭矩增大、进尺减少或无进尺等，表明井下有落物碰伤齿或者钻遇研石层或硬夹层而导致齿崩落、掉片等事故。

resulting in accidents such as tooth collapse and falling pieces.

(4) An operational error that causes slipping or stuttering of the drill bit, resulting in slow or no drilling, indicates that the drill bit with severe tooth loss is scrapped.

### 5.1.3 Prevention and Handling of Drill Bit Accidents

#### 5.1.3.1 Inspection of Drill Bit before Entering the Well

Firstly, choose a drill bit that meets the quality requirements. Secondly, carefully inspect the drill bit, including the manufacturer, number, model, production date, and thread of the drill bit. For a tooth-wheel drill bit, check the clearance and activity of the tooth wheel, palm, bearings, water eye groove sealing, as well as whether the toothwheel collide, and whether there are cracks, pores, etc. in the welding process; The PDC drill bit needs to check whether the thread, steps, and cutting teeth of the drill bit are intact, and whether the water hole is unobstructed. In addition, it is necessary to ensure that the wellbore is unobstructed, the bottom of the well is clean, and there are no falling objects before entering the well.

#### 5.1.3.2 Proper Use of Drill Bit

According to the instructions of the drill bit manufacturer, use the drill bit correctly. The thread specifications of the matching joint of the drill bit must be consistent with the specifications of the connecting thread of the drill bit to prevent the entire drill bit from slipping into the well.

#### 5.1.3.3 Choose a Reasonable Drill Bit Loader and Unloader

PDC drill bits need to be equipped with

（4）操作失误造成溜钻、顿钻导致进尺慢或无进尺，表明齿损失严重的钻头报废。

### 5.1.3 钻头事故的预防与处理

#### 5.1.3.1 钻头入井前的检查

首先要选择质量符合要求的钻头，再者要认真检查钻头，具体检查内容包括钻头厂家、编号、型号、生产日期和螺纹等，牙轮钻头还要检查牙轮、巴掌、轴承间隙及活动情况，水眼槽密封情况，以及牙轮是否相碰，焊接有无裂缝、气孔等。PDC钻头要检查钻头螺纹、台阶和切削齿是否完好，水眼是否畅通。此外，还要求井眼畅通，井底干净，没有落物才能入井。

#### 5.1.3.2 正确使用钻头

按照钻头出厂使用说明书要求正确使用钻头，钻头的配合接头螺纹规范必须与钻头连接螺纹的规范一致，防止整个钻头脱扣入井。

#### 5.1.3.3 选用合理的钻头装卸器

PDC钻头需选择专用的钻头装卸器，要避免上扣时碰伤牙轮、牙齿；按规定上

dedicated drill bit loaders to avoid damaging the tooth wheel and teeth during screwing up; Tighten the torque according to regulations to avoid being too loose or too tight.

### 5.1.3.4 Smooth Tripping is Necessary to Prevent Stuttering and Slipping During Drilling

The diameter of the open hole section is irregular and there are multiple wall steps present. When drilling in this section, it is necessary to lower it slowly. For the tooth-wheel drilling bit, it is necessary to prevent one of the cones from coming into contact with the wall step under excessive force and falling off. When the wall step is obstructed, only one tooth wheel is subjected to force, which can easily lead to a stuttering tooth wheel accident. Moreover, the stuttering tooth does not fall to the bottom of the well, but falls on the wall step, which is very unstable. Once it falls, it can cause malignant jamming of the drill bit. For PDC drill bits, it is also necessary to control the speed during drilling to prevent damage to the teeth, which may cause accidents such as broken teeth and chips.

### 5.1.3.5 Reasonably Selecting Drilling Parameters During Drilling

Drill within the recommended range of drilling pressure and speed by the drill bit manufacturer, and ensure even drilling. In addition, it is necessary to optimize hydraulic parameters to avoid accidents such as water piercing and erosion.

### 5.1.3.6 Control the Usage Time of the Drill Bit

Control the usage time of the drill bit to avoid exceeding the time limit. Based on the usage time and wear situation of the same type of drill bit mentioned above, referring to adjacent well data, and considering the

扣扭矩紧扣，避免过松或过紧。

### 5.1.3.4 起下钻要平稳，严防顿钻、溜钻

裸眼井段井径不规则，有多个壁阶存在，下钻时在此井段要慢速下放，对牙轮钻头而言，防止某个牙轮接触壁阶受力过大而折落。壁阶遇阻只有一个牙轮受力，很容易发生顿落牙轮事故，而且顿落的牙轮不是掉到井底，而是落在壁阶上，状态很不稳定，一旦下落，便会造成恶性卡钻。对于PDC钻头下钻同样需要控制速度，以防碰坏齿，造成断齿、掉片等钻头事故。

### 5.1.3.5 钻进中合理优选钻井参数

在钻头厂家推荐的钻压、转速范围内钻进，同时要求送钻均匀。除此之外，还需优选水力参数，避免刺水眼、冲蚀等事故。

### 5.1.3.6 控制钻头使用时间

控制钻头使用时间，避免超时使用。可以根据上只同类型钻头的使用时间及磨损情况，参考邻井资料，综合钻头特点和地层岩性，初步确定钻头的使用时间。但还要具体考虑本只钻头的具体井下工作状况，最后对

characteristics of the drill bit and formation lithology, the usage time of the drill bit can be preliminarily determined. But it is also necessary to consider the specific underground working conditions of the drill bit, and finally make a clear judgment on the usage time of the drill bit.

### 5.1.3.7 Timely Drilling

The use of drill bits should be based on the actual situation underground. If there are abnormal situations underground, such as broken drilling, jumping drilling, or increased torque, the cause should be carefully judged. According to the following situations, timely drilling should be carried out:

(1) There are falling objects underground, and adjusting the drilling pressure and speed is ineffective, resulting in a sharp decrease in footage or no footage;

(2) When drilling into special formations, if encountering gravel layers, the drill will jump, and when encountering certain mud shale layers, the drill will also fail. However, there is footage and the drilling speed is basically uniform. After adjusting the drilling pressure and speed, significant results can be seen;

(3) Early wear of tooth-wheel drill bits, such as bearing displacement, interlocking and jamming of tooth wheels, can lead to poor drilling and jumping, but with a decrease in drilling speed due to footage. When encountering these situations, if it is difficult to make a clear judgment at the moment, even if the drill bit has not been used for enough time, it is still necessary to get up early and not indulge in small profits to cause great trouble.

### 5.1.3.8 Abnormal Underground Conditions

When the underground situation is abnormal, such as a decrease in suspended weight or pump pressure, and the cause can not be found on the ground, it is absolutely prohibited to insert any objects from the water hole of the drill string, such as inclinometers, pressure holding steel balls, etc., to worsen the accident.

### 5.1.3.9 Select High-quality Drilling Fluid to Avoid "Mud package" on Drill Bit

Taking PDC drill bits as an example to illustrate engineering measures to avoid drill bit "mud package":

(1) Firstly, the selection of PDC drill bits should be done well, and the design of drill water holes and flow channels should be conducive to chip removal;

(2) Before inserting the PDC drill bit, the drilling fluid should be thoroughly circulated and the wellbore cleaned to prevent the drilling cuttings remaining in the wellbore from further hydration and dispersion after drilling;

(3) Before inserting the PDC drill bit, short trip the hole and scrape and squeeze the filter cake on the wellbore wall. Thin and compact the thick filter cake to ensure smooth wellbore flow and eliminate blockages as much as possible; In areas with a high incidence of drill bit mud pockets, if all methods are used but cannot avoid PDC drill bit "mud package", it is necessary to first use a tooth-wheel drill bit to clear the well;

(4) When drilling, apply butter to the surface of the drill bit flow channel to form a protective film, reducing the time it takes for the drill bit to directly contact the low-quality solid phase in the drilling fluid, or wrap the drill bit. This can have a certain effect even in deep wells;

(5) When drilling, the drill bit continuously scrapes the wellbore wall, and the filter cake on the wellbore wall or the drilling debris stuck

柱水眼内投入任何物件，如测斜仪、憋压钢球等，导致事故恶化。

### 5.1.3.9 选用优质钻井液，避免钻头"泥包"

以PDC钻头为例，说明避免钻头"泥包"的工程措施：

（1）首先做好PDC钻头的选型工作，钻头水眼、流道设计应利于排屑；

（2）下入PDC钻头之前，应充分循环钻井液，清洗井眼，防止起钻后滞留在井眼内的钻屑继续水化分散；

（3）下入PDC钻头前先短起下钻，对井壁滤饼进行刮削、挤压，将厚滤饼拉薄、压实，尽量保证井眼畅通、消除阻卡；在钻头泥包高发区，如果采用了所有方法也无法避免PDC钻头"泥包"，那么先使用牙轮钻头通一趟井就成了必要手段；

（4）下钻时在钻头流道表面涂满黄油，形成一层保护膜，减少钻头与钻井液中的劣质固相直接接触的时间，或者把钻头包起来，这样做即便在深井中也是有一定作用的；

（5）下钻时钻头不断刮削井壁，井壁上的滤饼或滞留于井内的钻屑会在钻头下堆积到一定程度便会压实在钻头上，那么下钻中途进行循环，将钻头冲洗干净也是有其必要的；

in the well will accumulate under the drill bit to a certain extent and compact on the drill bit. Therefore, it is necessary to circulate during drilling to flush the drill bit clean;

(6) During the drilling process, the speed should also be appropriately controlled to prevent the drill bit from suddenly rushing into the "sand bridge" and drilling into a pile of mud. In addition, if the speed is appropriate, the PDC drill bit will travel along the spiral wellbore trajectory drilled by the previous drill bit, instead of pulling a large amount of filter cake on the wellbore wall;

(7) When drilling to the bottom each time, the pump must be turned on first, and the displacement should be increased as much as possible. The bottom of the well and the drill bit should be thoroughly flushed. After the displacement meets the requirements, the PDC drill bit should be lightly pressed and rotated to drill 0.5~1.0m. This is also a requirement for the PDC drill bit design. Try to use high displacement drilling to ensure sufficient cleaning and cooling of PDC drill bits;

(8) When drilling in soft mudstone, low drilling pressure, high rotation speed, and large displacement should be used as much as possible;

(9) The operation should be precise, and the drilling pressure must be evenly applied, not fluctuating.

### 5.1.4 Handling of General Drill Bit Accidents

Dealing with drill bit accidents involves salvaging drill bits or cutting teeth and other components that have fallen into the well. Small irregular falling objects can be salvaged; Large falling objects can be salvaged by grinding or exploding methods. In short,

（6）下钻过程中还应适当控制速度，防止钻头突然冲入"砂桥"，钻进一堆烂泥中。另外，如果速度合适，PDC 钻头会顺着上一只钻头所钻的螺旋形井眼轨道行进，而不是在井壁上划拉下大量滤饼；

（7）每次下钻到底时必须先开泵，尽量提高排量，充分冲洗井底和钻头，等排量满足要求后再轻压旋转钻进 0.5~1.0m，这也是 PDC 钻头造型的要求。尽量采用大排量钻进，保证 PDC 钻头的充分清洗与冷却；

（8）在软泥岩中钻进，应尽量采用低钻压、高转速、大排量；

（9）操作要精细，送钻加压一定要均匀，不能忽大忽小。

### 5.1.4 一般钻头事故的处理

处理钻头事故就是打捞井下落入的钻头或钻头上的切削齿等部件。比较小的不规则落物可以采取打捞方式；较大落物，可以采取磨碎或者爆炸等方法处理。总之，要根据钻头落物的形状和大小选用不同的方法和工具进行打捞作业。

different methods and tools should be selected for salvage operations based on the shape and size of the falling objects on the drill bit.

If the overall drill bit falls into the well, although the thread has been damaged and it is not possible to use a joint to catch it, as long as there is a catch area, it should be salvaged using a male and female cone.

Precautions for handling drill bit falling into the well:

(1) If the overall drill bit falls into the well due to reverse threading or joint wear, it is best to use a male cone for fishing with internal threads, and a female cone for fishing with external threads;

(2) It is best to connect a stabilizer with a diameter similar to that of the drill bit at the upper part of the male and female cones. The main reason for salvage failure is due to the short and easy inclination of the drill bit body. With a stabilizer, the drill bit can be prevented from deviating, allowing for accurate and accurate alignment of the fish head;

(3) The matching between the thread diameter of the fishing tool and the diameter of the fishing area for falling objects should be very precise;

(4) If the thread is broken, and there is no possibility of salvage, you have to grind it with grinding shoes or use a salvage tool to salvage it.

If some parts of the drill bit break or fall off, causing irregular objects to fall into the well, various fishing tools should be used according to the specific shape and location of the falling objects.

如果整体钻头落井，虽然螺纹已经损坏，不可能用接头对扣打捞，但只要有抓捞的部位，就应该用公锥和母锥进行打捞。

钻头落井处理应注意的事项：

（1）因倒扣或接头磨损而使整体钻头落井，鱼头为内螺纹者最好用公锥打捞，鱼头为外螺纹者最好用母锥打捞；

（2）在公锥、母锥上部最好接一个与落井钻头直径相近的扶正器。因钻头体短易斜是打捞失败的主要原因，有了扶正器，可防止钻头偏斜，这样就可以准确无误地对正鱼头；

（3）打捞工具螺纹直径与落物打捞部位直径的配合要十分精细；

（4）如果螺纹断裂，没有打捞的可能，就只好用磨鞋磨碎或用打捞器打捞。

如果出现钻头某些部位断裂、掉落，造成不规则落物入井，就要根据落物的具体形态和具体落井位置，使用多种打捞工具进行处理。

## 5.2 On-site Accident Cases

### 5.2.1 Q well Sand Jam Accident

#### 5.2.1.1 Basic Information

(1) Well number: Q.

(2) Well category: evaluation well.

(3) Well type: vertical well.

(4) Oil testing date: September 10th to October 8th, 2011.

Date of occurrence: September 18th, 2011.
Termination date: January 1st, 2015.
Lost time: 14 days.

(5) Falling fish structure: 99mm vibrator lower joint × 0.36m+variable buckle joint× 0.22m+73mm lifting short joint × 1.45m+109mm hydraulic pump working cylinder× 0.34m+125mm the supporting sand skin bowl assembly × 0.58m+73mm lifting short joint× 1.0m+109mm single flow valve × 0.3m+114mm hydraulic anchor × 0.32m+Y221-114 packer× 1.38m+73mm oil pipe × 9.52m+100mm horn mouth × 0.2m.

#### 5.2.1.2 Process of Occurrence

On September 10th, 2011, during the fracturing process, the structure of the fracturing string was as follows: oil pipe hanger+140 oil pipes(73mm)+73mm thickened lifting short joints+109mm hydraulic fracturing pump working cylinder+125mm supporting sand skin bowl +73mm thickened lifting short joints+109mm single flow valve+114mm hydraulic laying+Y221-114 packer+1 thickened oil pipe(73mm)+100mm horn mouth.

After fracturing, use a 3mm oil nozzle to spray. On September 12th, open the sliding sleeve of the fracturing pump barrel, insert

## 5.2 现场事故案例

### 5.2.1 Q井砂卡事故

#### 5.2.1.1 基本情况

（1）井号：Q。

（2）井别：评价井。

（3）井型：直井。

（4）试油日期：2011年9月10日—10月8日。

发生时间：2011年9月18日。
解除时间：2015年1月1日。
损失时间：14d。

（5）落鱼结构：99mm震击器下接头×0.36m+变扣接头×0.22m+73mm提升短节×1.45m+109mm水力泵工作筒×0.34m+125mm托砂皮碗总成×0.58m+73mm提升短节×1.0m+109mm单流阀×0.3m+114mm水力锚×0.32m+Y221-114封隔器×1.38m+73mm油管×9.52m+100mm喇叭口×0.2m。

#### 5.2.1.2 发生经过

2011年9月10日，压裂，压裂管柱结构为：油管挂+73mm油管140根+73mm加厚提升短节+109mm压裂水力泵工作筒+125mm托砂皮碗+73mm加厚提升短节+109mm单流阀+114mm水力铺+Y221-114封隔器+73mm加厚油管1根+100mm喇叭口。

压裂后用3mm油嘴放喷，9月12日，打开压裂泵筒滑套，投泵芯到位，水力泵排液生产。

the pump core in place, and discharge the hydraulic pump for production.

From September 13th to 16th, the hydraulic pump discharged liquid with a pump pressure of 8~12MPa. On September 17th, the pump was stopped due to rain. On September 18th, the hydraulic pump discharged liquid and the ground pump was not working properly. When replacing the power fluid, it was found that there was 0.8m³ Taoli.

On September 19th, the hydraulic pump discharge pipe column was lifted up to 400kN, but the packer was not unsealed and the pipe column encountered jamming. After backwashing the well, the lifting of the pipe column was 480kN, and the release of the stuck pipe was ineffective. Backwash out the pump core, with a bottomhole flow pressure of 2.03MPa and a dynamic liquid level of 1146.54m.

5.2.1.3 Processing Process

(1) Phase 1 (September 20th to 22nd, 2011): Rinse, lift and release the card.

On September 21st, the well was flushed with sand, with a pump pressure of 5-6MPa, a displacement of 300L/min, and a flushing fluid of 30m of clean water ³, No sand particles return.

From September 21st to 22nd, after washing the well, the load was lifted up to 600kN and forcefully pulled out 5 times, but it was ineffective.

After research, it has been decided to re install the pump core for hydraulic pump discharge.

On September 22nd to 30th, the hydraulic pump was produced, and on September 30th, the pump core was backwashed out. The bottomhole flow pressure was 0.42MPa, and

9月13日—16日，水力泵排液，泵压为8~12MPa，9月17日因下雨停泵。9月18日，水力泵排液，地面泵工作不正常，更换动力液时发现方箱中有0.8m³陶粒。

9月19日，起水力泵排液管柱，上提400kN，封隔器未解封，管柱遇卡。反洗井后，上提管柱480kN，解卡无效。反洗出泵芯，井底流压为2.03MPa，动液面为1146.54m。

5.2.1.3 处理过程

（1）第一阶段（2011年9月20日—22日）：冲洗、上提解卡。

9月21日，正洗井冲砂，泵压为5~6MPa，排量为300L/min，冲砂液清水30m³，无砂粒返出。

9月21日—22日，洗井后上提600kN，强拔5次，无效。

经过研究决定，重新投泵芯进行水力泵排液。

9月22日—30日，水力泵生产，9月30日反洗出泵芯，井底流压为0.42MPa，动液面为1307.54m。

the dynamic liquid level was 1307.54m.

(2) Phase 2 (October 1st to 8th, 2011): Reversal and shockwave.

From October 1st to 4th, the well was washed and sandblasted in a regular cycle, and the repeated movement of the pipe column was ineffective. Invert 140 oil pipes. The remaining column structure is: 73mm lifting short section ×1.45m+109m hydraulic pump working cylinder×0.34m+125mm the supporting sand skin bowl assembly×0.58m+73mm lifting short joint×1.0m+109mm single flow valve×0.3m+114mm hydraulic anchor× 0.32m+Y221-114 packer×1.38m+73mm oil pipe×9.52m+100mm horn mouth×0.2m.

From October 5th to 8th, the bottom of the pipe was salvaged by a downward shock, with a shock absorber at the bottom. The pipe was lifted up by 280~320kN, but repeated shocks were ineffective.

On October 8th, with a reverse buckle, a shock-absorbing pipe column was launched, and the pipe column was reversed from the joint under the shock-absorbing device. Downhole fish falling: 99mm vibrator lower joint× 0.36m+variable buckle joint×0.22m+73mm lifting short joint × 1.45m+109mm working cylinder 5×0.34m+125mm the supporting sand skin bowl assembly×0.58m+73mm lifting short joint×1.0m+109mm single flow valve× 0.3m+114mm hydraulic anchor×0.32m+Y221-114 packer×1.38m+73mm oil pipe×9.52m+100mm horn mouth×0.2m. The top position of the fish is 1345.96m.

Research decision to undergo major repairs.

(3) The third stage (December 29th, 2014 to January 1st, 2015): buckle making and salvage.

（2）第二阶段（2011年10月1日—8日）：倒扣、震击。

10月1日—4日，正循环洗井冲砂，反复活动管柱无效。倒扣起出油管140根。剩余管柱结构为：73mm提升短节×1.45m+109m水力泵工作筒×0.34m+125mm托砂皮碗总成×0.58m+73mm提升短节×1.0m+109mm单流阀×0.3m+114mm水力锚×0.32m+Y221-114封隔器×1.38m+73mm油管×9.52m+100mm喇叭口×0.2m。

10月5日—8日，下震击打捞管柱，底部带震击器，上提管柱280~320kN，反复震击无效。

10月8日，倒扣，起出震击管柱，管柱从震击器下接头处倒开。井下落鱼为：99mm震击器下接头×0.36m+变扣接头×0.22m+73mm提升短节×1.45m+109mm工作筒5×0.34m+125mm托砂皮碗总成×0.58m+73mm提升短节×1.0m+109mm单流阀×0.3m+114mm水力锚×0.32m+Y221-114封隔器×1.38m+73mm油管×9.52m+100mm喇叭口×0.2m。鱼顶位置为1345.96m。

研究决定，上大修处理。

（3）第三阶段（2014年12月29日—2015年1月1日）：造扣打捞。

From December 29th, 2014 to January 1st, 2015, a 73mm reverse drilling rod + a male cone were lowered, and a buckle was made and salvaged. The 99mm shockhammer lower joint+variable buckle joint+73mm lifting short joint+109mm hydraulic pump working cylinder+125mm supporting sand skin bowl assembly+73mm lifting short joint+109mm single flow net+114mm hydraulic dislocation+Y221-114 packer+73mm oil pipe+100mm bell mouth, all removed from the well. This accident has been resolved after multiple positive and negative washings, 5 forced oil pipe withdrawals 20 shockwaves, 1 lead imprint, 1 male cone buckle fishing, and 1 well trip, a total of 18 trips of drilling tools, with a total of 14 days of processing.

#### 5.2.1.4 Cause Analysis

(1) Direct cause.

Sand production in the formation is the direct cause of pipe sticking.

(2) Indirect causes.

① The distance between the supporting sand skin bowl and the packer is 2.53m, and the water hole is 2.78m away from the packer rubber cylinder. A positive flushing well was carried out for sand flushing, but due to the distance from the packer, the impact force is small and cannot achieve the sand flushing effect.

② The unreasonable work system and excessive drainage intensity have led to serious sand production in the formation.

(3) Management reasons.

① Unreasonable tool selection. The inner diameter of the oil layer casing in this well is 121.36~125.74mm, while the outer diameter of the supporting sand skin bowl is 125mm. It may have been damaged during the process of reducing the diameter of the casing, resulting

2014年12月29日—2015年1月1日下73mm反扣钻杆+公锥，造扣打捞，打捞出99mm震击器下接头+变扣接头+73mm提升短节+109mm水力泵工作筒+125mm托砂皮碗总成+73mm提升短节+109mm单流网+114mm水力错+Y221-114封隔器+73mm油管1根+100mm喇叭口，井下落物全部打捞出，至此事故解除本次事故经过多次正反冲洗、5次油管强拔、20次震击器震击、1次打铅印1次公锥造扣打捞、1次通井，共计起下钻具18次，累计处理14d。

#### 5.2.1.4 原因分析

（1）直接原因。

地层出砂是导致管柱遇卡的直接原因。

（2）间接原因。

① 托砂皮碗和封隔器的间距为2.53m，水眼距离封隔器胶筒2.78m行了正洗井冲砂，但是由于距离封隔器较远，冲击力小，起不到冲砂效果。

② 工作制度不合理，排液强度过大，导致地层出砂严重。

（3）管理原因。

① 工具选择不合理。本井油层套管内径为121.36~125.74mm，而托砂皮碗外径为125mm，在下至套管变径过程中可能已经损坏，导致排液时排出的压裂砂沉积在封隔器上，形成砂卡。

in the deposition of fracturing sand discharged during fluid discharge on the packer, forming a sand jam.

② Insufficient on-site liquid storage capacity, inability to produce continuously on rainy days, deposition of sand particles in the annulus, resulting in sand jamming.

### 5.2.1.5 Lessons Learned

(1) The hydraulic pump discharge pipe column should minimize the distance between the sand cup and the packer as much as possible to ensure effective impact force on the deposited sand during circulation.

(2) Use sand proof packers instead of sand carrying cups to prevent damage to the sand carrying cups during the well lowering process, which can not provide sand prevention.

(3) Continuous operation is required for hydraulic pump drainage to reduce the settlement and accumulation of fracturing sand in the annulus. For wells with blocked roads on rainy days that affect the transportation of produced liquid from the well, on-site liquid storage capacity should be strengthened to avoid pump shutdown.

(4) Reasonably develop work systems based on specific geological conditions to prevent excessive negative pressure from causing sand production in the formation.

## 5.2.2 The Complex Detachment of the Pumping and Weighting Rod in S Well

### 5.2.2.1 Basic Information

(1) Well number: S Well.

(2) Well category: Evaluation well.

(3) Well type: high inclination well.

(4) Oil testing date: May 25th to August 6th, 2017.

Date of occurrence: July 11th, 2017.

② 现场储液能力不足，雨天生产不能够连续，环空砂粒沉积，造成砂卡。

### 5.2.1.5 经验教训

（1）水力泵排液管柱要尽可能减小托砂皮碗和封隔器之间的距离，保证循环时对沉积砂的有效冲击力。

（2）使用防砂封隔器取代托砂皮碗，防止下井过程中托砂皮碗损坏，起不到防砂作用。

（3）水力泵排液时要连续作业，以减少环空压裂砂的沉降、堆积。对于雨天道路不通、影响井内产出液外运的井，加强现场储液能力，避免停泵。

（4）根据具体地层情况合理制订工作制度，防止负压过大导致地层出砂。

## 5.2.2 井抽汲加重杆脱扣复杂

### 5.2.2.1 基本情况

（1）井号：S 井。

（2）井别：评价井。

（3）井型：大斜度井。

（4）试油日期：2017 年 5 月 25 日—8 月 6 日。

发生时间：2017 年 7 月 11 日。

Termination date: July 12th, 2017.

Lost time: 1 day.

#### 5.2.2.2 Process of Occurrence

On July 11th, 2017, at 20:00, the night shift personnel carried out a pumping and drainage operation. After the first pumping, they lifted the weighting rod to the wellhead to check and tighten the pumping rope cap. At 23:00, the second pumping was carried out. When the pumping wire rope reached 1150m, it encountered resistance. During the inspection, it was found that the weighting rod and the pumping rope cap were detached, and the weighting rod fell into the oil pipe.

#### 5.2.2.3 Processing Process

On July 12th, 2017, the KQ35/65 Christmas tree was unloaded and a blowout preventer was installed. When the oil pipe was lifted to the 114th one, liquid was found in the oil pipe. At the 121st one, it was found that the weight rod was stuck inside the oil pipe, and the suction cup broke and detached from the bottom steel wire frame. It was moved up and stuck on the weight rod.

The research decision was made, and the exploration company notified to remove the perforated string, lower the swab string, and re swab.

This accident involved one oil pipe salvage and a total of one trip, with a total of one day of handling.

#### 5.2.2.4 Cause Analysis

(1) Direct cause.

During the descent of the swabbing steel wire rope, the swabbing rubber cup broke and detached from the bottom steel wire frame, causing it to move up and get stuck on the weighting rod, resulting in resistance when lowering the swabbing steel wire rope. Due to sudden obstruction, the steel wire rope

解除时间：2017 年 7 月 12 日。

损失时间：1d。

#### 5.2.2.2 发生经过

2017 年 7 月 11 日 20:00，夜班人员进行抽汲排液作业，在第一次抽汲后将加重杆起至井口对抽汲绳帽进行检查紧扣，于 23:00 第二次进行抽汲，抽汲钢丝绳下至 1150m 时遇阻，活动起出检查发现加重杆与抽汲绳帽之间脱扣，加重杆落入油管内。

#### 5.2.2.3 处理过程

2017 年 7 月 12 日，卸提 KQ35/65 采油树，安装防喷器，起油管至第 114 根时油管内见液，第 121 根时发现加重杆卡死在油管内，抽汲胶杯从底端钢丝骨架处断裂脱开，上移卡死在加重杆上。

研究决定，勘探公司通知起出射孔管柱，下抽汲管柱，重新抽汲。

本次事故经过 1 次起油管打捞，累计 1 次起下钻，共计处理 1d。

#### 5.2.2.4 原因分析

（1）直接原因。

在抽汲钢丝绳下行过程中抽汲胶杯从底端钢丝骨架处断裂脱开，上移卡死在加重杆上，导致下放抽汲钢丝绳遇阻。由于突然遇阻，钢丝绳瞬间卸力松劲反向转动，导致加重杆脱扣。

suddenly released its force and turned in reverse, causing the weight rod to trip.

(2) Indirect causes.

① The driller did not pay attention to the condition of the wire rope when lowering it, resulting in sudden obstruction and loosening of the wire rope unloading force.

② The staff did not carefully inspect the suction cup and did not replace the worn suction cup in a timely manner.

(3) Management reasons.

The on-site follow-up cadres did not conduct sufficient supervision and did not properly supervise the operations of on-site personnel.

#### 5.2.2.5 Lessons Learned

(1) Before pumping, safety technical briefing should be conducted, comprehensive risk identification should be carried out, risk prevention measures should be taken, and all systems of the well repair machine should be checked and started to operate normally. The pumping wire rope should meet the standards.

(2) Before preparing for pumping, check the anti blowout device for pumping, check that the threaded connections of the weighting rod are securely fastened, and confirm that the well control facilities are in good condition.

(3) During pumping, check the rope cap, weighting rod, swab, and swab rubber cup every 2~3 times. When the swab rubber cup is severely deformed and cannot be inserted into the oil pipe or the support frame is worn, replace the rubber cup.

(4) The driller operates the sand removal drum of the well repair machine, and the wellhead worker stands next to the sand removal drum, staring at the marking of the

（2）间接原因。

① 司钻在下放钢丝绳时没有注意观察钢丝绳情况，导致突然遇阻，钢丝绳卸力松劲。

② 岗位人员对抽汲胶杯未进行仔细检查，对磨损的抽汲胶杯未及时进行更换。

（3）管理原因。

现场跟班干部督查不到位，对现场岗位人员操作没有监护到位。

#### 5.2.2.5 经验教训

（1）抽汲前做好安全技术交底，风险识别全面，做好风险防范措施，检查、启动修井机各系统运转正常，抽汲钢丝绳符合标准。

（2）准备抽汲前检查抽汲防喷装置，检查加重杆各螺纹连接处紧固完好，检查确认井控设施完好。

（3）抽汲时每抽2~3次对绳帽、加重杆、抽子和抽汲胶杯进行检查，当抽汲胶杯变形严重不能放入油管或支撑骨架磨损时，更换胶杯。

（4）司钻操作修井机捞砂滚筒，井口工站在捞砂滚筒旁边盯好大绳记号。抽汲下放速度要慢而匀速，防止钢丝绳打扭。上起时排好滚筒大绳，不骑绳不压绳，盯好大绳记号，及时提醒司钻操作。

drilling line. The pumping and lowering speed should be slow and uniform to prevent the steel wire rope from twisting. When lifting up, arrange the drum rope properly, do not ride or press the rope, pay attention to the rope markings, and promptly remind the driller to operate.

(5) On site follow-up cadres should strengthen supervision and inspection, supervise the operation of on-site personnel, replace worn suction cups in a timely manner, provide timely feedback on quality problems, and personally supervise and inspect the rope cap and weighting rod every time they are inspected.

### 5.2.3 D Well RD Safety Circulation Valve Failure Complex

#### 5.2.3.1 Basic Information

(1) Well number: D.

(2) Well category: Pre exploration well.

(3) Well type: vertical well.

(4) Oil testing date: April 3rd to 18th, 2019.

Date of occurrence: April 6th, 2019.
Termination date: April 6th, 2019.
Lost time: 1 day.

#### 5.2.3.2 Process of Occurrence

On April 3rd, 2019 at 23:45, the testing tool manhole was opened. On April 6th at 2:37, the reverse fluid construction began; At 9:05 on April 6th, when the person inside the wellbore is $10m^3$ transition slurry+$10m^3$ the isolation fluid is mixed with 77m of organic salt, the drilling team needs to reverse the process with the surface team. The casing pressure rises rapidly from 37.6MPa, and the downhole tool party calls the surface team with a telephone to open the power oil nozzle valve

（5）现场跟班干部加强监督检查，对现场岗位人员操作监督到位，对磨损的抽汲胶杯及时进行更换，对存在质量问题的应及时反馈，每次检查绳帽、加重杆时要亲自监督检查。

### 5.2.3 D井RD安全循环阀失效复杂

#### 5.2.3.1 基本情况

（1）井号：D。

（2）井别：预探井。

（3）井型：直井。

（4）试油日期：2019年4月3日—18日。
发生时间：2019年4月6日。
解除时间：2019年4月6日。
损失时间：1d。

#### 5.2.3.2 发生经过

2019年4月3日23：45，测试工具入井；4月6日2：37，开始反替液施工；4月6日9：05，当井筒内注入10m³过渡浆+10m³隔离液+77m³有机盐时，井队需要和地面队倒流程，套压从37.6MPa急速上升，井下工具方用报话机呼叫地面队，要求开动力油嘴阀门控制出口压力。喊了两次都没人应答（后与地面队核实，反馈为报话机未接收到信息，分析原因为两台报话机一起按下通话键时两台报话机都没有声音），此时套压升至41.0MPa，数采房内的朱某立即跑

to control the outlet pressure. After shouting twice but no one answered (after verifying with the ground team, it was reported that the phone did not receive any information, and the reason was analyzed to be that both phones pressed the call button together and there was no sound from both phones). At this time, the set pressure rose to 41.0MPa, and Zhu in the data collection room immediately ran to the ground manifold to open the power oil nozzle valve. Zhang and Zhu from the ground team opened the power oil nozzle valve together, Due to the fact that the two individuals did not wear a telephone (Li from the ground team carried a telephone to cooperate with the drilling team to reverse the process), Guo ran out of the data collection room to convey instructions to the two individuals. The two quickly opened the power oil nozzle valve for more than 10 turns, but the casing pressure was still rising (at this time, the casing pressure rose to 44.0MPa). Wang did not take emergency measures or report the pressure situation at this time in the data collection room. When the casing pressure rose to 48.0MPa, Wang saw no hope of the casing pressure falling back, Guo, who was running out of the data collection room and returning to the data collection room, reported the pressure and notified the pump truck to stop pumping through a telephone. However, when the pump truck stopped pumping, the casing pressure had already risen to a maximum of 49.8 MPa.

Afterwards, Party A notified to continue replacing the liquid. After adding the remaining organic salts, a 48mm steel ball was put in at 11:55 to prepare the packer for setting. After 90 minutes of waiting for the ball to seat, it is pressurized and sealed, with

向地面管汇开动力油嘴阀门，地面队张某和朱某一起开动力油嘴阀门，因两人没有佩戴报话机（地面队李某携带报话机配合井队倒流程），郭某跑出数采房向两人传达指令，两人迅速开动力油嘴阀门10圈以上，但套压仍在上升（此时套压升至44.0MPa），数采房内王某未进行应急处置也未通报此时压力情况，当套压升至48.0MPa，王某看套压回落无望，跑出数采房外向正往数采房返回的郭某汇报压力后用报话机通知泵车停泵，但泵车停泵时套压已最高升至49.8MPa。

之后甲方通知继续替液，在替入剩余有机盐后，于11:55投入48mm钢球准备封隔器坐封。候球入座90min后，正打压坐封，油压为31.0~32.9MPa，套压为30.9~32.7MPa，说明油套连通。考虑可能候球时间不够，钢球没有落入球座，之后直到16:49，共打压坐封6次，均显示油套连通无法坐封。分析为在替液套压升至49.8MPa时已将RD安全循环阀破裂盘打开，RD安全循环阀井下关井。

an oil pressure of 31.0~32.9 MPa and a sleeve pressure of 30.9~32.7 MPa, indicating that the oil sleeve is connected. Considering the possibility of insufficient waiting time for the ball, the steel ball did not fall into the ball seat. Afterwards, until 16:49, a total of 6 times of pressure sealing were applied, all of which showed that the oil sleeve was connected and could not be sealed. Analysis shows that the rupture disc of the RD safety circulation valve was opened when the pressure of the liquid replacement sleeve increased to 49.8 MPa, and the RD safety circulation valve was shut down.

#### 5.2.3.3 Processing Process

(1) Phase 1 (April 6th, 2019): Pressure release

Due to the abnormal opening of the RD safety circulation valve, the packer was unable to set. On April 6th, 2019, at 20:40, upon receiving instructions from Party A, the pipe column was not replaced and ground level pressure control and release were adopted.

At 20:50, surface wells were opened to release pressure, with oil pressure ranging from 23.1~41.2 MPa and casing pressure ranging from 23.0~41.0MPa. After closing the well for 30 minutes, the formation pressure was restored, and a cumulative organic salt production of 0.34 cubic meters was achieved.

At 21:26, the well was opened on the ground to release pressure, with an oil pressure of 15.2~23.1MPa and a casing pressure of 15.1~23.0MPa. After closing the well for 30 minutes, the formation pressure was restored, and a cumulative organic salt production of $0.51m^3$ was achieved。

At 22:16, the well was opened on the ground to release pressure, with an oil pressure of 5.1~15.2MPa and a casing pressure of

#### 5.2.3.3 处理过程

（1）第一阶段（2019年4月6日）：放压

由于RD安全循环阀异常打开，封隔器未能坐封，2019年4月6日20:40接甲方指令，不更换管柱，采取地面分级控压放压。

地面于20:50开井放压，油压为23.1~41.2MPa，套压为23.0~41.0MPa，关井30min，恢复地层压力，累计出有机盐$0.34m^3$。

地面于21:26开井放压，油压为15.2~23.1MPa，套压为15.1~23.0MPa，关井30min，恢复地层压力，累计出有机盐$0.51m^3$。

地面于22:16开井放压，油压为5.1~15.2MPa，套压为5.0~15.1MPa，关井30min，恢复地层压力，累计出有机盐$0.85m^3$。

5.0~15.1MPa. The well was shut in for 30 minutes to restore formation pressure, and a cumulative organic salt production of 0.85m³ was achieved。

At 22:55, the well was opened on the ground to release pressure, with an oil pressure of 0~5.1MPa and a casing pressure of 0~5.0MPa. The well was left open for 7 minutes without being shut in, and the formation pressure was restored. A total of 1.19m³ of organic salt was produced.

(2) Phase 2 (April 6th to 8th, 2019): lifting the pipe column.

Due to the lack of production in the formation, the direct testing operation was completed. After the tool inspection, the RD safety circulation valve ruptured, the circulation hole opened, the ball valve closed, and the 48mm sealing steel ball was placed on the upper part of the ball valve.

#### 5.2.3.4 Cause Analysis

(1) Direct cause.

During the fluid replacement process, improper emergency response was taken after an abnormal increase in casing pressure. The pump truck was not notified to stop the pump in a timely manner, resulting in the rupture of the RD safety valve disc, opening of the circulation hole, downhole shutdown, inability of the packer to set, and unsuccessful testing.

(2) Indirect causes.

Ground team operator Li left his original position with a telephone to cooperate with the drilling team in process switching. There was only Zhang at the ground manifold (without a telephone), and the ground team operator did not receive the instruction to turn on the power nozzle in a timely manner, missing the best time for pressure control.

地面于22:55开井放压，油压为0~5.1MPa，套压为0~5.0MPa，敞放7min不出关井，恢复地层压力，累计出有机盐1.19m³。

（2）第二阶段（2019年4月6日—8日）：提管柱。

因地层无产出，直接性测试作业结束，后提出工具检查，RD安全循环阀破裂盘破裂，循环孔打开，球阀关闭，48mm坐封钢球在球阀上部。

#### 5.2.3.4 原因分析

（1）直接原因。

替液过程中，套压异常升高后应急处置不当，未及时通知泵车停泵，导致RD安全阀破裂盘打破，循环孔打开，井下关井，封隔器无法坐封，测试未成功。

（2）间接原因。

地面队操作人员李某携带报话机离开原岗位配合钻井队进行流程切换，地面管汇处只有张某一人（未携带报话机），地面队操作人员没有第一时间接到开动力油嘴指令，错失了压力控制的最佳时间。

(3) Management reasons.

① The on-site technical leader did not conduct secondary technical briefing with the relevant operators, informing them of the risks and control measures during the operation process, and did not implement job responsibilities for key operators.

② The on-site team leader lacks emergency response capabilities and has not taken measures to reduce the displacement or stop the pump in response to abnormal pump pressure situations.

### 5.2.3.5 Lessons Learned

(1) Organize employees to carefully study and analyze this incident, conduct comparative self-examination based on actual situations, sort out the implementation of various management regulations and standards in the team, and prevent similar incidents from happening again.

(2) Prior to the construction of key wells and processes, technical briefing and job division should be carried out to inform relevant operators of the risks present during the operation process, and clarify the job responsibilities and emergency response measures of all parties.

(3) When encountering emergency situations during construction, construction personnel should strictly follow the testing construction operation procedures and emergency plans for construction, timely report any abnormal situations that occur during the operation to superiors, and do a good job in initial emergency response work.

## 5.2.4 T Well Perforation Misinjection Acciden

### 5.2.4.1 Basic Information

(1) Well number: T.

(2) Well category: Pre exploration well.

（3）管理原因。

① 现场带班技术干部未与相关作业方进行二次技术交底，告知其作业过程中存在的风险及控制措施，未对关键岗位操作人员进行岗位职责落实。

② 现场带班干部应急处置能力不足，并且未针对异常泵压情况采取降低替液排量或停泵的措施。

### 5.2.3.5 经验教训

（1）组织员工认真学习、分析本次事件，结合实际开展对照自查，梳理本队各项管理规定、标准的执行、落实情况，杜绝类似事件再次发生。

（2）针对重点井、重点工序施工前，开展技术交底与岗位分工，告知相关作业方作业过程中存在的风险，明确各方的岗位职责与应急处置措施。

（3）施工中遇到应急情况，施工人员要严格按照测试施工操作规程和应急预案进行施工，及时向上级汇报作业过程中出现的异常情况，并做好初期应急处置工作。

## 5.2.4 T井射孔误射事故

### 5.2.4.1 基本情况

（1）井号：T。

（2）井别：预探井。

(3) Well type: vertical well.

(4) Oil testing date: June 20th to November 25th, 2008.

Time of occurrence: October 11th, 2008.
Termination date: November 6th, 2008.
Lost time: 25 days.

#### 5.2.4.2 Process of Occurrence

From September 26th to October 1st, 2008, sand filling and cement injection were carried out. Column structure: 88.9mm externally thickened oil pipes (61 pieces)+variable buckle joints+73mm externally thickened oil pipes (180 pieces), the actual sand surface position is 2845.00m. Adjust to 2310.60m oil pipe filled with sand $2.0m^3$, During the sand filling process, sand blockage occurs and the well is being cleaned to remove the blockage. Cement injection, $0.46m^3$ cement slurry injection, Replacement volume is $7.8m^3$, The depth of the backwash well is 2272.35m, and the actual location of the cement surface on October 3rd was 2270.50m.

On October 5th to 7th, 2008, perforation testing was carried out in collaboration. The structure of the pipe column was as follows: oil pipe hanger+73mm external oil pipe 232 pieces+99mm RD safety circulation valve+95mm pressure gauge support cylinder+99mm LPR-N valve+95mm pressure gauge support cylinder+114mm RTTS packer+73mm screen tube+73mm detonator+102mm perforation gun. Open the LPR-N valve by pressurizing the annulus to open the well, and insert a rod into the oil pipe for perforation. After inserting the rod, there is no display at the wellhead, indicating that the injection has failed. The perforating team encountered obstacles at 1870.00m and 1893.99m, respectively, when throwing the fishing rod. On October 10th, the testing tool

（3）井型：直井。

（4）试油日期：2008年6月20日—11月25日。

发生时间：2008年10月11日。
解除时间：2008年11月6日。
损失时间：25d。

#### 5.2.4.2 发生经过

2008年9月26日—10月1日，填砂、注水泥。管柱结构：88.9mm外加厚油管（61根）+变扣接头+73mm外加厚油管180根，实探砂面位置为2845.00m。调整至2310.6m油管填砂$2.0m^3$，填砂过程中发生砂堵，正洗井解堵。注水泥，注入水泥浆$0.46m^3$，替量为$7.8m^3$，反洗井深度为2272.35m，10月3日实探水泥面位置为2270.50m。

2008年10月5日—7日，射孔测试联作，管柱结构：油管挂+73mm外加享油管232根+99mmRD安全循环阀+95mm压力计托筒+99mmLPR-N阀+95mm压力计托筒+114mmRTTS封隔器+73mm筛管+73mm起爆器+102mm射孔枪。环空加压打开LPR-N阀开井，从油管内投棒射孔，投棒后井口无显示，判断投棒射孔失败。射孔队下投杆打捞器分别在1870.00m、1893.99m遇阻。10月10日，起测试工人到井口（未起出射孔枪），发现射孔枪减振油管内盛满水泥浆，射孔队再次打捞投棒失败，考虑到施工安全，经请示决定把枪身重新下至井底，再用连续油管冲洗后进行打捞投棒。10月11日，在下入传输射孔管柱至379.0~383.3m时，在井听到射孔枪起爆声响，判断发生误射孔。

arrived at the wellhead (without the perforation gun), and it was found that the vibration reducing oil pipe of the perforation gun was filled with cement slurry. The perforation team failed to salvage and drop the rod again. Considering construction safety, after consultation, it was decided to lower the gun body back to the bottom of the well, wash it with continuous oil pipe, and then salvage and drop the rod. On October 11th, when the transmission perforation string was lowered to 379.0~383.3m, the sound of the perforation gun detonating was heard in the well, indicating that a false perforation had occurred.

5.2.4.3 Processing Process

(1) The first stage (October 12th to 15th, 2008): sand flushing, fishing, perforation, and rod casting.

On October 12th, a 73mm thick oil pipe and a perforation gun were installed to the bottom of the well. From October 13th to 15th, the continuous tubing was sandblasted to 2236.00m, and after sandblasting, the well was backwashed. The perforating team successfully salvaged the perforating rod and then lifted out all the pipe columns and perforating guns.

(2) The second stage (October 16th to November 6th, 2008): Bridge plugging and cement squeezing were used to seal the erroneous injection section.

On October 16th, log the diameter and inspect the casing. From October 17-19, bridge plugs will be installed at a location of 450.00m. Oil pipe sand filling $0.5m^3$, The position of the sand exploration surface is 395.00m. On October 20th, squeeze cement $10m^3$ inside。 From October 23rd to 25th, the plug was cleaned and the pressure test passed. On October 26th, sand flushing and removal of bridge plugs. From October 27th to 28th, sweep the cement plug

5.2.4.3 处理过程

（1）第一阶段（2008年10月12日—15日）：冲砂、捞射孔投杆。

10月12日，下73mm外加厚油管+射孔枪至井底。10月13日—15日，连续油管冲砂至2236.00m，冲砂后进行反洗井，射孔队打捞射孔投杆成功，然后起出全部管柱及射孔枪。

（2）第二阶段（2008年10月16日—11月6日）：打桥塞、挤水泥封堵误射井段。

10月16日、测井径，检查套管。10月17日—19日，打桥塞，位置为450.00m。油管填砂$0.5m^3$，探砂面位置为395.00m。10月20日，内挤水泥$10\ m^3$。10月23日—25日，扫塞，试压合格。10月26日，冲砂、捞出桥塞。10月27日—28日，扫水泥塞（2270.50~2348.59m）。10月29日，重新注水泥。10月31日，探水泥面位置为2317.70m。11月1日—3日，套管刮削，刮削至2310.00m。11月5日—6日，再次进行射孔测试联作，至此本井事故解除。

(2270.50~2348.59m). On October 29th, re inject cement. On October 31st, the position of the cement exploration surface was 2317.70m. From November 1st to 3rd, the casing was scraped to 2310.00m. From November 5th to 6th, perforation testing was conducted again, and the accident in this well was resolved.

This accident went through 1 sand filling, 1 logging diameter, 1 bridge plug installation, 1 casing squeezing cement, 1 continuous tubing sand flushing, 3 logging perforation and rod insertion, 1 cement injection, 2 plug cleaning, 1 sand flushing, and 1 bridge plug removal, for a total of 25 days of handling.

#### 5.2.4.4 Cause Analysis

(1) Direct cause.

① The vibration reducing oil pipe above the perforation gun was filled with cement slurry, causing the rod to encounter resistance and failing to trigger the detonator, resulting in perforation failure.

② During the process of lowering the oil pipe, the cement slurry above the detonator is gradually soaked and washed away by the liquid in the well until a critical point, and the plunger falls again due to gravity, triggering the detonator and causing misfiring.

(2) Indirect causes.

① The sand filling and cement injection in the pipe column is unreasonable, and a sand bridge is formed at the diameter change, causing sand blockage in the oil pipe. The sand particles in the washing well enter the annulus and do not completely return to the wellbore. After the cement injection is backwashed, the sand particles in the wellbore deposit, forming a large amount of mixed slurry zone. The subsequent construction cement mixed slurry enters the testing pipe column.

② During backwashing, the on duty personnel mistakenly judged that the cement slurry had returned, resulting in incomplete backwashing and cement slurry falling back in the oil pipe, causing sedimentation at the bottom of the well.

③ There is a 3m screen tube short joint on the vibration reducing oil pipe above the perforation detonator. During the process of combining the lower perforation test with the pipe string, cement slurry enters the vibration reducing pipe.

(3) Management reasons.

① The on-site technical personnel have weak quality awareness and have made unauthorized changes to the original sealing design. They have changed the design cement surface from 2295.00m to 2272.00m and raised the cement surface by 23.00m.

② Using a combination string without authorization during sand filling and cement injection can form a "sand bridge" in the oil pipe, resulting in incomplete cement backwashing and sedimentation of sand particles and excessive mixing.

③ The on-site operators did not strictly follow the cement injection design, and did not isolate clean water before and after replacing the cement slurry, resulting in direct contact between the cement slurry and the well washing fluid, affecting the cement setting effect and causing excessive mixing.

5.2.4.5 Lessons Learned

(1) It is strictly prohibited to use a combination string for sand filling inside the oil pipe.

(2) When the liquid in the well is not clear water, it is necessary to replace it with clear water isolation liquid when injecting cement

② 反洗时，坐岗人员误判水泥浆返出，造成反洗不彻底，油管内水泥浆回落导致沉积在井底。

③ 射孔起爆器以上减振油管上有3m筛管短节，下射孔测试联作管柱过程中造成水泥混浆进入减振管内。

（3）管理原因。

① 现场技术人员质量意识淡薄，私自改动原封堵设计，将设计水泥面由2295.00m改为2272.00m，上提水泥面23.00m。

② 填砂注水泥管柱私自采用组合管柱，填砂时在油管内形成"砂桥"，注水泥反洗井不彻底，造成砂粒沉积，混浆量过大。

③ 现场操作人员没有严格执行注水泥设计，替水泥浆前后没有进行清水隔离，造成水泥浆和洗井液直接接触发生反应，影响水泥凝结效果，造成混浆带过多。

5.2.4.5 经验教训

（1）严禁采用组合管柱进行油管内填砂。

（2）井内液体不是清水时，注水泥塞时一定要替入清水隔离液，以免造成接触带反应，使水泥速凝或不凝。

plugs to avoid causing contact zone reactions and causing cement to quickly set or not set.

(3) The combination of pressure detonation and rod throwing detonation should be used for oil pipe transportation perforation. If one detonation method fails, it can be changed to another.

## 5.2.5 Case Analysis of a Well in Tarim Oilfield

### 5.2.5.1 Basic Information

The well was drilled to a depth of 5527.00m on June 15th, 1996, and was completed by running a $\phi$177.8mm casing to 5525.00m for cementing. After electrical logging, it was found that the cementing quality was unqualified, so perforation and cement squeezing operations were carried out. Afterwards, overflow occurred during the process of replacing crude oil, causing the wellhead to lose control.

### 5.2.5.2 Accident Process

On July 17th, 1996, $\phi$177.8m scraper, with a pipe string structure of $\phi$177.8mm scraper+331mm×310mm joint+$\phi$18×121.8mm drill collars+$\phi$88.9mm drill pipe 184 columns+331mm×410mm joint+411×$\phi$50.8mm live joint+50.8mm rubber hose (to the anti overflow pipe for drilling fluid recovery) and then replace the crude oil by 100m$^3$, Before drilling out, a small amount of crude oil was found to have overflowed. It was determined on site that it was caused by high temperature and uneven pressure in the well. Further observation is needed. Later, it was discovered that crude oil flowed into the surface sewage tank from the connection between tanks 1 and 2, and the flow rate increased. Then, the drilling team unloaded 411×$\phi$50.8mm live joint was prepared to be connected to a single flow valve and shut in the well. However, due to the increase in injection potential and pressure in the well, only 2 buckles of the joint were removed, and the rubber hose was broken from the connection. The wellhead lost control, and crude oil sprayed up to about 15m from the wellhead.

#### 5.2.5.3 Accident Handling Process

Immediately carry out rescue work after a blowout occurs. First, turn off the blowout preventer (after turning off the annular blowout preventer) $\phi$88.9mm gate, then perform reverse circulation well killing, first inject clean water for 30m$^3$, The casing pressure decreased from 20MPa to 16MPa, followed by an injection density of 1.50g/cm$^3$ and 1.20g/m$^3$ 117m$^3$ of mixed drilling fluid, reduce the spray potential inside the drilling tool to the drilling surface, and seize the opportunity to connect Lower the rotary plug with a $\phi$127mm square drill pipe, shut in the well, continue throttling and circulating well pressure, and finally achieve a density of 1.50g/cm$^3$ The drilling of the hydraulic well was successful. At this point, the blowout accident has been resolved.

#### 5.2.5.4 Analysis of Accident Causes

(1) The quality of cementing is not up to standard, and the successful squeezing of cement through perforation provides conditions for oil and gas to enter the well.

(2) Improper well control measures resulted in uncontrolled overflow due to the lack of internal blowout prevention tools in the drilling rods during the construction process.

(3) When an overflow occurs, the on-site person in charge made an error in judgment, which delayed the timing and caused the wellhead to lose control.

(4) The design of perforation completion

#### 5.2.5.3 事故处理经过

井喷发生后，立即进行抢险工作。首先关防喷器（关环形防喷器后再关$\phi$88.9mm闸板），然后进行反循环压井，先注清水30m$^3$，套管压力由20MPa降至16MPa，接着注密度为1.50g/cm$^3$和1.20g/m$^3$的混合钻井液117m$^3$压井，钻具内喷势减弱至钻台面，抢接$\phi$127mm方钻杆下旋塞，关井，继续进行节流循环压井，最终以密度为1.50g/cm$^3$的钻井液压井成功。至此，井喷事故解除。

#### 5.2.5.4 事故原因分析

（1）固井质量不合格，通过射孔挤水泥不成功，为油气进入井内提供了条件。

（2）井控措施不当，在组织施工过程中，钻杆无内防喷工具，致使溢流无法控制。

（3）发生溢流时，现场负责人判断出现错误，以至延误时机，造成井口失控。

（4）射孔完井工程设计不完善，对井下已射开的油气层段和不挤水泥情况未做说明，更没有井控要求。

engineering is not perfect, and there is no explanation for the oil and gas intervals that have been shot underground and the situation of not squeezing cement, let alone well control requirements.

(5) During the construction period, the supervision of underground operations and drilling representatives have unclear division of labor and responsibilities, and their implementation of relevant rules and regulations is not effective. After the underground operation supervisor went to the well, he did not have a clear understanding of the underground situation with the drilling representative and the drilling team. Before construction, he did not carefully explain the construction tasks and techniques of the drilling team, and there were no well control measures or safety requirements during the process of replacing crude oil, which ultimately led to the occurrence of a blowout accident.

### 5.2.5.5 Lessons Learned and Preventive Measures

The occurrence of the blowout accident exposed the problems in various aspects such as design, construction, and emergency response, and also reflected that the safety management principle of "who is in charge, who is responsible" has not been implemented in practice. The production management personnel and drilling team employees of Party A have poor well control training effects and lack practical skills. To learn from lessons, the following preventive measures should be taken:

(1) According to the safety management principle of "who is in charge, who is responsible", it is necessary to effectively implement the well control system for

（5）井下作业监督、钻井代表在施工期间，各自分工职责不清，对有关的规章制度执行不力。井下作业监督上井后，未向钻井代表、钻井队了解清楚井下情况，施工前又未认真对井队施工任务、施工工艺交底，并且在替原油过程中无任何的井控措施和安全要求，最终导致了井喷事故。

### 5.2.5.5 教训及防范措施

该井井喷事故的发生，暴露了设计、施工、应急等环节存在的问题，也反映出"谁主管、谁负责"的安全管理原则没有落到实处。甲方生产管理人员及井队职工井控培训效果不好，缺乏实战能力。为吸取教训，应采取以下防范措施：

（1）按照"谁主管、谁负责"的安全管理原则，要切实落实完井、试油和井下作业井控制度；

completion, oil testing, and downhole operations;

(2) The drilling department should work together with the oil and gas development company and the oil testing department to develop practical and feasible well control systems and measures for completion testing and downhole operations. On site operators strictly adhere to various operating procedures and systems.

### 5.2.6　Accident of Wire Tool Breakage and Detachment in Y Well

#### 5.2.6.1　Basic Information

(1) Well number: Y.

(2) Well type: oil production well.

(3) Well type: horizontal well.

(4) Oil testing date: September 14th to November 14th, 2016.

Date of occurrence: October 22th, 2016.
Termination date: October 30th, 2016.
Lost time: 8.5 days.

(5) Falling fish structure: steel wire×45m+25.40mm steel wire rope cap×0.20m+38mm universal joint×0.20m+38mm tungsten steel weighting rod×4.00m+38mm universal joint×0.20m+38mm mechanical shock absorber×2.20m+67mm stabilizer×0.10m+47mm JDC×0.40m+68mm balanced plug×0.40m.

#### 5.2.6.2　Process of Occurrence

On October 19th, 2016, at 0∶00, a 200.03mm PHP-3 packer was inserted to complete the well string. The string structure was: pipe shoes×0.15m+88.90mm oil pipe 1 piece+fishing plug×Two 0.29m+88.90mm oil pipes+200.03mm PHP-3 packers×1.74m+88.90mm oil pipe 3 pieces+88.90mm expansion pipe 2 pieces+88.9mm lifting short joint×100 oil pipes with a diameter of 0.52m+88.90mm were

（2）钻井处要会同油气开发公司、试油处制定切实可行的完井试油、井下作业井控制度和措施。现场作业人员严格执行各项作业规程、制度。

### 5.2.6　Y井钢丝工具断脱落井事故

#### 5.2.6.1　基本情况

（1）井号：Y。

（2）井别：采油井。

（3）井型：水平井。

（4）试油日期：2016年9月14日—11月14日。

发生时间：2016年10月22日。
解除时间：2016年10月30日。
损失时间：8.5d。

（5）落鱼结构：钢丝×45m+25.40mm钢丝绳帽×0.20m+38mm万向节×0.20m+38mm钨钢加重杆×4.00m+38mm万向节×0.20m+38mm机械震击器×2.20m+67mm扶正器×0.10m+47mmJDC×0.40m+68mm可平衡式堵塞器×0.40m。

#### 5.2.6.2　发生经过

2016年10月19日0∶00，下入200.03mm PHP-3封隔器完井管柱，管串结构：管鞋×0.15m+88.90mm油管1根+投捞式堵塞器×0.29m+88.90mm油管2根+200.03mmPHP-3封隔器×1.74m+88.90mm油管3根+88.90mm伸缩管2根+88.90mm提升短节×0.52m+88.90mm油管100根，管柱下到位后，坐油管挂，换装采油树，替液。

installed. After the pipe column was lowered into place, the oil pipe was hung, replaced with an oil tree, and the liquid was replaced.

On October 21st, at 18:00, the steel wire operation began with the deployment of school darts. After the darts were in place, they were graded and positively pressurized to seat the packer, slowly releasing the oil pressure to 8MPa. The darts were lifted and the annulus was pressurized, increasing the pressure from 9MPa to 24MPa. The pressure remained stable for 30 minutes, and the packer passed the sealing inspection. Slowly release the sleeve pressure to 3MPa. On October 22nd, at 8:10 am, a dart tool was launched and dropped into the well.

### 5.2.6.3 Processing Process

On October 23rd from 8:00 to 3:00, lower the fishing tool (68mm inner spear× 0.56m) not caught. Then insert the steel wire head tool (68mm)×At a depth of 0.20m, a steel wire head was detected at 5559.00~5566.00m, and a fishing tool (65mm multiple internal fishing spears) was inserted×0.30m) Remove the steel wire and tool string (steel wire) from the well×45m+25.40mm steel wire rope cap× 0.20m+38mm universal joint × 0.20m+38mm tungsten steel weighting rod × 4.00m+38mm universal joint×0.20m+38mm mechanical shock absorber×2.20m+67mm stabilizer× 0.10+47mm JDC×0.40m), dart 68mm×0.4m) not taken out.

On October 24th, the fishing tool was used for salvage, and the tool string structure was a 38mm steel wire rope cap×0.20m+38mm universal joint×0.29m+50mm tungsten steel weighting rod×1.59m+38mm universal joint × 0.29m+38mm mechanical shock absorber× 0.21m+63mm stabilizer×0.10m+47mm JDC × 0.40m, encountered obstruction at 5587m, unable to reach the position of the spear.

10月21日18:00开始钢丝作业投放校镖,梭镖到位后,分级正打压坐封封隔器,缓慢泄油压至8MPa,提开梭镖,环空打压,压力由9MPa升至24MPa,稳压30min不降,封隔器验封合格,缓慢放套压至3MPa。10月22日8:10,起出投放梭镖工具,工具串落井。

### 5.2.6.3 处理过程

10月23日8:00—15:00,下打捞工具(68mm内捞矛×0.56m)未捞获。然后下入找钢丝头工具(68mm×0.20m)在5559.00~5566.00m探得钢丝头,又下入打捞工具(65mm多片内捞矛×0.30m)捞出落井钢丝及工具串(钢丝×45m+25.40mm钢丝绳帽×0.20m+38mm万向节×0.20m+38mm钨钢加重杆×4.00m+38mm万向节×0.20m+38mm机械震击器×2.20m+67mm扶正器×0.10+47mm JDC×0.40m),梭镖68mm×0.4m)未带出。

10月24日,下打捞工具打捞,工具串结构:38mm钢丝绳帽×0.20m+38mm万向节×0.29m+50mm钨钢加重杆×1.59m+38mm万向节×0.29m+38mm机械震击器×0.21m+63mm扶正器×0.10+47mmJDC×0.40m,在5587m遇阻,无法到达梭镖位置。

From October 27th to 29th, install the continuous tubing blowout preventer, spray pipe, and injection head, and pass the pressure test as required. Flush the continuous oil pipe string to a depth of 5600.10m at the top of the fish and encounter a resistance of 15kN (pipe structure: 43mm)×16.5cm jet flushing tool+43mm×65cm motor head assembly+43mm×15cm continuous oil pipe connector), retested three times with the same position, pump pressure of 33.9~35.8MPa, wellhead control pressure of 8.4~9.6MPa, casing pressure of 5.92MPa, displacement of 200L/min, sit and wash until sunrise with clear water and no debris, sit and wash at a depth of 5599.50m, lift the continuous oil pipe, lower the continuous oil pipe fishing tool at the wellhead, encounter obstacles, multiple attempts cannot pass, and close the well.

On October 30th, the continuous oil pipe injection head was dismantled and connected to the steel wire anti spray equipment. After passing the pressure test, the fishing tool was lowered. The tool string structure was a 38mm steel wire rope cap×0.13m+38mm universal joint×0.2m+45mm tungsten steel weighting rod×1.47m+38mm universal joint ×0.29m+38mm mechanical shock absorber× 1.60m+63mm stabilizer×0.12m+47mm JDC× 0.36m, catch the spear, the salvage is over.

### 5.2.6.4 Cause Analysis

(1) Direct cause.

After the steel wire operation is completed and the dart is placed in place, the graded positive pressure is applied to seal the packer. The packer passes the sealing inspection. During the process of lifting and dropping the dart tool, the tool string causes the steel wire to twist and then fall, causing the steel wire to break.

10月27日—29日，安装连续油管防喷器、防喷管、注入头并按要求试压合格。下连续油管冲洗管柱，冲洗至鱼顶深度5600.10m遇阻15kN（管柱结构：43mm×16.5cm射流冲洗工具+43mm×65cm马达头总成+43mm×15cm连续油管连接器），复探3次，位置不变、泵压为33.9~35.8MPa，井口控压8.4~9.6MPa，套压为5.92MPa，排量为200L/min，坐洗至出口见清水，无杂物，坐洗深度为5599.50m，起连续油管，下连续油管打捞工具在井口遇阻，多次尝试无法通过，关井。

10月30日，拆用连续油管注入头，连接钢丝防喷设备，试压合格，下打捞工具，工具串结构：38mm钢丝绳帽×0.13m+38mm万向节×0.2m+45mm钨钢加重杆×1.47m+38mm万向节×0.29m+38mm机械震击器×1.60m+63mm扶正器×0.12m+47mm JDC×0.36m，捞获梭镖，打捞结束。

### 5.2.6.4 原因分析

（1）直接原因。

钢丝作业投放梭镖到位后，分级正打压坐封封隔器，封隔器验封合格，起投放梭镖工具过程中，工具串在井下发生上冲导致钢丝打扭，然后工具串下落，造成钢丝扭断。

(2) Indirect causes.

① When the tool is inserted, impurities or drilling fluid sediment adhering to the pipe wall are scraped off, and after a certain period of time, they deposit on the upper and surrounding parts of the blocker. When pressing and sealing, the sediment is compacted, causing an increase in friction between the blocker and the pipe wall. When the wellhead is depressurized, the pressure difference energy accumulated below the blocker is suddenly released, causing the tool to top drill.

② After the tool is lowered to the working cylinder position, the pressure below the plug slowly recovers. After the first pressure setting, the pressure is released to the original pressure of 8MPa. It is possible that the drilling fluid below the plug will return under the driving force of the restored pressure, and then the second pressure will reach 36MPa, which cannot stabilize the pressure. There may be drilling fluid support on the sealing surface of the plug, causing poor sealing. The returning drilling fluid is compacted under pressure and blocks the pressure balance hole. Finally, during the pressure relief, the pressure difference energy below the plug is suddenly released, causing top drilling.

(3)Management reasons.

Insufficient risk identification and ineffective risk reduction measures for wire plug operations have not been developed and implemented in response to the well conditions.

### 5.2.6.5　Lessons Learned

(1) Strictly follow the operating procedures for fishing the plug. After the pressure setting is completed, lift the plug 50

meters up and check the seal again. If a second pressure setting is required, it can be lowered to the position of the plug again and re pressed.

(2) Before construction, thoroughly understand the well condition information, such as the properties of drilling fluid and protective fluid, formation pressure, etc. Based on specific well conditions, develop corresponding reduction and control measures.

(3) Increase the reserved pressure value of the pump truck after pressure relief, use a needle valve to finely control the pressure relief, and extend the pressure balance time.

(4) Summarize experience, learn from lessons, improve the safety control awareness of operators.

### 5.2.7 Fracture Accident of J Well Oil Testing Joint Pipe Joint

#### 5.2.7.1 Basic Information

(1) Well number: J.

(2) Well category: Risk exploration well.

(3) Well type: vertical well.

(4) Oil testing date: December 28th, 2019 to June 6th, 2020.

Date of occurrence: January 20th, 2020.
Termination date: April 5th, 2020.
Lost time: 74.8 days.

(5) Falling fish structure (below the buckle joint): with a perforated gun tail×0.13m+pressure delay detonation device×0.71m+tail air×0.26m+perforation gun×3.0m+sandwich gun×3.0m+perforating gun×22.0m+sandwich gun×2.0m+perforation gun×16.0m+sandwich gun×7.0m+perforation gun×20.0m+sandwich gun×7.0m+perforation gun×19.0m+sandwich gun×51.0m+perforation gun×16.0m+empty gun×1.91m+internal threaded joint×0.11m+pressure delay

再次下至堵塞器位置，重新打压。

（2）施工前详细了解井况信息，如钻井液、保护液性质、地层压力情况等针对具体井况，制订相应消减控制措施。

（3）提高泵车泄压后的预留压力值，用针形阀精细控制泄压，延长压力平衡时间。

（4）总结经验、吸取教训，提高操作人员安全管控意识。

### 5.2.7　J井试油联作管柱接头断裂事故

#### 5.2.7.1　基本情况

（1）井号：J。

（2）井别：风险探井。

（3）井型：直井。

（4）试油日期：2019年12月28日—2020年6月6日。

发生时间：2020年1月20日。
解除时间：2020年4月5日。
损失时间：74.8d。

（5）落鱼结构（变扣接头之下）：带孔枪尾×0.13m+压力延时起爆装置×0.71m+尾空×0.26m+射孔枪×3.0m+夹层枪×3.0m+射孔枪×22.0m+夹层枪×2.0m+射孔枪×16.0m+夹层枪×7.0m+射孔枪×20.0m+夹层枪×7.0m+射孔枪×19.0m+夹层枪×51.0m+射孔枪×16.0m+空枪×1.91m+内螺纹接头×0.11m+压力延时起爆装置×0.71m+长槽筛管×0.55m+减振器×1.12m+减振器×1.12m+长槽筛管×0.55m+变扣接头×0.12m+接球筛管×2.06m+调整短节2根×5.11m+73.0mm

detonation device×0.71m+long groove sieve tube×0.55m+shock absorber×1.12m+shock absorber×1.12m+long groove sieve tube×0.55m+variable buckle joint×0.12m+ball screen tube×2.06m+2 adjustable short joints ×22 5.11m+73.0mm oil pipes×210.86m, total length 391.32m.

### 5.2.7.2 Process of Occurrence

JT1 downhole perforation acidification test combined with tubing to 7591.10m, and carried out joint oil testing construction on the Dengying group (7424.00~7590.00m).

On January 7th, 2020, a low displacement of 4m$^3$ was achieved through the liquid replacement valve Isolation fluid, 23.9m$^3$ turned to acid, and closed the replacement valve. Positive pressure perforation and detonation were successful (oil pressure decreased from 18.90MPa to 0, and after 2 hours, oil pressure increased from 0 to 8.60MPa). Subsequently, a high-pressure turning acid preparation acidizing fracturing was carried out, with a high pump pressure. A total of 8 oscillation squeezing tests were conducted (pump pressure of 75.00~98.00MPa), with a cumulative injection of 3m$^3$ of turning acid No displacement established. When the 8th high extrusion occurs, the oil sleeve is blown through, and the oil pressure suddenly drops from 89.32MPa to 52.06MPa, while the balance sleeve pressure rises from 28.96MPa to 33.57MPa. After 0.5 hours, the oil pressure was 46.00 MPa and the casing pressure was 27.00 MPa, indicating that the sealing of the test packer had failed. After research, it was decided to kill the well, pull out the test string, connect the reservoir casing back, and then carry out reservoir renovation.

油管 22 根 ×210.86m，总长 391.32m。

### 5.2.7.2　发生经过

JT1 井下入射孔酸化测试联作管柱至 7591.10m，对灯影组（7424.00~7590.00m）进行联作试油施工。

2020 年 1 月 7 日，通过替液阀低替 4m$^3$ 隔离液、23.9m$^3$ 转向酸后关闭替液阀正加压射孔起爆成功（油压由 18.90MPa 降至 0，2h 后油压由 0 升至 8.60MPa）。随后进行高挤转向酸准备酸化压裂，泵压较高，采用振荡试挤共计 8 次（泵压为 75.00~98.00MPa），累计挤注入转向酸 3m$^3$ 未建立排量。当第 8 次高挤时油套窜通，油压突然由 89.32MPa 降至 52.06MPa，平衡套压由 28.96MPa 升至 33.57MPa。0.5h 后，油压为 46.00MPa，套压为 27.00MPa，判断测试封隔器密封失效。经研究决定压井，起出测试管柱，回接油层套管后再开展储层改造。

At 5:10 on January 20th, after the completion of the joint operation of the oil pipe column, it was found that the external thread end of the safety joint of the packer tail pipe and the variable buckle joint of the lower oil pipe connection was broken, with a relatively regular fracture. The remaining pipe columns were broken and fell into the well, with a fish length of 391.32 meters and a theoretical fish top of 7268.08 meters.

5.2.7.3　Processing Process

(1) Phase 1: The main method of fishing is to use male and female cone inverted fishing, and catch 279.43m of fallen fish.

On January 23th, 2020, the retrievable fishing spear reached the top of the fish at 7211.20m (tail pipe hanging well depth of 7258.80m), and after pressurized fishing, multiple lifting and releasing activities were carried out, with a maximum lifting force of 450kN. The caught fish did not get stuck. This drilling tool combination: LM73-105 fishing spear+variable buckle joint+88.9mm drill collar+variable buckle joint+101.6mm drill pipe+backpressure valve+bypass valve+101.6mm drill pipe.

After research, it has been decided to withdraw from the salvage spear and handle it in an inverted manner. Due to the fact that the main method of fishing is the 73mm flat oil pipe, which has a much lower up torque than the up torque of the fishing tool, it has been decided to mainly use the forward buckle fishing tool for reverse buckle fishing. The drilling tool combination (from bottom to top) is: reverse buckle male cone, female cone+variable buckle joint+73mm reverse buckle drill pipe+forward and reverse buckle change joint+101.6mm forward buckle drill pipe.

1月20日5:10，起联作油管柱完，发现封隔器尾管安全接头与下部油管连接的变扣接头外螺纹端断裂，断口较规则，其余管柱断落入井，落鱼长391.32m，理论鱼顶7268.08m。

5.2.7.3　处理过程

（1）第一阶段：以公母锥倒扣打捞为主进行打捞，捞获落鱼279.43m。

2020年1月23日，下可退打捞矛至7211.20m探得鱼顶（尾管悬挂井深7258.80m），加压打捞后，多次反复提放活动，最高过提450kN，捞获的落鱼未解卡。本次钻具组合：LM73-105打捞矛+变扣接头+88.9mm钻铤+变扣接头+101.6mm钻杆+回压阀+旁通阀+101.6mm钻杆。

经研究决定，退出打捞矛，后续以倒扣方式处理。由于落鱼以73mm平式油管为主，其上扣扭矩远低于打捞钻具上扣扭矩，因此决定主要以正扣打捞钻具进行倒扣的方式打捞，钻具组合（从下至上）：反扣公锥、母锥+变扣接头+73mm反扣钻杆+正反转换接头+101.6mm正扣钻杆。

From January 26th to February 18th, tools such as reverse threaded female cone, reverse threaded male cone, and reverse threaded large end male cone were successively inserted for reverse fishing. The depth was processed to 7483.95m, and a total of 10 drills were carried out. A total of 279.43m of 73mm oil pipes and their lower part of the perforating gun were retrieved. Among them, the 18th single oil pipe was found to have a fracture at a distance of 10mm above the tool retraction buckle, with a neat fracture surface. The remaining length of falling fish is 117.89m.

Remaining fish structure: gun tail ×0.13m+delayed detonation device×0.71m+tail air×0.26m+perforation gun× 3.0m+sandwich gun×3.0m+perforating gun× 22.0m+sandwich gun×2.0m+perforation gun ×16.0m+sandwich gun×7.0m+perforation gun×20.0m+sandwich gun×7.0m+perforation gun×19.0m+sandwich gun×16.8m.

(2) Phase 2: During the process of continuing the reverse fishing, the drill pipe tripped twice, resulting in an additional fish drop of 234.85m+30.65m.

On February 20th, 2020, put salvaged drilling tools, combination: reverse buckle large head male cone×0.4m+conversion joint×24 pieces of 0.53m+73.00mm reverse drilling rods×234.09m+forward and reverse conversion joint×One 0.47m+101.6mm positive buckle drill pipe×9.73m+adapter×0.47m+bypass valve×0.38m+adapter×281 pieces of 0.50m+ 101.6mm positive buckle drill rods×2731.14m.

Drilling down to a depth of 7483.95m, a fish head was detected and successfully salvaged. The hanging weight of the reverse buckle increased from 2050kN to 2350kN and

1月26日—2月18日，先后下入反扣母锥、反扣公锥、反扣大头公锥等工具进行倒扣打捞，处理至深度7483.95m，共计10趟钻，捞获73mm全部油管及其下部部分射孔枪共计279.43m，其中捞获的第18根单根油管外螺纹退刀扣以上10mm处断裂，断口整齐。剩余落鱼长度117.89m。

剩余落鱼结构：枪尾×0.13m+延时起爆装置×0.71m+尾空×0.26m+射孔枪×3.0m+夹层枪×3.0m+射孔枪×22.0m+夹层枪×2.0m+射孔枪×16.0m+夹层枪×7.0m+射孔枪×20.0m+夹层枪×7.0m+射孔枪×19.0m+夹层枪×16.8m。

（2）第二阶段：继续倒扣打捞过程中，两次钻杆脱扣，新增落鱼234.85m+30.65m。

2020年2月20日，下公打捞钻具，组合：反扣大头公锥×0.4m+转换接头×0.53m+73.00mm反扣钻杆24根×234.09m+正反转换接头×0.47m+101.6mm正扣钻杆1根×9.73m+转换接头×0.47m+旁通阀×0.38m+转换接头×0.50m+101.6mm正扣钻杆281根×2731.14m。

下钻至井深7483.95m探得鱼头，打捞造扣成功，反转倒扣悬重由2050kN升至2350kN再降至2030kN，开泵泵压为4.1MPa，低于正常值，下放钻具多次正转对扣、反转倒扣处理无明显效果，起钻发现73.00mm反扣钻24根及下部打捞管柱脱落扣，新增落鱼234.85m（大头公锥×0.0m+转换接头×0.33m+73mm反扣钻杆24根×234.12m）。

then decreased to 2030kN. The pump pressure was 4.1MPa, which was lower than the normal value. The drilling tool was lowered multiple times for forward and reverse buckles, but there was no significant effect. When drilling out, 24 73.00mm reverse buckles were found, and the lower fishing pipe column fell off. A new fish drop of 234.85m was added (large head male cone) × 0.0m+conversion joint × 24 pieces of 0.33m+73mm reverse drilling rods × 234.12m).

On February 22nd, a large head male cone was lowered to a depth of 7255.75m and a new fish head was found after fishing. The hanging weight of the drilling tool was raised from 2080kN to 2300kN, and the hook was successfully made. The drilling tool was repeatedly moved up and down, with a hanging weight of 2080~2300kN. Suddenly, the hanging weight decreased from 2300kN to 2080kN. After drilling, it was found that the fish had tripped from the lower end of the positive buckle bypass valve, and another fish drop of 30.65m was added (reverse buckle large head male cone)×0.45m+variable buckle joint×0.51m+73mm reverse drilling rod× 9.73m+front and back buckle joints×2 drill rods with a diameter of 0.51m+101.6mm× 19.45m).

Falling fish structure: gun tail× 0.13m+ delayed detonation device×0.71m+tail air×0.26m+ perforation gun×3.0m+sandwich gun×3.0m+ perforating gun×22.0m+sandwich gun×2.0m+ perforation gun×16.0m+sandwich gun×7.0m+ perforation gun×20.0m+sandwich gun×7.0m+ perforation gun×19.0m+sandwich gun×11.8m+ large male cone×0.40m+adapter×24 pieces of 0.33m+73mm reverse drilling rods×234.12m+ reverse buckle large head male cone×0.45m+

2月22日，下入大头公锥至井深7255.75m探得鱼头后打捞新增落鱼，上提钻具悬重由2080kN升至2300kN，造扣成功，反复上下活动钻具，悬重为2080~2300kN，悬重突然由2300kN降至2080kN，起钻完，发现从正扣旁通阀下端脱扣，再次新增落鱼30.65m（反扣大头公锥×0.45m+变扣接头×0.51m+73mm反扣钻杆×9.73m+正反扣接头×0.51m+101.6mm钻杆2根×19.45m）。

落鱼结构：枪尾×0.13m+延时起爆装置×0.71m+尾空×0.26m+射孔枪×3.0m+夹层枪×3.0m+射孔枪×22.0m+夹层枪×2.0m+射孔枪×16.0m+夹层枪×7.0m+射孔枪×20.0m+夹层枪×7.0m+射孔枪×19.0m+夹层枪×11.8m+大头公锥×0.40m+转换接头×0.33m+73mm反扣钻杆24根×234.12m+反扣大头公锥×0.45m+变扣接头×0.51m+73mm反扣钻杆×9.73m+正反扣接头×0.51m+101.6mm钻杆2根×19.45m总长377.4m。

variable buckle joint×0.51m+73mm reverse drilling rod×9.73m+front and back buckle joints ×2 drill rods with a diameter of 0.51m+101.6mm ×19.45m in total length 377.4m.

(3) Phase 3: During the process of dealing with newly added fish falling, accidents such as drill collar breakage, drill pipe detachment, and joint fracture occur, mainly through the use of male cone inverted fishing. As a result, the number of fish falling gradually increases.

From February 22nd to March 5th, a total of 5 drilling trips were carried out to catch 206.16m of newly added fish by using buckle and male cone tools to catch inverted fish. When salvaging the reverse buckle at 7431.20m from the top of the fish using the 19th large head male cone, the salvaging tool broke from the outer thread of the drill collar, resulting in an additional 28.64m of fish falling.

Falling fish structure: gun tail × 0.13m+ delayed detonation device × 0.71m+tail air× 0.26m+perforation gun×3.0m+sandwich gun× 3.0m+perforating gun×22.0m+sandwich gun× 2.0m+perforation gun×16.0m+sandwich gun× 7.0m+perforation gun×20.0m+sandwich gun× 7.0m+perforation gun×19.0m+sandwich gun ×11.79m+large male cone×0.40m+adapter ×Six 0.33m+73mm reverse drilling rods× 58.62m+large male cone × 0.45m+safety joint ×0.5m+variable buckle joint×0.52m+3 drill collars×27.17m, total length 199.88m, fish top well depth 7402.56m.

On March 7th, one drill collar was retrieved by inserting a reverse buckle cone; On March 9th, during the fishing of the reverse buckle big head male cone, the drill pipe detached from the middle of the 73.00mm drill pipe, and another 49.73m of fish fell (reverse buckle big head male cone) ×

（3）第三阶段：以公锥倒扣打捞为主处理新增落鱼过程中，发生钻铤断落、钻杆脱扣、接头断裂事故，落鱼逐渐增多。

2月22日—3月5日，采用对扣、公锥工具打捞倒扣处理新增落鱼，共计5趟钻捞获落鱼206.16m。在第19下大头公锥打捞钻具至鱼顶7431.20m打捞倒扣时，打捞钻具从钻铤外螺纹断裂，新增落鱼28.64m。

落鱼结构：枪尾×0.13m+延时起爆装置×0.71m+尾空×0.26m+射孔枪×3.0m+夹层枪×3.0m+射孔枪×22.0m+夹层枪×2.0m+射孔枪×16.0m+夹层枪×7.0m+射孔枪×20.0m+夹层枪×7.0m+射孔枪×19.0m+夹层枪×11.79m+大头公锥×0.40m+转换接头×0.33m+73mm反扣钻杆6根×58.62m+大头公锥×0.45m+安全接头×0.5m+变扣接头×0.52m+钻铤3根×27.17m，总长199.88m，鱼顶井深7402.56m。

3月7日，下入反扣母锥捞获钻铤1根；3月9日，下反扣大头公锥打捞时，钻杆从73.00mm钻杆中部脱扣，再次新增落鱼49.73m（反扣大头公×0.40m+转换接头×0.52m+73.00mm钻杆5根×48.81m），随后下入正扣大头公锥两次，捞取落鱼73.00mm钻杆5根×48.81m+转换接头×0.52m，再次下入正扣大头公锥打捞时，钻具安全接头之上的转换接头外螺纹断裂，井内新增落鱼0.97m（正扣大头公锥×0.45m+安全接头×0.52m）角顶井深7410.32m。

0.40m+adapter×five drill rods with a diameter of 0.52m+73.00mm×48.81m), then insert the front buckle big head male cone twice and fish out five 73.00mm drill rods × 48.81m+adapter ×0.52m, when fishing again with a positive buckle big head male cone, the external thread of the conversion joint above the safety joint of the drilling tool broke, and a new fish drop of 0.97m (positive buckle big head male cone) was added to the well×0.45m+safety joint× The depth of the corner top well is 7410.32m (0.52m).

Falling fish structure: gun tail× 0.13m+ delayed detonation device×0.71m+tail air× 0.26m+perforation gun×3.0m+sandwich gun× 3.0m+perforating gun×22.0m+sandwich gun× 2.0m+perforation gun×16.0m+sandwich gun× 7.0m+perforation gun×20.0m+sandwich gun× 7.0m+perforation gun×19.0m+sandwich gun× 11.79m+large male cone×0.40m+adapter ×Six 0.33m+73.00mm reverse drilling rods× 58.62m+ large male cone×0.45m+safety joint ×0.50m+ adapter×Two 0.52m+88.90mm drill collars× 18.01m+reverse buckle large head male cone× 0.40m+positive buckle large head male cone× 0.45m+safety joint×0.52m, total length 192.09m。

(4) Stage 4: Using 3 rounds of grinding and milling, 2 rounds of nesting, slow processing progress, and termination of salvage.

Due to handling stuck fish inside the small casing, the buckle type of the 73.00mm drill pipe, large end male cone, 88.90mm drill collar, and adapter used bears less torque. At present, there are two large male cones and safety joints on the fish head, and there are residual broken external threads in the water hole. The fishing method is limited, and drilling tool breakage accidents occur

落鱼结构：枪尾×0.13m+延时起爆装置×0.71m+尾空×0.26m+射孔枪×3.0m+夹层枪×3.0m+射孔枪×22.0m+夹层枪×2.0m+射孔枪×16.0m+夹层枪×7.0m+射孔枪×20.0m+夹层枪×7.0m+射孔枪×19.0m+夹层枪×11.79m+大头公锥×0.40m+转换接头×0.33m+73.00mm反扣钻杆6根×58.62m+大头公锥×0.45m+安全接头×0.50m+转换接头×0.52m+88.90mm钻铤2根×18.01m+反扣大头公锥×0.40m+正扣大头公锥×0.45m+安全接头×0.52m，总长192.09m.

（4）第四阶段：采用3次磨铣，2次套，处理进展缓慢，终止打捞。

由于在小套管内处理卡死的落鱼，所用的73.00mm钻杆、大头公锥、88.90mm钻铤及转换接头的扣型承受扭矩较小。目前鱼顶为2只大头公锥及安全接头，水眼内残留有断裂外螺纹，打捞方式受限，且钻具断落事故频现，决定采用钻磨、套铣处理鱼头。

frequently. Therefore, it has been decided to use drilling and milling to treat the fish head.

From March 12th to 22nd, drill and grind 113.00mm shoes three times (combination: 113.00mm shoes)×0.68m+88.9mm drill collar with 6 columns×162.11m+HT38× 211A conversion joint×92 drill rods with a diameter of 0.53m+101.60mm×2682.51m+410 ×HT38 male adapter×0.45m+127.00m drill 92 columns×475 drill pipes with a diameter of 2682.51m+127.00m×4576.65m), with no progress after a cumulative footage of 0.95m.

From March 26th to April 5th, 113mm/88mm and 115mm/86mm milling barrels were used to mill the fish twice. The milling was carried out to a depth of 7411.50m, but the progress was not significant. If the processing period cannot be predicted and the secondary complexity risk is high, it has been decided to terminate the salvage after research.

Remaining fish falling structure underground: gun tail×0.13m+delayed detonation device× 0.71m+tail air×0.26m+perforation gun×3.0m+ sandwich gun×3.0m+perforating gun×22.0m+ sandwich gun×2.0m+perforation gun×16.0m+ sandwich gun×7.0m+perforation gun×20.0m+ sandwich gun×7.0m+shooting gun×19.0m+ sandwich gun×11.79m+large male cone× 0.40m+adapter×Six 0.33m+73.00mm reverse drilling rods× 58.62m+large male cone×045m+ safety joint ×0.50m+adapter×Two 0.52m+ 88.90mm drill collars×18.01m+reverse deduction big head male×0.40m+positive buckle large head male cone×0.43m, total length 191.57m.

The complex handling of this accident took a total of 74.8 days, with a total of 23 salvages, 2 drilling fluid treatments, 3 grinding and milling operations, and 2 casing and milling operations. A total of 279.43m of

3月12日—22日，下入113.00mm磨鞋进行3次钻磨（组合：113.00mm磨鞋×0.68m +88.9mm钻铤6柱×162.11m+HT38×211A转换接头×0.53m+101.60mm钻杆92柱×2682.51m +410×HT38公转换接头×0.45m+127.00m钻92柱×2682.51m+127.00m钻杆475根× 4576.65m），累计进尺0.95m后无进展。

3月26日—4月5日，分别换用113mm/ 88mm、115mm/86mm的铣筒套铣落鱼两次，套铣至井深7411.50m，进展不明显，若继续处理周期无法预计，且次生复杂风险高，经研究决定终止打捞。

井下剩余落鱼结构：枪尾×0.13m+延时起爆装置×0.71m+尾空×0.26m+射孔枪×3.0m+夹层枪×3.0m+射孔枪×22.0m+夹层枪×2.0m+射孔枪×16.0m+夹层枪×7.0m+射孔枪×20.0m+夹层枪×7.0m+射枪×19.0m+夹层枪×11.79m+大头公锥×0.40m+转换接头×0.33m+73.00mm反扣钻杆6根×58.62m+大头公锥×045m+安全接头×0.50m+转换接头×0.52m+88.90mm钻铤2根×18.01m+反扣大头公×0.40m+正扣大头公锥×0.43m，总长191.57m。

本次事故复杂处理共耗时74.8天，累计打捞23趟次，处理钻井液2次，磨铣3趟次，套铣2趟次，累计捞获原始落鱼279.43m，累计新增落鱼长度344.92m，捞获新增落鱼264.70m，磨铣长度0.54m，剩余落鱼长度191.57m。接甲方通知于2020年4月5日1:00终止打捞，转入侧钻。

original fish were caught, 344.92m of new fish were added, 264.70m of new fish were caught, 0.54m of grinding and milling operations, and 191.57m of remaining fish were caught. Upon receiving notice from Party A, the salvage was terminated at 1:00 on April 5th, 2020, and the drilling was transferred to side drilling.

5.2.7.4 Cause Analysis

(1) Direct cause.

The comprehensive strength of the variable buckle joint and oil pipe below the packer is insufficient, and the transverse shear force generated by the perforation impact load exceeds the yield stress of the oil pipe and joint, which is the direct cause of the fracture of the oil pipe and variable buckle joint.

(2) Indirect causes.

① Insufficient understanding of the importance of the perforation string below the packer. It is generally believed that the force on the string above the packer is harsh and complex, and the force on the string below the packer has very little impact. The mechanical strength verification of the string in the oil testing design is only limited to the string above the packer, and the strength verification of the string below the packer has not been carried out.

② Insufficient understanding of perforation impact load. Regarding the phenomenon of bending or fracture of oil pipes caused by axial compression load and radial vibration generated by perforation, only Schlumberger Company has conducted partial research on related aspects abroad. At present, the impact load of perforation in China is mainly focused on the impact of axial tensile force on the perforation string, and there is insufficient understanding of the destructive effects of axial

5.2.7.4 原因分析

（1）直接原因。

①封隔器以下变扣接头和油管综合强度不足，射孔冲击载荷产生的横向剪切力超过了油管和接头的屈服应力，是造成油管和变扣接头断裂的直接原因。

（2）间接原因。

① 对封隔器以下的射孔管柱的重要程度认识不到位。通常认为封隔器以上管柱受力苛刻复杂，封隔器以下管柱受力影响非常小，试油设计中管柱力学强度校核也仅限于封隔器以上管柱，未对封隔器以下管柱强度进行校核。

② 对射孔冲击载荷认识不足。对于射孔产生的轴向压缩载荷和径向振动造成油管发生弯曲或断裂现象，国外仅斯伦贝谢公司对相关方面有部分研究。国内目前对射孔冲击载荷关注较多的为轴向拉伸力对射孔管柱造成的影响，对轴向压缩和径向摆动力的破坏性认识不足。

compression and radial oscillation force.

(3) Management reasons.

① The management of the use of wellbore joints is not standardized. The broken buckle joint was used for the second time in the well (not newly purchased), and its strength has decreased. No non-destructive testing was conducted before using.

② The technical specifications for the procurement of wellbore joints are incomplete. During procurement, only material and hardness requirements were proposed to the manufacturer, without any special requirements for heat treatment, yield strength, processing technology, etc; The inspection is mainly based on appearance judgment, without testing and reviewing key parameters such as material, hardness, and strength.

#### 5.2.7.5 Lessons Learned

(1) Deepen the research on the impact of perforation and detonation effects on the mechanics of the pipe string in ultra deep slim hole wells, optimize the structure of the pipe string, and ensure the successful implementation of the perforation and testing combined process.

(2) Improve the design of oil testing engineering, conduct strength verification on the pipe column below the packer in mechanical analysis, and make requirements for the triaxial stress strength of the pipe column below the packer.

(3) Strengthen the procurement, acceptance, and usage management of drilling tools. Improve procurement technical specifications, provide standardized processing drawings for all variable buckle joints, and conduct non-destructive testing after processing is completed; The variable buckle joint used in

（3）管理原因。

① 入井接头使用管理不规范。断裂的变扣接头为第二次入井使用（非新购），强度有所降低，使用前也未进行探伤检测。

② 入井接头采购的技术规范不完善。采购时只向生产厂家提出了材质、硬度的要求，未对热处理、屈服强度、加工工艺等提出特殊要求；验货时主要依据外观判断，未对材质、硬度、强度等关键参数进行检测复核。

#### 5.2.7.5 经验教训

（1）深入开展超深小井眼井射孔爆轰效应对管柱力学影响的研究，优化管柱结构，保障射孔测试联作工艺的成功实施。

（2）完善试油工程设计，力学分析中应对封隔器以下管柱进行强度校核，对封隔器以下管柱的三轴应力强度做出要求。

（3）加强入井工具的采购、验收及使用管理。完善采购技术规范，所有变扣接头的加工，必须提供规范的加工图纸，加工完成后进行无损检测；射孔测试联作管柱使用的变扣接头必须为全新接头。

the perforation test joint string must be a brand new joint.

(4) Continuously carry out research and development on small wellbore salvage technology, reasonably select drilling tool combinations, optimize operating parameters, prevent secondary complications, and improve salvage efficiency.

### 5.2.8 M Well Completion Packer Stuck Accident

#### 5.2.8.1 Basic Information

(1) Well number: M.

(2) Well category: Evaluation well.

(3) Well type: high inclination well.

(4) Oil testing date: December 29th, 2019 to April 18th, 2020.

Date of occurrence: March 8th, 2020.

Termination date: March 20th, 2020.

(5) Falling fish structure (completion packer insertion string): 110.00mm insertion seal×0.22m+107.00mm short seat joint×One 0.22m+88.90mm oil pipe×9.58m+88.90mm variable buckle short joint×1.05m+146.00mm TNT completion packer×2.54m+88.90mm variable buckle short joint×Residual 1.05m+88.90mm oil pipe×2.47m, with a total length of 17.13m.

#### 5.2.8.2 Process of Occurrence

On March 7th, 2020, at 20:00, a 88.90mm airtight buckle oil pipe with TNT completion packer was inserted back into the pipe column (with a maximum outer diameter of 146.00mm), and a 50MPa airtight seal test was performed on each joint thread. At 13:18 on March 8th, when the casing was lowered to a depth of 1032.24m, the suspended weight suddenly decreased from 350kN to 230kN due to resistance, and the first oil pipe at the wellhead was bent.

（4）持续开展小井眼打捞工艺技术攻关，合理选择钻具组合、优化操作参数，防范次生复杂情况，提升打捞效率。

### 5.2.8　M井完井封隔器遇卡事故

#### 5.2.8.1　基本情况

（1）井号：M。

（2）井别：评价井。

（3）井型：大斜度井。

（4）试油日期：2019年12月29日—2020年4月18日。

发生时间：2020年3月8日。

解除时间：2020年3月20日。

（5）落鱼结构（完井封隔器回插管柱）：110.00mm回插密封×0.22m+107.00mm坐放短节×0.22m+88.90mm油管1根×9.58m+88.90mm变扣短节×1.05m+146.00mmTNT完井封隔器×2.54m+88.90mm变扣短节×1.05m+88.90mm油管残体×2.47m，总长17.13m。

#### 5.2.8.2　发生经过

2020年3月7日20：00，下入88.90mm气密封扣油管带TNT完井封隔器回插管柱（封隔器最大外径146.00mm），并对每个接头螺纹做50MPa气密封检测。3月8日13：18，下管柱至井深1032.24m时，突然遇阻悬重由350kN降至230kN，井口第一根油管被压弯。

On March 8th from 13:18 to 14:05, during the first lifting of the hanging weight to 400kN, it was not released: it was lowered to the original hanging weight of 350KN, and during the second lifting to 440kN, it was not released. This confirmed that the TNT completion packer had been set in advance, resulting in a stuck drill pipe accident during the completion reinsertion.

#### 5.2.8.3 Processing Process

(1) Phase 1: Reinforce and seal the stuck packer.

On March 14th from 8:00 to 13:10, a 50.80mm continuous tubing with a setting ball was inserted into the 88.90mm return tubing, and the setting short joint was inserted into the well. The depth of the well was 1020.66m, and the pressure was gradually increased into the continuous tubing. According to the normal procedure, the TNT completion packer was set and inspected to be qualified. The purpose was to prevent the packer from rotating or sliding down and damaging the bare string return tube during the next milling process of the packer.

(2) Phase 2: Place a liquid rubber plug under the stuck packer for isolation.

At 11:50—20:00, on March 15th, the 50.80mm coiled tubing was lowered with head washing to the well depth of 3000.00m, and the gel fluid was pumped in. When it was lifted to 1613.00m, 27.5m$^3$ of gel fluid was pumped in。 The purpose is to form a gel slug in the 1500.00~3000.00m well section and build a line of defense to prevent the iron chips and fragments from falling into the bare sealing string when sharpening the packer, which will affect the back plugging or acidizing steering in the future.

(3) Stage 3: Cut the oil pipe above the packer.

On March 16th, the RCT chemical cutting tool under the cable was successfully used to cut the 88.90mm return insertion oil pipe column at a depth of 1012.50m. On March 17th, the cut oil pipe string was removed, with neat and no deformation on the fracture surface.

(4) Stage 4: Grinding and milling to salvage the packer.

March 18th, 148mm lower × The 70mm lead eye was ground to 1011.89m, and the actual exploration of the fish top began milling. The drilling pressure was 10~35kN, the pump pressure was 5~8MPa, the displacement was 13~15L/s, the torque was 0~2.5kN·m, and the milling depth was 3.70m. Theoretically, the remaining oil pipe, variable buckle short joint, and the small end of the joint on the TNT completion packer have been ground. During drilling inspection, the fishing cup was filled with iron filings, and the grinding shoes were severely worn with obvious circular grooves. Based on the wear area, it was determined that the joint on the packer had been in contact.

March 19th, 148mm lower×76mm milling and fishing integrated tool was used to explore the depth of the fish top at 1015.57m. Pressure was applied to mill the upper part of the packer and above the main body. After milling to a depth of 1015.89m, the lifting pipe encountered a jam and the lifting tonnage was 300~500kN, but it could not be released. The rotation was normal. Continue grinding and milling with a depth of 0.41m, and after reaching a depth of 1015.98m, there is no obvious depth. Theoretically, the slips on the packer have been ground. The suspended weight was lifted

（3）第三阶段：切割封隔器之上油管。

3月16日，电缆下RCT化学切割工具至井深1012.50m，对88.90mm回插油管柱切割成功。3月17日，起出被切割的油管串，断口整齐、无变形。

（4）第四阶段：磨铣打捞封隔器。

3月18日，下148mm×70mm领眼磨鞋至1011.89m，实探鱼顶开始磨铣，钻压为10~35kN，泵压为5~8MPa，排量为13~15L/s，扭矩为0~2.5kN·m，磨铣进尺3.70m，理论上已磨掉油管残体、变扣短节及TNT完井封隔器上接头小端部分。起钻检查，捞杯装满铁屑，磨鞋磨损严重，有明显环形凹槽，根据磨损面积判断已接触封隔器上接头。

3月19日，下148mm×76mm磨铣打捞一体工具、实探鱼顶深度1015.57m，开始加压磨铣封隔器上卡瓦及其以上的本体部分，磨铣至井深1015.89m后，上提管柱遇卡，过提吨位300~500kN未能解卡，旋转正常。继续磨铣进尺0.41m，至井深1015.98m后无明显进尺，理论上已经磨完封隔器上卡瓦。上提悬重由470kN升至550kN再降至470kN，解卡成功，继续上提管柱有轻微挂卡。3月20日，起出磨铣打捞一体化管柱，捞获封隔器残体及下部油管串全部落鱼。

from 470kN to 550kN and then lowered to 470kN. The lifting was successful, and there was a slight hanging of the pipe column when it continued to be lifted. On March 20th, the integrated milling and salvage pipe string was launched, and the remaining packer and lower oil pipe string were all caught.

This accident went through a total of 11.5 days of handling, including one set of coiled tubing, one injection of lower liquid segment plug into the coiled tubing, one chemical cutting of steel wire, one lifting of the upper string, one drilling and grinding of the leading hole, and one grinding and salvaging.

#### 5.2.8.4　Cause Analysis

(1) Direct cause.

During the drilling process of the TINT completion packer, the cone slip undergoes relative displacement and sets in advance.

(2) Indirect causes.

The inner diameter of the oil reservoir casing is 152.50mm, and the outer diameter of the TNT packer is 146.00mm. The annular space gap is too small, and the impact force of the friction resistance during drilling causes the setting pin to shear.

Retrieve logging curves and confirm that the drilling speed is 27m/min. Although there is no violation of the rule that the drilling speed should not exceed 30m/min, it is no longer suitable for the drilling technology requirements of this specification of packer.

(3) Management reasons.

The TNT packer was used for the first time in Sichuan, but the management of the "four new" was not in place, and the risk analysis and identification of the application of new tools were not in place.

本次事故经过1次连续油管坐封、1次连续油管注下部液体段塞、1次钢丝化学切割、1次起上部管柱、1次领眼钻磨、1次磨捞一体，共计处理11.5d。

#### 5.2.8.4　原因分析

（1）直接原因。

TINT完井封隔器在下钻过程中，锥体卡瓦发生相对位移提前坐封。

（2）间接原因。

油层套管内径为152.50mm，TNT封隔器外径为146.00mm，环空间隙过小，下钻摩擦阻力瞬间遇阻的冲击力致使坐封销钉剪断。

调取录井曲线，证实下钻速度为27m/min，虽然没有违反"下钻速度不超过30m/min的规定"，但已不适合本规格封隔器的下钻技术要求。

（3）管理原因。

TNT封隔器首次在川内使用，"四新"管理不到位，对新工具应用的风险分析和识别不到位。

### 5.2.8.5 Lessons Learned

(1) Strengthen the "four new" management, conduct pre use demonstration of new tools, carefully study the compatibility and risk identification of new tools entering the well with the wellbore, and formulate preventive measures.

(2) Carry out on-site supervision and guidance for key processes, strictly follow regulations, operate smoothly, and especially reduce and control speed when lifting special tools.

(3) Grinding and milling the suspended drilling packer, reinforcing the packer to prevent rotation, and adding liquid rubber plugs at the bottom to prevent secondary complications in the later stage, ensuring the success rate and efficiency of complex processing.

## Post-class exercises

Well Completion and Stimulation

Completion and stimulation activities enable the flow of natural gas out of the formation and up to the surface. Completion essentially involves preparing the bottom of the well hole to specifications designed to maximize gas production, installing production tubing and related downhole tools, perforating and stimulating the well (collectively called "lower completion") and installation of wellhead equipment. Stimulation is a treatment performed on hydrocarbon-bearing formations to improve the flow of gas from the formation to the wellbore. While stimulation treatments vary depending on formation type, hydraulic fracturing is the type stimulation treatment used to open up the shale formation, enabling the efficient flow of oil and gas from the rock.

### 5.2.8.5 经验教训

（1）加强"四新"管理，开展新工具使用前论证，认真研究入井新工具与井筒的匹配性及风险识别，制订好防范措施。

（2）做好关键工序的旁站监督和指导，严格执行规程，平稳操作，尤其起下特殊工具应进一步降低和控制速度。

（3）磨铣处理悬空卡钻封隔器，加固封隔器防止转动，下部加垫液体胶塞防止后期的次生复杂，确保复杂处理的成功率及效率。

## 课后练习

完井与增产

完井和增产活动使天然气能够从地层流出并上升到地表。完井应按照设计规格准备井眼的底部，以最大限度地提高天然气产量，安装生产管柱和相关井下工具，射孔和增产（统称为"下部完井"）以及安装井口设备。增产是对含油气地层进行的一种处理，以改善从地层到井筒的天然气流动。根据地层类型，增产处理方式各不相同，而水力压裂是用于打开页岩地层的一种增产处理方式，可使油气能够从岩石中高效流出。

Completion Activities

There are several types of completion techniques used in oil and gas development. Selection of the completion technology is largely dictated by the technical and economic viability of the approach to effectively intake gas given the location and characteristics of the hydrocarbon formation that the well is accessing. Generally, one of the two techniques is used in shale gas well completions: One is cased-hole, perforated completion, where production casing and cement extend to the toe (bottom tip) of the well, the walls of which are perforated in the production zone using a shaped explosive charge prior to hydraulic fracturing; the other one is open-hole completion, in which no casing is cemented into place within the reservoir, the end of the piping is left open and gas is collected through the production liner.

完井活动

在石油和天然气开发中使用了多种类型的完井技术。完井技术的选择主要受所采用方法的技术和经济可行性影响，以有效地从井所在的油气藏的位置和特性中吸收气体。一般来说，在页岩气井完井中使用了两种技术中的一种：一是套管完井，射孔完成，其中生产套管和水泥延伸到井底的（底部尖端），该井壁在压裂前使用成型炸药进行射孔；二是开放式完井，在这种方式下，在储层内没有固化套管，管道末端保持开放，气体通过生产管柱收集。

# Chapter 6 Specifications and Standards Related to Oil and Gas Well Engineering

# 第六章 油气井工程相关的规范与标准

## 6.1 Safety Technical Regulations for Drilling Well Site Equipment Operation (SY/T 5974-2020)(Part)

## 6.1 《钻井井场设备作业安全技术规程》(SY/T 5974—2020)(部分)

### 6.1.1 Well Control Signal and Emergency assembly signal

### 6.1.1 井控信号及紧急集合信号

The well control signals are shown in Table 6.1.

井控信号见表6.1。

Table 6.1 The Well Control Signals

| Operation content | voice signal | Hand signals (on the drill floor, facing the remote console) |
|---|---|---|
| Overflow alarm signal | Sound the siren for 15~30s | |
| Open the poppet valve (throttle). | | Left arm flat to the left |
| Off Ring Blowout Preventer | | Raise your arms flat to the sides in a straight line, with your five fingers in a half arc, then swing them upward at the same time and close them above your head |
| Half-sealed shutter | | Raise your arms flat to the sides in a straight line, five fingers outstretched, palms forward then simultaneously swing forward flat and close in front of your chest |
| Closing the Pilot Operated Discharge Valve (Throttle Valve) | | Extend your left arm flat and draw a flat circle with your right hand clockwise downwards. |
| Open the annular blowout preventer | | Palms outstretched, palms facing outward, arms spread sideways and upward in the air |
| Open the half-sealed gate | | Palms outstretched, palms facing outward, arms spread flat in front of chest |

表 6.1 井控信号

| 操作内容 | 声音信号 | 手势信号<br>（在钻台上，面对远程控制台） |
|---|---|---|
| 溢流警报信号 | 鸣汽笛 15~30s | |
| 打开液动放喷阀（节流阀） | | 左臂向左平伸 |
| 关环形防喷器 | | 双臂向两侧平举呈一直线，五指呈半弧状，然后同时向上摆，合拢于头顶 |
| 关半封闸板 | | 双臂向两侧平举呈一直线，五指伸开，手心向前然后同时向前平摆，合拢于胸前 |
| 关闭液动放喷阀（节流阀） | | 左臂平伸，右手向下顺时针划平圆 |
| 打开环形防喷器 | | 手掌伸开，掌心向外，双臂侧上方高举展开 |
| 打开半封闸板 | | 手掌伸开，掌心向外，双臂胸前平举展开 |

When an emergency muster is required, the emergency muster shall be signaled as specified in the emergency plan.

## 6.1.2 Well Control Design and Installation of Well Control Devices, Pressure Testing and Well Control Operations

### 6.1.2.1 Well Control Design

Drilling well control design, well control device installation, pressure testing and well control operations should comply with the provisions of GB/T 31033.

According to the information provided by the geology, the drilling fluid density design is based on the highest formation pore pressure equivalent drilling fluid density value SY/T 5974—2020 in each bare borehole section, with an additional safety value of:

(1) 0.05~0.10g/cm$^3$ for oil wells and water wells or control bottomhole differential pressure 15.0~35.0MPa;

(2) 0.07~0.15g/cm$^3$ for gas wells or control bottomhole differential pressure of 3.0~5.0MPa;

需要紧急集合时，应按应急方案中的规定发出紧急集合信号。

## 6.1.2 井控设计和井控装置安装、试压及井控作业

### 6.1.2.1 井控设计

钻井井控设计、井控装置安装、试压及井控作业应符合 GB/T 31033 的规定。

根据地质提供的资料，钻井液密度设计以各裸眼井段中的最高地层孔隙压力当量钻井液密度值 SY/T 5974—2020 为基准，另附加一个安全值：

（1）油井、水井为 0.05~0.10g/cm$^3$ 或控制井底压差 15.0~35.0MPa；

（2）气井为 0.07~0.15g/cm$^3$ 或控制井底压差 3.0~5.0MPa；

(3) 0.02~0.15g/cm³ for coalbed methane wells.

For the design of drilling fluid density in oil and gas formations containing hydrogen sulfide, carbon dioxide and other toxic and harmful gases, the additional safety value or additional pressure should be taken as the upper limit. When specifically selecting the additional safety value of drilling fluid density, the following influencing factors should be considered according to the actual situation.

(1) Accuracy of formation pore pressure prediction;

(2) Capacity and burial depth of oil, gas, and water formations;

(3) Hydrogen sulfide content in formation hydrocarbons;

(4) Geostress and formation rupture pressure;

(5) Well control device package.

According to the formation pore pressure gradient, formation fracture pressure gradient, lithologic profile and the need to protect the oil and gas reservoirs, design a reasonable well structure and casing procedure and meet the following requirements:

(1) Exploratory wells, ultra-deep wells and complex wells should be structured with a layer of spare casing to allow for adequate estimation of unpredictability;

(2) When drilling a well in an underground mineral extraction area, the distance between the wellbore and the extraction pit, the mine access road shall be not less than 100m, and the surface casing or technical casing shall be lowered to a depth that seals the extraction layer and exceeds the extraction section by 100m;

(3) The casing down depth takes into account the highest drilling fluid density in the

（3）煤层气井为 0.02~0.15g/cm³。

含硫化氢、二氧化碳等有毒有害气体的油气层钻井液密度设计，其附加安全值或附加压力应取上限。具体选择钻井液密度安全附加值时，应根据实际情况考虑下列影响因素：

（1）地层孔隙压力预测精度；

（2）油层、气层、水层的产能和埋藏深度；

（3）地层油气中硫化氢的含量；

（4）地应力和地层破裂压力；

（5）井控装置配套情况。

根据地层孔隙压力梯度、地层破裂压力梯度、岩性剖面及保护油气层的需要，设计合理的井身结构和套管程序，并满足如下要求：

（1）探井、超深井、复杂井的井身结构应充分估计不可预测因素，留有一层备用套管；

（2）在地下矿产采掘区钻井，井筒与采掘坑道、矿井通道之间的距离不少于100m，表层套管或技术套管下深应封住开采层并超过开采段100m；

（3）套管下深要考虑下部钻井最高钻井液密度和溢流关井时的井口安全关井余量；

lower drilled well and the safe shut-in margin at the wellhead in case of overflow shut-in;

(4) The technical casing and oil casing cement of high sulfur oil and gas wells and high pressure oil and gas wells should be returned to the upper casing or to the surface.

After each layer of casing cementing and drilling, the rupture pressure of the first leakage-prone layer under the casing shoe (the length of the new borehole should not be more than 100m) shall be measured according to the requirements of SY/T 5623.

The control unit package includes:

(1) Drilled wells should be fitted with blowout preventers or blowout deflectors;

(2) The blowout preventer pressure rating should be matched to the highest formation pressure in the openhole section, and the size series and combination of the blowout preventers for each subdrilling should be selected according to the different downhole conditions;

(3) Regional exploratory wells, high-pressure oil and gas wells, sulfur-containing oil and gas wells, gas wells, deep wells, and complex wells should use a standard casing head with a pressure rating that matches the highest formation pressure in the corresponding well section;

(4)The pressure rating of the throttle manifold should match the wellhead blowout preventer pressure rating;

(5) The pressure rating of the pressurized tubing manifold shall match the pressure rating of the blowout preventer at the wellhead;

(6) Draw the installation schematic of wellhead device and parallel control manifold for each sub-drilling, and install and test pressure according to SY/T 5964;

（4）高含硫油气井和高压油气井的技术套管、油层套管水泥应返至上一级套管内或地面。

每层套管固井开钻后，按 SY/T 5623 的要求测定套管鞋下第一个易漏层（新井眼长度不宜大于 100m）的破裂压力。

控装置配套包括：

（1）钻井应装防喷器或防喷导流器；

（2）防喷器压力等级应与裸眼井段中最高地层压力相匹配，并根据不同的井下情况选用各次开钻防喷器的尺寸系列和组合形式；

（3）区域探井、高压油气井、含硫油气井、气井、深井和复杂井应使用标准套管头，其压力等级与相应井段的最高地层压力相匹配；

（4）节流管汇的压力等级应与井口防喷器压力等级相匹配；

（5）压井管汇的压力等级应与井口防喷器压力等级相匹配；

（6）绘制各次开钻井口装置及并控管汇安装示意图，按 SY/T 5964 的规定安装和试压；

(7) For wellhead devices and well control manifolds with sulfur resistance requirements, their metal materials shall meet the corresponding requirements of GB/T 20972, and their non-metallic materials shall have the performance of being used in the environment of hydrogen sulfide without failing;

(8) Shear gates should be installed during drilling operations of regional exploratory wells, high-pressure oil and gas wells, and drilling operations in the purpose section of high sulfur oil and gas wells.

The blowout prevention tools, well control monitoring instruments, meters and drilling fluid treatment devices and irrigation devices in the drilling tools shall meet the requirements of well control technology.

Sulfur-containing oil and gas wells and wells with high gas-to-oil ratios should be equipped with gas detection equipment.

Exploratory wells, gas wells, sulfur-containing oil and gas wells and wells with high gas-to-oil ratios should be equipped with a drilling fluid liquid-gas separator.

According to the content of sulfide and carbon dioxide in the formation fluid and the maximum shut-in pressure after completion, and taking into account that it can meet the needs of further measures to increase production and the later water injection and well workover operations, the model, pressure level and size series of wellhead device for well completion are selected according to the provisions of GB/T 22513.

The design document of the drilling project should specify the reserve quantity of weighted drilling fluid and weighted materials before drilling open the oil and gas layer, as well as the main technical measures for

（7）有抗硫要求的井口装置及井控管汇，其金属材质应符合 GB/T 20972 的相应要求，其非金属材料应具有在硫化氢环境下使用而不失效的性能；

（8）区域探井、高压油气井、高含硫油气井目的层段钻井作业中，应安装剪切闸板。

钻具内防喷工具、井控监测仪器、仪表及钻井液处理装置和灌注装置，应满足井控技术的要求。

含硫油气井、气油比高的油井应配置气体检测设备。

探井、气井、含硫油气井及气油比高的油井应配备钻井液液气分离器。

根据地层流体中硫化和二氧化碳含量及完井后最大关井压力值，并考虑能满足进一步采取增产措施和后期注水、修井作业的需要，按 GB/T 22513 的规定选用完井井口装置的型号、压力等级和尺寸系列。

钻井工程设计书中应明确钻开油气层前加重钻井液和加重材料的储备量，以及油气井压力控制的主要技术措施，并对同一区域曾发生的井喷、溢流、井漏等情况进行描述和风险提示。

pressure control of oil and gas wells, and provide descriptions of blowouts, overflows and well leaks that have occurred in the same area and risk tips.

When drilling wells in areas that may contain hydrogen sulfide, the location, depth of burial and content of the layers should be predicted, and the corresponding safety and technical measures to be taken should be specified in the design and comply with the requirements of SY/T 5087.

The technology of formation pressure checking (monitoring) with drilling should be adopted for exploratory wells, pre-exploratory wells and data wells; the predicted formation pressure gradient curve of the well, the design drilling fluid density curve, the $dc$-indexed drilling with drilling to monitor the formation pressure gradient curve and the actual drilling fluid density curve should be drawn up, and the density of drilling fluid should be adjusted promptly according to the results of the monitoring and the actual drilling.

Drilling in the developed adjustment area, the oil and gas field development department should promptly find out the distribution of water injection, gas injection (steam) wells and water injection, gas injection (steam) situation, to provide layered dynamic pressure data, drilling open the oil and gas layer should be taken before the corresponding measures such as stopping the injection, pressure relief and pumping, until the corresponding layers of casing cementing waiting to be condensed until the completion of the casing.

### 6.1.2.2 Installation of Well Control Devices
#### 6.1.2.2.1 Wellhead Devices

The blowout preventer and blowout deflector should be corrected and fixed firmly

在可能含硫化氢地区钻井，应对其层位、埋藏深度及含量进行预测，并在设计中明确应采取的相应安全和技术措施，且符合 SY/T 5087 的要求。

对探井、预探井、资料井应采用地层压力随钻检（监）测技术；绘制本井预测地层压力梯度曲线、设计钻井液密度曲线、$dc$ 指数随钻监测地层压力梯度曲线和实际钻井液密度曲线，根据监测和实钻结果，及时调整钻井液密度。

在已开发调整区钻井，油气田开发部门要及时查清注水、注气（汽）井分布及注水、注气（汽）情况，提供分层动态压力数据，钻开油气层之前应采取相应的停注、泄压和停抽等措施，直到相应层位套管固井候凝完为止。

### 6.1.2.2 井控装置的安装
#### 6.1.2.2.1 井口装置

防喷器和防喷导流器安装完毕后应校正井口、转盘、天车中心并固定牢固。

at the center of the wellhead, turntable and overhead crane after installation.

With a manual locking mechanism of the gate blowout preventer should be equipped with manual operation lever, against the handwheel end should be supported firmly, easy to operate, and hang a sign indicating the direction of opening and closing and the number of turns in the end.

The installation of the drilling tee should comply with the relevant provisions of SY/T 5964.

The installation of casing head shall comply with the relevant provisions of SY/T 6789.

6.1.2.2.2　Blowout Preventer Control Device

Installation requirements for blowout preventer remote console:

(1) Installed in the face of the derrick door on the left side, the entrance to the well is not less than 25m of the special room, around 10m shall not be stacked flammable, explosive, corrosive substances;

(2) A certain distance should be maintained between the pipe rack and the spray prevention pipeline and the spray discharge pipeline, and the pipe rack should be protected by guards at the crossing of the automobile road and the sidewalk. It is not allowed to pile up debris on the pipe rack, use it as welding grounding line or carry out welding and cutting operations on it;

(3) The main gas supply should be connected separately from the gas supply room, separate from the driller's console gas supply, and equipped with a gas drain separator; the gas hose bundle should not be forcibly bent or crimped;

(4) The power supply should be led directly from the main switch of the switchboard with

具有手动锁紧机构的闸板防喷器应装齐手动操作杆，靠手轮端应支撑牢固，便于操作，并挂牌标明开、关方向和到底的圈数。

钻井四通的安装应符合SY/T 5964的相关规定。

套管头的安装应符合SY/T 6789的相关规定。

6.1.2.2.2　防喷器控制装置

防喷器远程控制台安装要求：

（1）安装在面对井架大门左侧、井口不少于25m的专用活动房内，周围10m内不得堆放易燃、易爆、腐蚀物品；

（2）管排架与防喷管线、放喷管线之间应保持一定距离，在穿越汽车道、人行道处用防护装置保护。不允许在管排架上堆放杂物和以其作为电焊接地线或在其上进行焊割作业；

（3）总气源应从气源房单独接出，与司钻控制台气源分开连接，并配置气源排水分离器；不应强行弯曲和压折气管束；

（4）电源应从配电板总开关处用专线直接引出，并用单独的开关控制；

a dedicated line and controlled by a separate switch;

(5) Accumulators are intact, pressurized to specified values and always at operating pressure;

(6) Fully sealed gate directional valves shall be protected by a cover, and shear gate control directional valves shall be protected by a limit.

The driller's console should be mounted in a location conducive to the driller's operation and securely fastened.

It is desirable to install a blowout preventer with a brake linkage anti-lift safety device for the rig hoisting system, whose air line is connected in parallel with the air line of the anti-bumping overhead crane.

6.1.2.2.3　Well Control System

The blowout prevention pipeline, blowout release pipeline and drilling fluid recovery pipeline should use pipes qualified by flaw detection. The blowout prevention pipeline with predicted formation pressure greater than 35MPa should be made of metal material, and the blowout prevention pipeline and drilling fluid recovery pipeline matched with blowout preventer of 35MPa and the following pressure level can use high-pressure fire-resistant flexible pipeline of the same pressure level. The wellhead pipelines and pipe sinks of sulfur-containing oil and gas wells should adopt special pipes with sulfur resistance.

Anti-spray pipeline should be connected by standard flange, should not be welded on site, the pressure level is matched with the pressure level of the blowout preventer, and the length of more than 7m should be fixed firmly.

（5）蓄能器完好，压力达到规定值，并始终处于工作压力状态；

（6）全封闸板换向阀应装罩保护，剪切闸板控制换向阀应限位保护。

司钻控制台应安装在有利于司钻操作的位置并固定牢固。

宜安装防喷器与钻机提升系统刹车联动防提安全装置，其气路与防碰天车气路并联。

6.1.2.2.3　井控管汇

防喷管线、放喷管线和钻井液回收管线应使用经探伤合格的管材，预测地层压力大于35MPa的防喷管线应采用金属材料，35MPa及以下压力等级防喷器所配套的防喷管线及钻井液回收管线可以使用同一压力等级的高压耐火软管线。含硫油气井的井口管线及管汇应采用抗硫的专用管材。

防喷管线应采用标准法兰连接，不应现场焊接，压力等级与防喷器压力等级匹配，长度超过7m应固定牢固。

The outlet of the drilling fluid recovery line should be connected to the drilling fluid tank and fixed securely, with the angle of the turn greater than 120°, and its diameter is not smaller than the outlet diameter of the throttling manifold.

Installation requirements for discharge piping:

(1) The diameter of the discharge pipe line is not less than $\phi$78 mm;

(2) Discharge piping should not be welded in the field;

(3) The layout should take into account local seasonal winds, residential areas, roads, tank farms, power lines and various facilities;

(4) When two pipelines are in the same direction, they should be spaced more than 0.3m apart and fixed separately, and their outlets should be in the same direction;

(5) The pipeline should be flat and straight out of the well field, and there should be a bridge cover at the traveling place, and the pipeline under it should have no joints; the cast (forged) steel elbow not less than 120° or 90° elbow with anti-erosion function should be used at the turning place;

(6) Pipeline every 10~15m turn ends, the exit should be fixed firmly; if across more than 10m wide ditches, ponds and other obstacles, should be supported firmly.

Flat plate valves for well control manifolds should comply with the corresponding provisions of GB/T 22513.

The wellhead tee shall be connected to blowout preventer lines on both sides of the wellhead tee, and each blowout preventer line shall be fitted with two gate valves each, one of which shall be directly connected to the tee and shall be in a normally open condition.

钻井液回收管线出口应接至钻井液罐并固定牢靠，转弯处角度大于120°，其通径不小于节流管汇出口通径。

放喷管线安装要求：

（1）放喷管线通径不小于$\phi$78mm；

（2）放喷管线不应在现场焊接；

（3）布局要考虑当地季节风向、居民区、道路、油罐区、电力线及各种设施等情况；

（4）两条管线走向一致时，应保持间距大于0.3m，并分别固定，其出口应朝同一方向；

（5）管线宜平直接出井场，行车处应有过桥盖板，其下的管线应无接头；转弯处应使用不小于120°的铸（锻）钢弯头或90°带抗冲蚀功能的弯头；

（6）管线每隔10~15m转处两端、出口处应固定牢靠；若跨越10m宽以上的河沟、水塘等障碍，应支撑牢固。

井控管汇所配置的平板阀应符合GB/T 22513的相应规定。

井口四通的两侧应接防喷管线，每条防喷管线应各装两个闸阀，其中一只应直接与四通相连，宜处于常开状态。

Requirements for installation and use of pressure gauges on blowout preventer lines, throttling manifolds and pressure well manifolds:

(1) Matching installation of shut-off valves;

(2) Use high and low range seismic pressure gauges with the low pressure range gauge in the normally closed position;

(3) Pressure gauges are regularly calibrated and have test certificates;

(4) The flow control box is placed on the side of the drilling table against the riser, and the required air supply should be accessed from the special air drain separator;

(5) Backpressure well lines should be securely fastened.

6.1.2.2.4  Spray Prevention Tools in Drilling Tools

The pressure rating of the blowout preventer tool within the drilling tool matches the blowout preventer pressure rating required in the design.

Plug valves should be periodically active, plug valves should not be used as splash valves, and drilling rigs should be equipped with rig check valves or plug valves that match the buckling type of the rig, and special tools for snagging.

In oil and gas well drilling operations, a blowout preventer should be prepared on the concurrent drilling table using rotary drilling, and a column of blowout preventer risers should be prepared for wells using top drive drilling.

Operations in high sulfur oil and gas formations shall be equipped with a near-bit drilling tool check valve on the drilling tool, except in the following special cases:

(1) Plugging drilling tool combinations;

防喷管线、节流管汇和压井管汇上压力表安装、使用要求：

（1）配套安装截止阀；

（2）使用高、低量程抗震压力表，低压量程表处于常关状态；

（3）压力表定期检定，并有检测合格证；

（4）节流控制箱摆放在钻台上靠立管一侧，所需气源应从专用气源排水分离器上接入；

（5）反压井管线应固定牢固。

6.1.2.2.4  钻具内防喷工具

钻具内防喷工具的额定压力与设计中要求的防喷器额定压力相匹配。

旋塞阀应定期活动，旋塞阀不应作为防溅阀，钻台上配备与钻具扣型相符的钻具止回阀或旋塞阀，并配备抢装专用工具。

油气井钻作业中，使用转盘钻进的并钻台上应准备一根防喷钻，使用顶驱钻的井应准备一柱防喷立柱。

高含硫油气层作业应在钻具上加装近钻头钻具止回阀，但下列特殊情况除外：

（1）堵漏钻具组合；

(2) Weighing rig assemblies before lowering the tailpipe;

(3) Handling explosive loose-buckle drilling tool assemblies in jammed drilling accidents;

(4) Core-piercing and salvaging logging cable and instrumentation drilling tool assemblies;

(5) Transmission logging drilling tool assemblies.

6.1.2.2.5　Liquid-Gas Separator

It is installed on the side of the throttle manifold manifold and connected to the throttle manifold with special piping.

The outlet of the safety relief valve should be directed toward the outside of the well site and should not be connected to a relief line.

The discharge line is connected to the distribution box of the vibrating screen on the recirculation tank, and should be supported and fixed if the overhang length is more than 6 m. The nozzle should not be buried in the liquid inside the tank. The nozzle should not be buried in the liquid inside the box.

The exhaust line is configured according to the design diameter, connected to the safety zone in the downwind direction along the wind direction of the local seasonal wind, and the outlet is fixed firmly and equipped with ignition devices.

6.1.2.3　Pressure Test of Well Control Devices

6.1.2.3.1　Test Pressure Values

(1) In the well control workshop, the annular blowout preventer (sealing the drill pipe, not sealing the empty well), the gate preventer, the four-way, the blowout preventer line, the throttling manifold pressure well manifold should do the rated pressure sealing test, and the gate preventer should also do the

（2）下尾管前的称重钻具组合；

（3）处理卡钻事故中的爆炸松扣钻具组合；

（4）穿心打捞测井电缆及仪器钻具组合；

（5）传输测井钻具组合。

6.1.2.2.5　液气分离器

安装在节流管汇汇流管一侧，与节流管汇之间用专用管线连接。

安全泄压阀出口应朝向井场外侧，不应接泄压管线。

排液管线接至循环罐上振动筛的分配箱，悬空长度超过 6m 应支撑固定。管口不应埋在箱内液体中。

排气管线按设计通径配置，沿当地季节风风向接至下风方向安全地带，出口处固定牢固并配置点火装置。

6.1.2.3　井控装置的试压

6.1.2.3.1　试压值

（1）在井控车间，环形防喷器（封钻杆，不封空井）、闸板防喷器、四通、防喷管线、节流管汇压井管汇应做额定压力密封试验，闸板防喷器还应做 1.4~2.1MPa 低压密封试验；

low pressure sealing test of 1.4~2.1MPa;

(2) After installation at the drilling site, the test pressure is not greater than 80% of the casing's internal pressure strength, and the test pressure of the annular blowout preventer to close the drill pipe is 70% of the rated pressure; the gates of the blowout preventer, the pressure well tubing sinks, and the blowout preventer piping should do the sealing test of the rated pressure, and the throttling tubing sinks should be tested according to the rated pressure rating of the parts and components: the test pressure of the blowout preventer piping should not be less than 10MPa;

(3) The wellhead unit shall be pressure seal tested before drilling through the formation and after replacing well control unit components;

(4) The blowout preventer control system does a reliability test at its rated pressure.

6.1.2.3.2　Pressure Test Rules

(1) In addition to the sprayer control system using the prescribed pressure oil test pressure, the rest are clear water, cold areas in winter can be added antifreeze;

(2) Test pressure stabilization time of not less than 10min, high-pressure test pressure drop does not exceed 0.7MPa, low-pressure test pressure drop does not exceed 0.07MPa, sealing parts of no leakage is qualified.

6.1.2.4　Use of Well Control Devices

Annular blowout preventers should not be used to shut down wells for long periods of time, and should not be used to seal off empty wells except under special circumstances.

If the casing pressure is not more than 7MPa, when the annular blowout preventer is used for up and down drilling of unpressurized wells, drilling tools with 18° slope joints

（2）在钻井现场安装好后，试验压力在不大于套管抗内压强度80%的前提下，环形防喷器封闭钻杆试验压力为额定压力的70%；闸板防喷器、压井管汇、防喷管线应做额定压力密封试验，节流管汇按零部件额定压力分级试压：放喷管线试验压力不低于10MPa；

（3）钻开油气层前及更换井控装置部件后，井口装置应进行压力密封试验；

（4）防喷器控制系统按其额定压力做一次可靠性试压。

6.1.2.3.2　试压规则

（1）除防喷器控制系统采用规定压力油试压外，其余均为清水，寒冷地区冬季可加防冻剂；

（2）试压稳压时间不少于10min，高压试验压降不超过0.7MPa，低压试验压降不超过0.07MPa，密封部位无渗漏为合格。

6.1.2.4　井控装置的使用

环形防喷器不应长时间关井，非特殊情况下不能用来封闭空井。

套压不大于7MPa的情况下，用环形防喷器进行不压井起下钻作业时，应使用18°斜坡接头的钻具，起下钻速度不得大于0.2m/s。

should be used, and the up and down drilling speed should not be more than 0.2m/s.

After closing the well, the gate preventer with manual locking mechanism should be locked manually. Before opening the gate, it should be manually unlocked, and both locking and unlocking should be to the bottom first, and then turned back 1/4 ~ 1/2 turn.

After the annular blowout preventer or gate blowout preventer is closed, it is permitted to move the drilling tools up and down at a speed of not more than 0.2m/s, but it is not permitted to rotate the drilling tools or pass the joints of the drilling tools if the pressure of the shut-in sleeve does not exceed 14MPa.

The full seal gate blowout preventer should not be closed when drilling tools are present in the well.

The pressure in the well should not be relieved by opening the blowout preventer.

When field servicing a gate blowout preventer equipped with a more chained side gate or replacing its gate, both sides of the gate should not be opened at the same time.

The blowout preventer and valve opening and closing activities should be performed periodically during oil and gas reservoir operations.

The following procedure is recommended for cutting drilling tools/tubing with a shear gate blowout preventer:

(1) Lock the rig brake system after ensuring that the drill tool/tubing fitting is not in the shear gate position;

(2) Close the ring blowout preventer above the shear gate preventer;

(3) Open the open spray line gate valve to relieve pressure;

(4) Install the appropriate deadbolt at the

具有手动锁紧机构的闸板防喷器关井后，应手动锁紧闸板。打开闸板前，应先手动解锁，锁紧和解锁都应先到底，然后回转 1/4 ~ 1/2 圈。

环形防喷器或闸板防喷器关闭后，在关井套压不超过 14MPa 情况下，允许以不大于 0.2m/s 的速度上下活动钻具，但不准许转动钻具或过钻具接头。

当井内有钻具时，不应关闭全封闸板防喷器。

不应用打开防喷器的方式来泄井内压力。

现场检修装有铰链侧门的闸板防喷器或更换其闸板时，两侧门不应同时打开。

油气层作业期间，应定期对防喷器和阀门开、关活动。

用剪切闸板防喷器剪断钻具/油管宜按以下程序操作：

（1）确保钻具/油管接头不在剪切闸板位置后，锁定钻机刹车系统；

（2）关闭剪切闸板防喷器以上的环形防喷器；

（3）打开放喷管线闸阀泄压；

（4）在转盘面上的钻具/油管上适当位置安装相应的死卡，并与钻机底座连接固定；

appropriate location on the drilling tool/tubing on the face of the turntable and attach and secure it to the rig base;

(5) Open the half-sealed gate blowout preventer above and below the shear gate blowout preventer;

(6) Close the shear gate blowout preventer with remote console reservoir pressure until the well drilling tool/tubing is sheared;

(7) Close the full seal gate blowout preventer and control the wellhead;

(8) Manually locking full-seal gate blowout preventer and shear gate blowout preventer;

(9) Close the remote console reservoir bypass valve and adjust the blowout preventer remote console manifold pressure to normal.

The wellsite shall have half-seal gates of the same specifications as the half-seal gates in use and the corresponding seals and their removal and pressure testing tools.

The maintenance of the blowout preventer and its control device is carried out according to the corresponding provisions of SY/T 5964.

Gate blowout preventers and flat valves with secondary seals can only be used in emergencies where their seals have failed to the point of serious leakage, and stopping leakage is sufficient; when the emergency is lifted, the secondary seals are cleaned and replaced immediately.

After the manual plate valve is opened and closed to the bottom, the one with labor-saving mechanism should be turned back $3/4 \sim 1\frac{1}{2}$ turns. Its opening and closing should be in place at once, and should not be half-open and half-closed or used as a throttle valve.

Pressure well manifolds should not be used for daily filling of drilling fluids. Anti-

（5）打开剪切闸板防喷器上面和下面的半封闸板防喷器；

（6）用远程控制台储能器压力关闭剪切闸板防喷器，直至剪断井内钻具/油管；

（7）关闭全封闸板防喷器，控制井口；

（8）手动锁紧全封闸板防喷器和剪切闸板防喷器；

（9）关闭远程控制台储能器旁通阀，将防喷器远程控制台管汇压力调至正常值。

井场应备有与在用半封闸板同规格的半封闸板和相应的密封件及其拆装工具和试压工具。

防喷器及其控制装置的维护保养按SY/T 5964的相应规定执行。

有二次密封的闸板防喷器和平板阀，只能在其密封失效至严重漏失的紧急情况下才能使用其二次密封功能，且止漏即可，待紧急情况解除后，立即清洗更换二次密封件。

手动平板阀开、关到底后，带省力机构的应回转3/4～1½圈。其开、关应一次性到位，不应半开半闭或作节流阀用。

压井管汇不能用作日常灌注钻井液用，防喷管线、节流管汇和压井管汇应采取防堵、防漏防冻措施。最大允许关井套压值在节流管汇处用标示牌标示。

jet lines, throttling manifolds and pressure well manifolds should be protected against plugging, leakage and freezing. The maximum allowable shut-in sleeve pressure is indicated by a sign at the throttling manifold.

All valves on the well control manifold should be tagged and numbered and labeled with their open and closed status.

Oil (gas) well device and other well control devices should be inspected, test qualified before installation on the well oil (gas) port device in the well assembly should also be the overall test pressure, qualified before being put into use.

#### 6.1.2.5 Management of Well Control Devices

Well control devices should have a specialized agency responsible for management, maintenance and periodic on-site inspections.

The management and operation of well control devices of the drilling team should be implemented by specialized personnel with clear job responsibilities.

Specialized fittings warehouse and rubber parts air-conditioning warehouse should be set up, and the temperature of the warehouse should meet the requirements for storage of fittings and rubber parts.

Strengthen the stratum comparison with drilling, and provide timely and reliable geological forecasts 50 ~ 100m before entering the oil and gas layer, according to the designed maximum drilling fluid density value for the next drilling, conduct the pressure bearing capacity test on the bare stratum.

Wells drilled in adjustment blocks should be inspected for stoppages and pressure relief in neighboring water and gas (steam) injection wells.

Technical briefing on engineering,

井控管汇上所有阀门都应挂牌编号并标明其开、关状态。

采油（气）井装置等井控装置应经检验、试合格后方能上井安装采油（气）口装置在井上组装后还应整体试压，合格后方可投入使用。

#### 6.1.2.5 井控装置的管理

井控装置应有专门机构负责管理、维修和定期现场检查工作。

钻井队井控装置的管理、操作应落实专人负责，并明确岗位责任。

应设置专用配件库房和橡胶件空调库房，库房温度应满足配件及橡胶件储藏要求。

加强随钻地层对比，及时提出可靠的地质预报在进入油气层前 50 ~ 100m，按照下一步钻井的设计最高钻井液密度值，对裸眼地层进行承压能力试验。

在调整区块钻井，应检查邻近注水、注气（汽）井停注、泄压情况。

geology, drilling fluids, well control devices and well control measures shall be provided to the relevant staff at the drilling site.

The drilling team should implement the well control responsibility system; since the installation of well control equipment in the operation team, no less than one blowout prevention drill for different working conditions should be conducted every 30d. The wellhead should be controlled within 3 minutes for drilling operations and empty wells, and within 5 minutes for up and down drilling operations.

The drilling team should organize fire drills for all employees on site, and drilling in sulfur-containing areas should also conduct antihydrogen sulfide drills, and check and implement all aspects of safety prevention.

The cadres of the drilling team have implemented a 24h shift system at the production site, responsible for checking and supervising the strict implementation of the well control post responsibility system at each post, and organizing immediate rectification when problems are found.

Implement the "sitting watch" system, designate a person to observe and record the changes in the liquid level of the circulating tank and the situation of filling in or returning drilling fluid from the drill, so as to discover the overflow display in time.

Check whether the installation of drilling equipments, instruments, well control devices, protective equipments and special tools, fire-fighting equipments, explosion-proof circuits and gas circuits are in compliance with the regulations and function properly, and rectify the problems in time.

应向钻井现场有关工作人员进行工程、地质、钻井液、井控装置和井控措施等方面的技术交底。

钻井队应落实井控责任制；作业班自安装井控设备后，每30d应不少于一次不同工况的防喷演习。钻进作业和空井状态应在3min内控制住井口，起下钻作业状态应在5min内控制住井口。

钻井队应组织现场全体员工进行消防习，含硫地区钻井还应进行防硫化氢演习，并检查落实各方面安全预防工作。

实行钻井队干部在生产现场24h带班制度，负责检查、监督各岗位严格执行井控岗位责任制，发现问题立即组织整改。

实行"坐岗"制度，指定专人观察和记录循环罐液面变化和起下钻灌入或返出钻井液情况，及时发现溢流显示。

检查钻井设备、仪器仪表、井控装置、防护设备及专用工具、消防器材、防爆电路和气路的安装是否符合规定、功能是否正常，发现问题应及时整改。

The density and other major properties of drilling fluids meet the design requirements, and reserve weighted drilling fluids, weighting agents, plugging materials and other treatment agents according to the design, and circulate the reserve weighted drilling fluids on a regular basis to maintain their performance in accordance with the requirements.

Before drilling a sulfur-containing formation, the well site's hydrogen sulfide protection measures (including emergency plans and drills) should be checked.

The drilling team shall report the self-inspection to the construction unit and apply for inspection and acceptance after confirming the readiness through comprehensive self-inspection.

After the inspection and acceptance team passes the inspection and acceptance according to the requirements for drilling open the oil and gas layer, the oil and gas layer can be drilled open only with the approval of the construction unit.

### 6.1.2.6 Well Control Operations During Drilling in Oil and Gas Formations

The drilling team should select the type of drilling fluid and density value in strict accordance with the engineering design, and when it is found that the design is not in line with the actual situation, it should declare the change of design in accordance with the approval procedure in a timely manner, and can only be carried out after approval; in case of emergencies, the drilling team can deal with it first, and then report it to the authorities in a timely manner.

In the event of stuck drilling requiring oil soaking, oil mixing, or other reasons requiring appropriate adjustment of drilling fluid density, the wellbore column pressure should not be

钻井液密度及其他主要性能符合设计要求，并按设计储备加重钻井液、加重剂、堵漏材料和其他处理剂，对储备加重钻井液定期循环处理，保持其性能符合要求。

钻开含硫油气层前，应对井场的硫化氢防护措施（含应急预案及演练等）进行检查。

钻井队应通过全面自检，确认准备工作就绪后，向建设单位汇报自检情况，并申请检查验收。

检查验收组按钻开油气层的要求进行检查验收合格后，经建设单位批准方可钻开油气层。

### 6.1.2.6 油气层钻井过程中的井控作业

钻井队应严格按工程设计选择钻井液类型和密度值，当发现设计与实际不相符合时，应按审批程序及时申报更改设计，经批准后才能实施；若遇紧急情况，钻井队可先处理，再及时上报。

发生卡钻需泡油、混油或因其他原因需适当调整钻井液密度时，井筒液柱压力不应小于裸眼段中的最高地层压力。

less than the highest formation pressure in the barehole section.

Before each new well drill bit starts drilling and before the daily day shift starts drilling, the circulating pressure should be detected with 1/3 ~ 1/2 drilling flow rate and a good record of the number of pump strokes, flow rate, and circulating pressure should be made, and supplementary measurements should be made when the performance of the drilling fluid or the combination of the drilling tools has changed considerably.

A short trip up and down the drill should be performed to check for oil and gas intrusion and overflow in the following cases.

(1) Before the first start of drilling after drilling through an oil or gas formation;

(2) Severe oil and gas intrusion had occurred in drilling before starting drilling;

(3) Before starting drilling after drilling open oil and gas reservoir well leakage plugging;

(4) Before starting drilling after overflow pressure wells;

(5) Before starting drilling after the density of drilling fluid in the well is reduced;

(6) Before starting drilling when the cycle has to be stopped for an extended period of time for other operations (electrical survey, casing down, tubing down, midway test, etc.).

Two basic approaches to short-range up-and-down drilling.

(1) Generally speaking, try to start 10 ~ 15 drilling tools, and then go down to the bottom of the well for one week, if the drilling fluid is free of oil and gas invasion, then you can start drilling formally; otherwise, you should circulate to exclude the invasive contamination of the drilling fluid and adjust the density of

每只新入井钻头开始钻进前及每日白班开始钻进前,都要以 1/3~1/2 钻进流量检测循环压力并做好泵冲数、流量、循环压力记录,当钻井液性能或钻具组合发生较大变化时应补测。

下列情况应进行短程起下钻检查油气侵和溢流:

(1)钻开油气层后第一次起钻前;

(2)钻进中曾发生严重油气侵起钻前;

(3)钻开油气层井漏堵漏后起钻前;

(4)溢流压井后起钻前;

(5)井内钻井液密度降低后起钻前;

(6)需长时间停止循环进行其他作业(电测、下套管、下油管、中途测试等)起钻前。

短程起下钻的两种基本做法:

(1)一般情况下试起 10~15 个钻具,再下入井底循环一周,若钻井液无油气侵,则可正式起钻;否则,应循环排除受侵污钻井液并适当调整钻井液密度后再起钻:

the drilling fluid appropriately before you start drilling again;

(2) In special cases (when the circulation needs to be stopped for a long time or when the well is complicated), the drilling tool will be lifted to the casing shoe or the safe well section, the pump will be stopped to check the starting and stopping drilling cycle or the pump needs to be stopped for the working time, and then it will be lowered back to the bottom of the well to circulate for one week for observation.

Technical measures to prevent overflow and blowout during drilling and lowering:

(1) Maintain good walling and rheological properties of the drilling fluid;

(2) Fully circulate the drilling fluid in the well before drilling so that its performance is uniform and the difference between the import and export densities is not more than $0.02g/cm^3$;

(3) In the starting drilling, fill the well with drilling fluid in a timely manner in strict accordance with the regulations, and make records and calibrations to detect abnormalities in time;

(4) The drilling speed of the drill bit is not more than 0.5m/s in the oil and gas formation and in the well section of 300m above the top of the oil and gas formation;

(5) In loose formations, especially those with strong pulverization. Sufficient circulating flow should be maintained when encountering obstructed drilling to prevent the drill bit from "mud bagging";

(6) Drilling should be completed in a timely manner, and equipment should not be serviced while the well is empty;

（2）特殊情况时（需长时间停止循环或井下复杂时），将钻具起至套管鞋内或安全井段，停泵检查个起下钻周期或需停泵工作时间，再下回井底循环一周观察。

起下钻中防止溢流、井喷的技术措施：

（1）保持钻井液有良好的造壁性和流变性；

（2）起钻前充分循环井内钻井液，使其性能均匀，进出口密度差不大于 $0.02g/cm^3$；

（3）起钻中严格按规定及时向井内灌满钻井液，并做好记录、校核，及时发现异常情况；

（4）钻头在油气层中和油气层顶部以上300m 井段内起钻速度不大于 0.5m/s；

（5）在疏松地层，特别是造浆性强的地层，遇阻划眼时应保持足够的循环流量，防止钻头"泥包"；

（6）起钻完应及时下钻，不应在空井情况下进行设备检修；

(7) Down drilling should be controlled. If stationary or down drilling takes too long, the drilling fluid should be circulated in sections if necessary.

Improve the degassing performance of the drilling fluid, find gas intrusion should be timely excluded, gas-infiltrated drilling fluid without venting shall not be re-injected into the well.

If it is necessary to add weight to the gas-infiltrated drilling fluid, this should be done after the gas-infiltrated drilling fluid has been drained and drilling has been stopped, and should not be done while drilling.

Enhance the observation of overflow prognosis and overflow display, so as to detect overflow in time; "sitting Watch" workers should immediately report to the driller when they find overflow, well leakage, oil and gas display and other abnormal situations.

When well leakage occurs during drilling, the drilling tools should be lifted off the bottom of the well, the square drill pipe should be raised to the rotary table, and measures should be taken to backfill the drilling fluid at regular intervals to keep the pressure of the fluid column in the well in equilibrium with the pressure of the stratum, and then corresponding measures should be taken to deal with the well leakage.

Electrical testing cementing and midway testing should be done for well control and blowout prevention as follows.

(1) Conditions in the well should be normal and stable prior to electrical testing. If the electrical test takes a long time, consideration should be given to circulating the well halfway through the test before the electrical test;

(2) Before lowering the casing, the

（7）下钻应控制下钻速度。若静止或下钻时间过长，必要时应分段循环钻井液。

改善钻井液的脱气性能，发现气侵应及时排除，气侵钻井液未经排气不得重新注入井内。

若需对气侵钻井液加重，应在对气侵钻井液排完气后停止钻进的情况下进行，不应边钻进边加重。

加强溢流预兆及溢流显示的观察，做到及时发现溢流；"坐岗"人员发现溢流、井漏及油气显示等异常情况，应立即报告司钻。

钻进中发生井漏应将钻具提离井底、方钻杆提出转盘，采取定时定量反灌钻井液措施，保持井内液柱压力与地层压力平衡，其后采取相应措施处理井漏。

电测、固井、中途测试应做好以下井控防喷工作：

（1）电测前井内情况应正常、稳定。若电测时间长，应考虑中途通井循环再电测；

（2）下套管前，应换装与套管尺寸相同的防喷器闸板，固井全过程（起钻、下套管、固井）应保证井内压力平衡；

blowout preventer gate of the same size as the casing should be replaced, and the whole process of cementing (starting drilling, lowering casing and cementing) should ensure the pressure balance in the well;

(3) For midcourse testing and prior completion wells, observe one operating period time before proceeding with operations, and starting and lowering drill pipe or tubing shall be done provided the wellhead unit meets the installation and pressure test requirements;

(4) In sulfur-bearing formations it is generally not advisable to use conventional midway test tools for formation testing work, and if it is necessary to do so, the immersion time of the drill pipe in the hydrogen sulfide environment should be reduced and measures should be taken accordingly.

Shut down wells immediately when overflow is detected, and shut down wells for inspection when overflow is suspected.

The maximum allowable shut-in sleeve pressure shall not exceed the minimum of the rated pressure of the wellhead unit, 80% of the internal compressive strength of the casing, and the allowable shut-in sleeve pressure at the fracture pressure of the weak formation.

After shutting down the well, it should promptly obtain the shutdown vertical pressure, shutdown set pressure and overflow rate, and take corresponding measures according to the different situations of shutdown vertical pressure and set pressure.

Natural gas overflows should not be left unattended for long periods of time with the well shut in.

After shutting down an empty well with overflow, according to the severity of overflow, methods such as forced down drilling and

（3）中途测试和先期完成井，在进行作业以前观察一个作业期时间，起、下钻杆或油管应在井口装置符合安装、试压要求的前提下进行；

（4）在含硫地层一般情况下不宜使用常规中途测试工具进行地层测试工作，若需进行时，应减少钻柱在硫化氢环境中的浸泡时间，并采取相应措施。

发现溢流立即关井，疑似溢流关井检查。

最大允许关井套压不应超过井口装置额定压力、套管抗内压强度的80%和薄弱地层破裂压力所允许关井套压三者中的最小值。

关井后应及时求得关井立压、关井套压和溢流量，根据关井立压和套压的不同情况，分别采取相应处理措施。

发生天然气溢流时不宜长时间关井而不做处理。

空井溢流关井后，根据溢流严重程度，可采取强行下钻分段压井法、置换法、压回法等方法进行处理。含硫油气井发生溢流时，宜选用压回法进行处理。

segmented well pressure method, replacement method and pressure back method can be adopted to deal with it. When overflow occurs in sulfur-containing oil and gas wells, it is preferable to use the press-back method to deal with it.

### 6.1.3 Drilling and Ancillary Operations

#### 6.1.3.1 The First Drilling After Burying the Conduit and Before Lowering the Surface Casing

Conduit shoes should sit on hard strata and deepen the conduit for soft strata down.

A person should be in charge when drilling "rat holes" with power drilling tools.

The location and slope of the "rat hole" should facilitate the smooth raising and lowering of the square drill pipe.

The location of the "rat hole", the slope of the rathole pipe and the height of the exposed drilling platform shall be conducive to the easy operation of starting and placing the square drill pipe and taking off and hanging the tap.

The first drilling borehole should be straight, and the entry drilling tools should meet the quality standards required by SY/T 5369.

At the start of the first drilling, control the drilling pressure to be no more than 60% of the mass of the drillship column.

Drilling should be analyzed and judged according to changes in downhole conditions and changes in information collected by surface equipment and instruments, and corresponding measures should be taken in time to realize safe drilling.

#### 6.1.3.2 Drill Passes After Sealing Surface Casing

The wellhead unit shall be installed before each drilling and the overhead crane, turntable and wellhead center shall be corrected and

### 6.1.3 钻进及辅助作业

#### 6.1.3.1 埋设导管后，下表层套管前的第一次钻进

导管鞋应坐在硬地层上，对松软地层下加深导管。

用动力钻具钻"鼠洞"时应由专人指挥。

大"鼠洞"的位置和斜度应有利于方钻杆的顺利起下。

"鼠洞"的位置、鼠洞管的斜度与出露钻台高度，应有利于方钻杆的起放和摘挂水龙头操作方便。

第一次钻进井眼要直，入井钻具应符合 SY/T 5369 要求的质量标准。

第一次钻井开始，控制钻压不大于钻铤柱质量的 60%。

钻进中应根据井下情况变化和地面设备、仪表采集的信息变化分析判断，及时采取相应措施实现安全钻进。

#### 6.1.3.2 封固表层套管后的各次钻进

各次钻进前应先安装好井口装置，并校正天车、转盘和井口中心，固定牢固。

securely fastened.

After drilling the cemented plug and resuming drilling, the casing should be protected by.

(1) Before the drill chain is out of the casing shoe, the drilling pressure is not more than 60% of the drilling mass, and the rotary speed is preferred to be low;

(2) Measures to protect the casing should be taken in wells where the technical casing is lowered deeper and re-drilled into longer sections of the well.

The use of PDC bits and jetting bits in soft formations that are prone to shrinkage should be based on the actual situation, and each drilling footage of not more than 300~500m should be performed with a short take-up and take-down, and the take-up length should exceed the length of the newly drilled section.

Drilling fluid selection:

(1) For long sections of mudstone formations, the mineral components should be analyzed and the drilling fluid system with appropriate inhibition should be selected accordingly;

(2) The drilling fluid should be purified, and the solid phase content should be controlled according to the drilling design requirements, and the solid control equipment should be equipped with vibrating screen, desander centrifuge and mud remover (or cleaner);

(3) Drilling fluid performance should meet logging, logging and testing requirements.

Drilling should be based on changes in well conditions (drilling speed, drilling fluid performance, drill cuttings performance, drilling fluid volume and import and export

钻完固井水泥塞，再次恢复钻进，应对套管采取保护措施：

（1）在钻链未出套管鞋前，钻压不大于钻质量的60%，转盘速度宜采用低转速；

（2）技术套管下入较深、再次钻进井段较长的井，应采取保护套管的措施。

易缩径的软地层使用PDC钻头和喷射钻头应根据实际情况，每次钻进进尺不大于300~500m应进行短程起下钻，起出长度应超过新钻进井段。

钻井液的选择：

（1）对长段泥岩地层，应进行矿物组分分析，并依此选择具有相应抑制性的钻井液体系；

（2）钻井液应进行净化处理，按钻井设计要求控制固相含量，固控设备配备应有振动筛、除砂器离心机和除泥器（或清洁器）；

（3）钻井液性能应满足录井、测井和测试要求。

钻进中应根据井内情况变化（钻速、钻井液性能、钻屑性能、钻液体积和进出口流量等）和地面设备运转、仪表信息变化，判断分析异常情况，及时采取相应处理措施。

flow rate, etc.) and surface equipment operation, instrumentation information changes, judgment and analysis of anomalies, and timely take appropriate measures to deal with them.

New tine bit wells should start drilling with light pressure, appropriate speed drilling 0.2~1.0m, and then gradually increase to normal drilling pressure and speed, should not be pressurized to start the turntable.

When the new PDC drill bit enters the well and starts drilling, it should be used to clean the bottom of the well before the drill bit touches the bottom of the well by 0.5~1.0m, and then use the light pressure and appropriate rotational speed to drill 0.5~1.0m, and then gradually return to the normal drilling pressure and rotational speed.

Drilling should be terminated when one of the following conditions occurs.

(1) The drill bit is working abnormally at the bottom of the well, such as sudden whole jump drilling, sudden drop of drilling speed, increase of rotary torque, etc., which is ineffective after treatment;

(2) The drill bit is working properly at the bottom of the well, but the economic curve rate of the bit varies more than the allowable range;

(3) A sudden change in the pump pressure of the drilling pump has been determined to be a short circuit in circulation, a dislodged or clogged drill bit nozzle;

(4) Severe slip drilling occurs.

The bottom of the well should be free of metal fallout when using PDC drill bit, and the hole should not be scratched with PDC drill bit.

A milled teeth bit should be used for drilling or expanding holes in long sections of

新牙轮钻头井开始钻进时应采用轻压、适当转速钻进0.2~1.0m，再逐增至正常钻压和转速，不应加压启动转盘。

新PDC钻头入井开始钻进时，应在钻头接触井底前0.5~1.0m先开大排量清洁井底，然后采用轻压、适当转速钻进0.5~1.0m，再逐渐恢复到正常钻压和转速。

钻进中出现下列情况之一时应终止钻头使用：

（1）钻头在井底工作有异常，如突发性整跳钻、钻速突降、转盘扭矩增大等，经处理无效；

（2）钻头在井底工作正常，但钻头经济曲线率变化超过允许范围；

（3）钻井泵泵压突变，已判断为循环短路、钻头喷嘴脱落或堵塞；

（4）发生严重溜钻。

使用PDC钻头时井底应无金属落物，不能用PDC钻头划眼。

长井段的划眼或扩眼时应采用铣齿牙轮钻头。如用镶齿钻头划眼时，转速应控制在60r/min以下。

the well. In case of drilling with a tooth-set bit, the rotational speed should be kept below 60 r/min.

The drilling tools should not be left in the well for more than 3min to prevent sticking and jamming.

After safely drilling to the depth of the lower technical (oil) casing, other operations such as logging and cementing should be carried out in a timely manner in accordance with the requirements of the drilling design.

#### 6.1.3.3 Connecting a Single Root

Single roots and wellhead tools and materials should be prepared for inspection before connecting single roots.

No turntable shackles.

When using a small "rat hole" to connect a single, the connection threads should be tightened according to the specified torque, and care should be taken to prevent the connection threads of the single and square drill pipe from backing out during operation.

When connecting a single root, measures should be taken to prevent falling objects from entering the well.

After connecting the single and square drill pipe connection threads, the pump should be turned on to establish normal circulation before lowering the drill pipe to resume drilling.

#### 6.1.3.4 Tripping

Before drilling and tripping, you should be responsible for the division of labor in accordance with the operating position, and do a good job of checking the instruments, tools, equipment and safety protection facilities, and the well head the operation should be prevented from falling objects into the well measures.

钻具在井内静止时间不得超过 3min，防止黏附卡钻。

安全钻达下技术（油层）套管深度后，应根据钻井设计要求，及时进行测井、固井等其他作业。

#### 6.1.3.3 接单根

接单根前应做好单根、井口工具和材料的检查准备。

不能用转盘卸扣。

采用小"鼠洞"接单根时，应按规定力矩紧连接螺纹，操作时应注意防止单根和方钻杆的连接螺纹退松。

接单根时应有防落物入井措施。

接好单根和方钻杆连接螺纹后，应开泵建立正常循环，才能下放钻柱恢复钻进。

#### 6.1.3.4 起下钻

起下钻前应按照操作岗位负责分工，做好仪表、工具、器材和安全防护设施的检查，井口操作应有防落物入井措施。

Before starting drilling, the wellbore should be cleaned by adequate circulation washing according to the requirements of borehole conditions, mechanical drilling speed, drilling fluid performance and geologic recording and data.

According to the load of the drilling rig and the quality of the drilling tools and the borehole conditions, the tripping should be operated by double-hanging card or kawa, and when the depth of the well is more than 1000m or the load of the big hook is more than 300kN, the drill should be operated by double-hanging card with a small square filler or with a long drilling rod kawa.

The tripping lifting joint (or lifting connector) should be used at the same time for lifting and lowering the drilling chain, and the threads of kawa, safety lifting joint and drilling chain should be tightened with a clamp (or power clamp), and the safety kawa should be stuck at 0.05~0.10m from the upper part of the kawa, and the rotary disk should not be used to spin off the threads of the drilling chain.

Threads of drilling tools should be tightened according to the optimum torque value specified in SY/T 5369. It is advisable to use hydraulic pliers with direct reading torque meter to screw and unclamp threads of drilling tools.

The threads of connecting drilling tools should be lubricated with grease that meets the performance indexes specified in SY/T 5198.

Keep threads clean and intact before threading.

Speed limiting measures should be taken for down drilling. Auxiliary brakes should be used for large hook loads in excess of 300kN in the downdrill.

起钻前应根据井眼条件、机械钻速、钻井液性能和地质录井资料要求，充分循环洗井，清洁井筒。

起下钻应根据钻机载荷、钻具质量井眼条件，采用双吊卡或卡瓦操作，在井深大于 1000m 或大钩载荷大于 300kN 时用双吊卡加小方补心或用长钻杆卡瓦。

起下钻链应同时使用提升短节（或提升接头），卡瓦、安全提升短节和钻链连接螺纹应用吊钳（或动力吊钳）旋紧，安全卡瓦应卡在距卡瓦上部 0.05~0.10m 处，不应用转盘旋卸钻链螺纹。

钻具联结螺纹应按 SY/T 5369 规定的最佳扭矩值旋紧宜采用带有直读扭矩仪的液压大钳旋卸钻具螺纹。

连接钻具螺纹应采用符合 SY/T 5198 规定性能指标的润滑脂。

螺纹连接前应保持螺纹清洁完好。

下钻应采取限速措施。下钻大钩载荷超过 300kN 应使用辅助刹车。

When drilling with a check valve in the tool, fill the tool with drilling fluid every 20~30 strokes of the drill pipe.

If the jammed load exceeds 50~100kN of the suspended weight of the drilling tools (directional wells and horizontal wells should reduce the speed of lifting and lowering to take into account the influence of friction in complex jammed wells), timely measures should be taken to completely eliminate the jam before resuming normal operation.

After the drilling tool has been lowered to meet the square drill pipe, the pump will be turned on to circulate normally before transferring to normal operation.

### 6.1.3.5 Changing Drill Bit

The drill bit should be loaded and unloaded with a clamp and a special drill bit loader. The threads of the drill bit should be fastened manually and then tightened with the lifting pliers, and should not be pulled violently to prevent damage to the drill bit. Unload the drill bit by first loosening the threads with a hoist and then unloading the drill bit with a turntable at low speed (10~12r/min). Do not use the turntable to open the threads.

Connecting drill threads should be lubricated with standard thread grease and tightened to the specified thread torque value.

The type of entry bit and bit operating parameters should be selected based on the wear of the starting bit and the effectiveness of its use, in conjunction with the drilable nature of the rock to be drilled.

Before the tooth-wheel drill bit manhole should check the diameter of the drill bit, bearing clearance, tooth-wheel plane, teeth, connecting thread quality, weld quality, jet drill bit should check the quality of the nozzle installation.

钻具装有止回阀下钻时，每下20~30柱钻杆向钻具内灌满一次钻井液。

阻卡载荷超过当时钻具悬重（定向井、水平井起下钻在复杂卡阻井段应降低上提下放速度考虑摩阻影响）50~100kN时，要及时采取措施彻底消除阻卡后才能恢复正常作业。

钻具下完接方钻杆后，先开泵循环正常再转入正常作业。

### 6.1.3.5 换钻头

上卸钻头应用吊钳和专用钻头装卸器。钻头螺纹先用人工引扣，再用吊钳旋紧，不得猛拉猛绷，防止损坏钻头。卸钻头先用吊旋松螺纹，再用转盘低速（10~12r/min）卸开。不得用转盘绷开螺纹。

连接钻头螺纹应用标准螺纹润滑脂，并按规定螺纹扭矩值上紧。

应根据起出钻头磨损情况和使用效果，结合钻进岩石可钻性选择入井钻头类型和钻头工作参数。

牙轮钻头入井前应检查钻头直径、轴承间隙、牙轮平面、牙齿、连接螺纹质量、焊缝质量、喷射钻头应检查喷嘴安装质量。

Scraper bits should be checked for bit diameter, connection thread quality, blade height difference, alloy block and blade welding quality, nozzle quality, etc. before entering the well.

PDC drill bits should be checked for bit diameter, quality of welds between the carcass and the steel body, quality of sintering of the diamond or cutting block, quality of installation of the water eye sleeve and quality of the threaded connection before entering the well.

In and out of the well drill bits should be checked for bit diameter, and timely measures should be taken to scribe the eyes when the starting bit is badly worn.

### 6.1.3.6　Drill a Plug of Cement

It is suitable to use milled tooth bit for drilling cement plugs, and can use weighted drill pipe or add 1~2 columns of drill collars.

Drilling fluids for drilling cement plugs should be resistant to calcium contamination.

After drilling cement plugs out of the casing shoe, a casing shoe formation rupture pressure test should be performed in accordance with the drilling design requirements.

### 6.3.1.7　Extracted Rock

The following preparations should be made prior to the removal of the heart：

(1) Before coring, the relevant professionals will give a briefing to the drilling team, which should be clear about the requirements, basis, well depth, section length, lithology, structure of coring tools and inspection requirements of coring, and implement good coring technical measures;

(2) The diameter of the core bit should match the size of the drill bit for full-scale drilling. If, due to conditions, a small diameter

刮刀钻头入井前应检查钻头直径、连接螺纹质量、刀片高度差、合金块及刀片焊接质量、喷嘴质量等。

PDC 钻头入井前应检查钻头直径、胎体与钢体焊缝质量、金刚石或切削块烧结质量、水眼套安装质量和螺纹连接质量。

出入井钻头应进行钻头直径检查，起出钻头磨损严重时应及时采取划眼措施。

### 6.1.3.6　钻水泥塞

钻水泥塞宜用铣齿牙轮钻头，可采用加重钻杆或加 1~2 柱钻铤。

钻水泥塞的钻井液应具有抗钙化污染性能。

钻水泥塞出套管鞋后，应根据钻井设计要求，进行套管鞋地层破裂压力试验。

### 6.3.1.7　取心

取心前应做好以下准备：

（1）取心前由相关专业人员向钻井队交底，钻井队应清楚取心的要求、依据、井深、段长、岩性、取心工具结构及检查要求，执行好取心技术措施；

（2）取心钻头的直径，应与全面钻进的钻头尺寸相匹配，如因条件限制，要用小直径取心钻头，其小井眼段长应小于 50m；

core bit is to be used, the length of the small borehole section should be less than 50m. The diameter of the core bit should match the size of the drill bit for full-scale drilling;

(3) When starting and stopping the drilling of blocked or stuck sections of the well, measures such as paddling the eye through the well should be used to eliminate the blockage, and the core bit should not be used to paddle the eye;

(4) Where wells are to be cored immediately after cementing, the bottom of the well should be cleaned up before cored operations are carried out;

(5) Handle the drilling fluid well and keep its performance stable, which can ensure a smooth borehole, no collapse, no sand sinking, and smooth drilling down to the bottom of the well;

(6) Check the drilling equipment well;

(7) The outer barrel shall be probed, gauged and thoroughly inspected and a coring tool card filled out prior to delivery of the well;

(8) The coring tool should be prevented from being bent or flattened during loading and unloading.

The coring tool should meet the following requirements before entering the well:

(1) The inner cylinder rotates flexibly, and the suspension assembly should be unloaded and cleaned before each activation of the coring tool, and sufficient grease should be added, and the axial clearance should be adjusted before each change of the coring bit down the well;

(2) The check valve should drain freely and seal reliably;

(3) The water eye of the tap fitting should be clear;

（3）起下钻阻、卡井段，应采用划眼通井等措施消除阻卡，不能采用取心钻头划眼；

（4）凡固井后即需取心的井，应把井底处理干净后，才能进行取心作业；

（5）处理好钻井液，保持其性能稳定，能保证井眼畅通，无垮塌、无沉砂、能顺利下钻到井底；

（6）检查好钻井设备；

（7）送井前应对外筒进行探伤、测厚和全面检查并填写取心工具卡片；

（8）取心工具在装卸过程中，要防止摔弯、碰扁。

取心工具入井前应符合以下要求：

（1）内筒转动灵活，每次启用取心工具前悬挂总成均要卸开清洗干净，加足润滑脂，每次换取心钻头下井前应调整好轴向间隙；

（2）止回阀应排液畅通，密封可靠；

（3）分水接头水眼应畅通；

(4) The inner diameter of the inner core barrel is at least 5~6mm larger than the inner diameter of the core drill bit; the inner diameter of the free state of the card is 2~3mm smaller than the inner diameter of the core drill bit; the up and down sliding is flexible, and the sliding distance should be in accordance with the design requirements; the diameter of the cardboard core claw should be larger than the inner diameter of the core drill bit by 3~4mm;

(5) After passing all the inspections, the main dimensions such as the full length of the outer cylinder, the length of the inner cylinder, the free inner diameter of the clamp, the outer diameter of the drill bit, the diameter of the core inlet, etc. will be measured before assembly;

(6) The coring tool should be lifted on and off the drill table at both ends, operated smoothly, and wrapped and bundled with the drill bit;

(7) After hoisting to the wellhead, check the longitudinal clearance between the bottom of the core claw and the shoulder of the drill bit table, and that the inner core jane rotates flexibly.

The threads of each part should be intact, and the threads should be tightened during assembly, and the recommended values for the upper thread torque are shown in Table 6.2.

Table6.2  Recommended Torque for Buckling on Coring Cylinders

| Outer diameter×Inner diameter/<br>mm×mm | Torque/<br>N·m |
|---|---|
| 121×93 | 6000~7000 |
| 133×101 | 8000~9000 |
| 146×114 | 10000~12000 |
| 172×136 | 12000~13000 |
| 180×144 | 13000~16000 |
| 194×153 | 26000~31000 |

（4）内岩心筒的内径至少大于取心钻头内径 5~6mm 卡的自由状态内径比取心钻头内径小 2~3mm；上下滑动灵活，滑动距离应符合设计要求；卡板岩心爪的通径要大于取心钻头内径 3~4mm；

（5）经全部检查合格后，丈量外筒全长、内筒长、卡箍自由内径、钻头外径、岩心进口直径等主要尺寸，方能组装；

（6）取心工具上、下钻台，应两端吊，操作平稳，并包捆好钻头；

（7）吊到井口后，检查岩心爪底端与钻头台肩之间的纵向间隙，内岩心筒转动灵活。

各部分螺纹应完好无损，组装时应上紧螺纹，上螺纹扭矩推荐值见表 6.2。

表 6.2  取心筒上扣扭矩推荐值

| 外筒直径×内筒直径<br>mm×mm | 扭矩/<br>N·m |
|---|---|
| 121×93 | 6000~7000 |
| 133×101 | 8000~9000 |
| 146×114 | 10000~12000 |
| 172×136 | 12000~13000 |
| 180×144 | 13000~16000 |
| 194×153 | 26000~31000 |

When the coring cylinder enters and exits the well, it should be stuck with a safety kava.

The drill should be operated smoothly, and should not be lifted, released or braked violently, and the unthreading should be operated smoothly.

Core drilling meets the following requirements:

(1) Coring well sections shall be performed in accordance with design requirements and field geologic oversight instructions;

(2) Parametric fits for core drilling should be tailored to the different specifications of the tool;

(3) Coring tool down to 1.0m from the bottom of the well, first with a larger displacement of circulating drilling fluid to flush the bottom of the well, and then light pressure, slow turn the tree heart 0.3 ~ 0.5m, and then gradually add enough to the normal drilling pressure drilling, cutting the heart of the drilling, such as wells left in the heart of the next time to take the heart of the heart should be set before drilling;

(4) Coring drilling, should be adjusted beforehand square people, try to avoid midway to pick up a single, send drilling strive to be uniform, smooth, to prevent slippery drilling, found crappy drilling, skipping drilling or drilling time significantly higher, analyze the reasons for timely treatment, the reason is not clear should be up to check the drilling;

(5) During the coring operation, when the drilling tool should be lifted up due to unforeseen circumstances, the core should be cut first and the drilling tool should be lifted up; in case of overflow blowout, it should be handled according to the requirements of well

取心筒出入井时，应卡安全卡瓦。

起下钻操作平稳，不应猛提、猛放、猛刹，卸螺纹应平稳操作。

取心钻进符合以下要求：

（1）取心井段应按设计要求和现场地质监督指令执行；

（2）取心钻进的参数配合，应根据不同规范的工具具体制订；

（3）取心工具下到距井底1.0m时，先用较大排量循环钻井液冲洗井底，再以轻压、慢转树心0.3~0.5m后，逐步加够正常钻压钻进，割心起钻后，如井下留有余心，下次取心钻进前应套心；

（4）取心钻进，应事先调整好方可下入，尽量避免中途接单根，送钻力求均匀、平稳，防止溜钻，发现蹩钻、跳钻或钻时明显增高，分析原因及时处理，原因不明应起钻检查；

（5）取心作业时，因意外情况应上提钻具时，应先割断岩心，上提钻具；如遇溢流井喷，按井控要求处理；

control;

(6) Tree coring, coring drilling, cutting, casing and starting operations shall be operated by the head driller and deputy driller.

If the drilling encounters tunnels or pyrite strata, the coring drilling should be stopped and resumed after drilling with a lower tooth-wheel bit.

Directional centering shall be performed according to the requirements of SY/T 5347.

The quality of centering should be in accordance with the provisions of SY/T 5593.

The coring tools should be cleaned, maintained, assembled, and a card filled out by the well team drilling technician after the coring work is completed.

Hydrogen sulfide formation coring.

(1) Before lifting a core from a known or suspected hydrogen sulfide-bearing formation, be alert and wear a positive-pressure air respirator at least 10 risers before the core barrel reaches the surface, or as soon as the safety threshold concentration is reached;

(2) When the core barrel has been opened or when the core has been removed, the core barrel shall be checked with a portable hydrogen sulfide monitor; personnel shall continue to use positive pressure air respirators until it has been determined that the concentration of hydrogen sulfide in the atmosphere is below the safety threshold concentration;

(3) Vigilance should be exercised when handling and transporting core samples containing hydrogen sulphide; rock sample boxes should be made of materials resistant to hydrogen sulphide and labeled.

（6）树心、取心钻进、割心、套心和起钻作业，均应由正副司钻操作。

如钻遇隧石或夹有黄铁矿地层时，应停止取心钻进，改下牙轮钻头钻过后再恢复取心作业。

定向取心应按 SY/T 5347 的要求执行。

取心质量应符合 SY/T 5593 的规定。

取心工作结束后应将取心工具清洗、保养、组装好，并由井队钻井技术人员填好卡片。

硫化氢地层取心：

（1）在从已知或怀疑含硫化氢地层中起出岩心之前应提高警惕，在岩心筒到达地面以前至少 10 个立柱，或在达到安全临界浓度时，应立即戴上正压式空气呼吸器；

（2）当岩心筒已经打开或当岩心已移走后，应使用便携式硫化氢监测仪检查岩心筒；在确定大气中硫化氢浓度低于安全临界浓度之前，人员应继续使用正压式空气呼吸器；

（3）在搬运和运输含有硫化氢的岩心样品时，应提高警惕；岩样盒应采用抗硫化氢的材料制作并附上标签。

## 6.2 Safety Regulations for Underground Operations (SY 5727-2014) (Part)

### 6.2.1 Construction Operations

#### 6.2.1.1 General Requirements

Construction workers should be trained in the appropriate job skills and qualified for the work.

Workers entering the site should correctly wear and use labor protective equipment and other protective gear and do a good job of maintaining safety protection facilities High-altitude workers should wear safety belts, and will carry tools tied to the anti-drop rope, and before the operation of the safety belt in the derrick fastened firmly.

Workers going up and down the derrick should wear a safety belt and then hook up a fall arrest device.

Protective measures shall be taken for oil, gas and water pipelines and cables exposed on the ground when construction vehicles pass through the well site.

The construction design shall be prepared according to the geological design and engineering design of the underground operation, and shall be graded and approved according to the enterprise regulations.

Risk identification and assessment should be carried out according to changes in the construction environment, risk control measures should be taken and contingency plans should be formulated.

Make pre-construction preparations in accordance with the requirements of the

## 6.2 《井下作业安全规程》（SY 5727—2014）（部分）

### 6.2.1 施工作业

#### 6.2.1.1 一般要求

施工作业人员应参加相应的岗位技能培训，培训合格方可上岗。

进入现场人员应正确穿戴和使用劳动防护用品及其他防护用具，并做好安全防护设施的维护高处作业者应系安全带，并将随身携带的工具系上防掉绳，作业前将安全带在井架上系牢。

上下井架的人员应系好安全带后挂上防坠落装置。

施工车辆通过井场时，应对裸露在地面上的油、气、水管线及电缆采取保护措施。

应根据井下作业地质设计、工程设计编制施工设计，并按企业规定分级审批。

根据施工环境变化应进行风险识别、评估，采取风险控制措施并制订应急预案。

按施工设计要求做好施工前准备，经开工验收合格方可开工。

construction design, and start construction only after acceptance of the start of construction.

Conduct technical and safety briefing before construction, and insist on safety speech and post-shift evaluation every shift.

The construction process should implement the relevant operating procedures, quality standards and safety measures.

Strengthen the implementation of the post tour inspection system, timely rectification of the problems and hidden dangers found, can not be rectified to take control measures and immediately report to the superior.

6.2.1.2 Lowering Ramps

Before the operation and construction should check whether the tools used in the operation and construction are flexible and good for use, and the drilling platform and wellhead operating platform should not be stacked with other sundries except for the necessary tools.

Operators should have uniformly defined hand signals, movements and other means of communicating information for consistent and smooth operation.

When blocked or jammed, the uplift load should not exceed the safe load of the system.

The hydraulic power clamp jaws should not be touched by hand without cutting off the power source of the hydraulic power clamp.

There should be a person who directs the activity of unclipping, and a person observes the derrick, foundation, ground anchors, taut ropes, and finger weight gauges (tensiometers).

Construction shall be carried out in accordance with the construction design requirements to prevent blowouts.

When lifting and lowering the pipe column, you should always observe the finger weight

施工前进行技术、安全交底，每班坚持安全讲话和班后评价。

施工过程应执行相关操作规程、质量标准及安全措施规定。

加强岗位巡回检查制度的落实，及时整改发现的问题及隐患，不能整改的采取控制措施并立即向上级汇报。

6.2.1.2 起下管柱

作业施工前应检查作业施工中所用工具、用具是否灵活好用，钻台、井口操作台除必要的工具外不应堆放其他杂物。

操作人员应有统一规定的手势、动作和其他信息传递方式，配合一致、平稳操作。

当遇阻、遇卡，上提载荷不应超过系统的安全负荷。

在未切断液压动力钳动力源时，不应用手触碰液压动力钳钳牙。

活动解卡时应由专人指挥，专人观察井架、基础、地锚、绷绳、指重表（拉力计）等。

应按施工设计要求施工，防止井喷。

起下管柱时，应随时观察指重表（拉力计），密切注意井喷显示，发现异常及时采取有效措施。

gauges (tensiometer), pay close attention to the blowout display, and take effective measures in time when you find any abnormality.

Operations shall cease in the event of gusts of wind of magnitude 6 (or higher), dense fog with visibility less than the height of the derrick, heavy rain and thunderstorms, and when equipment is not operating normally.

Lifting and lowering a single pipe column on a tubing bridge should be done with the hitch opening facing upward, and precautions should be taken if the tubing support is lower than the face of the self-sealing flange.

Pulling and delivering tubing should be protected by threading measures, and the site operator should stand on one side of the tubing and should not ride across the tubing with both legs.

When raising and lowering the column, the threads should not be removed from the turntable.

Mechanical catheads should not be used to remove threads.

### 6.2.1.3 Perforation of the Wellhead.

During the injection process, a person should be assigned to observe the wellhead to prevent blowouts.

The power supply should be disconnected when the shot hole is connected.

### 6.2.1.4 Fracturing

Fracturing well selection and design should be in accordance with SY/T 5289 and should meet the following safety requirements:

(1) The casing elevated short section grouping matches the formation casing material and wall thickness;

(2) The use of non-toxic or low-toxic materials, the relevant chemical products should be provided with safe technical instructions for

遇有6级(含6级)以上大风、能见度小于井架高度的浓雾天气、暴雨雷电天气及设备运行不正常时,应停止作业。

在油管桥上提放单根管柱时应使吊卡开口朝上,如油管支架低于自封法兰面,应采取防范措施。

拉送油管应有保护螺纹措施,场地操作人员站在油管一侧,不应两腿跨骑油管。

起下立柱时,不应用转盘上卸螺纹。

不应使用机械猫头上卸螺纹。

### 6.2.1.3 射孔

射孔过程中,应设专人观察井口,防止井喷。

射孔连接时应切断电源。

### 6.2.1.4 压裂

压裂选井和设计应符合SY/T 5289的规定,并应达到下列安全要求:

(1)套管升高短节组配与油层套管材质、壁厚相符;

(2)使用无毒或低毒物资,相关化工产品应提供安全技术使用说明书和应急处置方法;

use and emergency disposal methods;

(3) The downhole tools and connections should be adequate for normal fracturing construction and facilitate other operations before and after fracturing;

(4) The road to the well site provides safe passage for construction vehicles;

(5) The wellsite area, foundation leveling, load-bearing capacity, and infrastructure should meet the requirements for large-scale fracking equipment placement, safe spacing and access, and emergency disposal.

The following safety-related elements should be included in the fracking design:

(1) There are issues that could affect fracking construction;

(2) Description and requirements for construction well sites and construction vehicle routes;

(3) Ground process connections, construction equipment inspection requirements;

(4) Test pressure and squeeze requirements;

(5) Construction handover and inspection requirements;

(6) Emergency plans and other technical safety requirements.

The following fracturing construction designs should have special safety technical requirements:

(1) Acid fracturing, foam fracturing, or the use of other toxic, hazardous, flammable or explosive substances;

(2) Possible blowout or other hazards and impacts;

(3) Fracturing of wells and gas wells containing toxic and hazardous substances such as $H_2S$;

(4) New technology trials, etc.

（3）下井工具、连接方式应满足正常压裂施工，并有利于压裂前后的其他作业；

（4）通往井场的道路能保证施工车辆安全通行；

（5）井场面积、地基平整度、承重能力及基础设施应满足大型压裂施工设备布置、安全间距与通道及应急处置要求。

压裂设计中应包括下列与安全有关的内容：

（1）存在可能影响压裂施工的问题；

（2）施工井场、施工车辆行驶路线说明及要求；

（3）地面流程连接、施工设备检查要求；

（4）试压、试挤要求；

（5）施工交接、检查要求；

（6）应急预案及其他安全技术要求。

下列压裂施工设计应提出特殊安全技术要求：

（1）酸化压裂、泡沫压裂或使用其他有毒有害、易燃易爆物质；

（2）可能出现井喷或其他危害及影响；

（3）含有 $H_2S$ 等有毒有害物质井、气井压裂；

（4）新技术试验等。

Fracturing design should be undertaken by an appropriately qualified organization.

Fracturing preparation work teams and fracturing construction teams should be trained and qualified.

Accessories for fracturing equipment should be selected as designed and meet the following safety requirements:

(1) Complete and free of leaks and other faults;

(2) The valves are flexible to open and close and the cue markings are clearly legible;

(3) Metering instruments and pressure-limiting protection and other indicating, alarming and control devices are in good condition;

(4) The safety valve is effective, the pressure relief pipe is clear, safe and secure, and the outlet is facing down;

(5) The protective facilities are complete.

Liquid storage tanks shall meet the following safety requirements:

(1) Smooth placement;

(2) The tank is clean, free of foreign matter and leaks;

(3) Inlet and outlet valves and fittings are intact;

(4) When storing flammable and explosive, toxic and hazardous substances, the tanks shall have appropriate warning signs;

(5) Mobile liquid storage tanks should have lids and changeover valves;

(6) Metal packaging shall be reliably earthed.

Other ancillary equipment and accessories shall meet the safety requirements.

Fittings for fracturing shall comply with SY/T 6270.

压裂设计应由具有相应资质的单位承担。

压裂准备作业队伍、压裂施工队伍应经培训，具有相应资质。

压裂用设备配件应按设计选配压裂设备，并达到下列安全要求：

（1）完整，无泄漏及其他故障；

（2）阀门开关灵活，提示标识清晰可辨；

（3）计量仪表和限压保护及其他指示、报警、控制装置完好；

（4）安全阀有效，泄压管畅通、安全牢靠、出口朝下；

（5）防护设施齐全。

储液罐应达到下列安全要求：

（1）摆放平稳；

（2）罐内清洁、无异物、无泄漏；

（3）进出口阀门、接头完好；

（4）储存易燃易爆、有毒有害物质时，罐体应有相应的警示标志；

（5）移动的储液罐应有罐盖、换气阀；

（6）金属容器应有可靠的接地。

其他附属设备、附件应达到安全要求。

压裂用配件应符合 SY/T 6270 的规定。

Before fracturing, the fracturing preparation work crew shall be organized in accordance with the requirements of the design document. Tools, tubing columns, and wellhead devices for fracturing should meet design requirements.

Fracturing tubing columns and fracturing tools should be inspected according to the design requirements before going down the well, the inspection results should be recorded, and the connections should be tightened to meet the design requirements when going down the well.

Wellheads are secured with special brackets or other means.

The casing shall be fitted with a pressure indicating device.

Install the blowout pipeline in the well, avoiding the fracking construction area. Fix the land (cement pier, etc.) at every 10m interval and at the exit, and keep the vent line free, and use steel elbow with angle not less than 120° or 90° elbow with anti-erosion function at the turning point of the vent line.

After the fracturing preparation operation is completed and the merge port and downhole pipe test pressure is qualified according to the design requirements, the valve of the oil (casing) pipe is closed, and the professional personnel carry out acceptance and signature.

Fracturing crews should be prepared according to design requirements prior to construction and should do the following:

(1) Analyze construction wells, identify, evaluate and develop measures for risks such as hazards and impacts;

(2) Inspect and maintain equipment, accessories and tools used;

(3) Provide technical and safety education to construction personnel.

压裂前，压裂准备作业施工队伍应按设计书要求组织施工。压裂用工具、管柱和井口装置应符合设计要求。

压裂管柱、压裂工具下井前应按设计要求进行检查，记录检查结果，下井时连接紧固达到设计要求。

井口用专用支架或其他方式固定。

套管应安装压力指示装置。

井安装放喷管线，朝避开压裂施工区域。每间隔10m和出口处用地（水泥基墩等）固定，放空管线畅通，放喷管线转弯处应使用角度不小于120°钢质弯头或90°带抗冲蚀功能的弯头。

压裂准备作业完毕，按设计要求对井口和井下管试压合格后，关闭油（套）管阀门，由专业人员进行验收、签字。

压裂施工队伍施工前应按设计要求进行准备，并应做好下列工作：

（1）分析施工井情况，识别、评估危害和影响等风险并制订措施；

（2）对所用设备、配件、工具进行检查、维护；

（3）对施工人员进行技术和安全教育。

For gas well fracturing, foam fracturing, fracturing with toxic, hazardous, flammable and explosive substances or other special fracturing, measures and contingency plans for preventing hazards and impacts should be formulated individually in advance according to the design requirements, and sufficient protective gears, special tools, and equipment and devices required for emergency response should be provided.

If necessary, fire engines, ambulances and other rescue equipment and facilities and specialized personnel should be provided. Foam fracturing should be equipped with sufficient air respirators and at the same time with cold-proof gloves, acid fracturing should be equipped with acid-proof protective gears, and the implementation of gas well fracturing and fracturing with flammable and explosive substances should be equipped with explosion-proof tools.

The use of radioactive substances should be subject to appropriate protective measures in accordance with the relevant regulations, and the activity of radioactive substances and the integrity of storage devices should be tested on a regular basis, and medical examinations should be conducted for related personnel.

Fracturing construction vehicles should form up and drive and stop in accordance with traffic regulations.

Fracturing vehicles should be placed upwind or to the side of the wellhead in accordance with the design requirements, at a distance from the wellhead in accordance with SY/T 5289, and with safe and emergency evacuation routes. The area where fracking vehicles are placed should be free of flammable materials.

实施气井压裂、泡沫压裂及使用有毒有害、易燃易爆物质压裂或其他特殊压裂，应按设计要求预先单独制订防止危害及影响的措施和应急预案，配备足够的防护用品、特殊工具及应急所需设备和器材。

必要时应备有消防车、救护车等抢险救护设备设施及专业人员。泡沫压裂应配备足够的空气呼吸器，同时应配备防寒手套；酸化压裂应同时配备防酸防护用品；实施气井压裂和使用易燃易爆物质压裂应配备防爆型工具。

使用放射性物质应按有关规定采取相应的防护措施，并定期对放射物质的活度、存储装置是否完好进行检测，对接触人员进行体检。

压裂施工车辆应编队并按交通法规行驶、停靠。

压裂施工车辆应按设计要求摆放在施工井井口上风或侧风方向，与井口距离符合 SY/T 5289 中的规定，并留有安全和应急疏散通道。压裂施工车辆摆放区域应无易燃物。

Prior to fracturing, the site director, wellhead operator, and the fracture preparation operator's designee should inspect the construction to confirm that the design requirements have been met.

The construction site should be equipped with 4~8 fire extinguishers with more than 8kg powder. When special needs, it is implemented according to the emergency plan.

Fracturing construction should be unified and coordinated by the on-site director.

Explosion-proof wireless radios are used at the fracking site to transmit command messages.

The on-site command should issue instructions in a standardized and accurate manner, repeating the instructions 1 to 2 times each time they are issued, and repeating them when the fiduciary party has not received or has not received them accurately. After the fiduciary party receives the instruction accurately, it shall reply to the issuing party with the instruction.

Above ground processes should be connected as designed. Tube manifold outlet to wellhead is connected by high pressure rigid piping.

Take 10m radius of the construction well as the boundary, and take 10m as the boundary along both sides of the ground flow from the exit of the pump truck to the entrance of the construction well as the high-pressure danger zone. The high-pressure hazardous area is fenced with special safety warning line (belt), the height of which is 0.8~1.2m, and the high-risk area should be set up with eye-catching safety signs and warnings. The safety signs shall comply with the provisions of Appendix B of SY/T 6355—2017.

压裂施工前，现场指挥、井口操作人员与压裂准备作业方指定人员应对施工进行检查，确认达到设计要求。

施工现场应配备 4~8 具 8kg 以上粉灭火器。特殊需要时按应急预案执行。

压裂施工应由现场指挥统一指挥、协调。

压裂施工现场采用防爆无线对讲机传递指令信息。

现场指挥发布指令应规范、准确无误，每次发布指令重复 1~2 次，在受信方未接到或未准确接到指令时，应重复发布指令。受信方准确接到指令后，应向发令方回复指令。

地面流程应按设计要求连接。管汇出口至井口采用高压硬质管线连接。

以施工井 10m 为半径，沿泵车出口至施工井口地面流程两侧 10m 为边界，设定为高压危险区。高压危险区使用专用安全警示线（带）围栏，高度宜为 0.8~1.2m。高危险区应设立醒目的安全标志和警句。安全标志应符合 SY/T 6355—2017 中附录 B 的规定。

The maximum pressure for construction operations should be no greater than the rated working pressure of the lowest pressure-bearing component, the downhole tool.

The wellhead unit shall be rated at a pressure greater than or equal to the maximum pressure of the construction design.

Fracturing operations should operate smoothly and in accordance with design requirements.

Acid fracturing, foam fracturing and fracturing with flammable and explosive substances should be zoned for containment, with no tampering of the pipeline flow.

On-site commanders should react to abnormal changes in construction in a timely manner. When a situation occurs that jeopardizes the safety of persons or equipment or affects construction, the following appropriate measures should be taken immediately:

(1) Fracturing equipment malfunctions that cannot be repaired, stop the pump immediately, and do not repair the attached pressurized equipment until it is safely relieved;

(2) In the event of a puncture, deformation, or fracture of the surface process and wellhead unit, stop the pump immediately and repair it after a safe pressure relief;

(3) The long line of sand plugs should be recirculated to replace the sand mixing fluid, and the recirculation pressure should not exceed the design pressure;

(4) Leaks and punctures in high- and low-pressure pipelines for acid fracturing, foam fracturing and fracturing with flammable and explosive substances should be dealt with in accordance with the emergency plan by immediately stopping the pumps or closing the fluid storage tanks and wellhead valves.

施工作业的最高压力应不大于承压最低部件、井下工具的额定工作压力。

井口装置的额定压力应大于或等于施工设计的最高压力。

压裂作业应按设计要求平稳操作。

酸化压裂、泡沫压裂和使用易燃易爆物质压裂应分区密闭、管汇流程无窜漏。

现场指挥对施工出现的异常变化应及时做出反应。当出现危及人身、设备安全或影响施工情况时，应立即采取以下相应的措施：

（1）压裂设备出现故障无法修复，立即停泵，附属带压设备未安全泄压前不得进行修理；

（2）地面流程和井口装置出现刺漏、变形、断裂，立即停泵，待安全泄压后进行修理；

（3）长线砂堵应反循环替出混砂液，反循环压力不应超过设计压力；

（4）酸化压裂、泡沫压裂和使用易燃易爆物质压裂的高低压管线出现泄漏、刺漏，应按应急预案立即停泵或关闭储液罐及井口阀门后进行处理。

When the ground process is pressurized, no personnel should enter the high-pressure hazardous area. When it is necessary to enter a high-pressure hazardous area, the following safety conditions shall be met:

(1) With permission from the field commander;

(2) Supervision is provided outside the danger zone;

(3) Leave quickly upon completion of your mission;

(4) Operations should not be changed without the operator leaving the hazardous area.

Acid fracturing, foam fracturing, and fracturing with flammable and explosive substances should be injected into the well with a replacement fluid for residual fluids within the surface process.

Upon completion of the fracturing work, the well should be shut in according to the design requirements, the internal cavity of the pump head should be cleaned using fresh water, and the fracturing lines should be disassembled.

Before the winter construction, the fracturing equipment and wellhead device should be preheated according to the actual needs, and after stopping the pump in the middle of the process, the fracturing equipment, the ground process and the wellhead device should be heated up before starting the pump again; at the end of the construction, all the equipment, pipelines and fittings should be antifreeze treated.

After fracturing construction the well should be shut in as designed to diffuse the pressure and observe pressure changes.Drain and perform subsequent construction as

地面流程承压时，任何人员不应进入高压危险区。因需要进入高压危险区时，应符合下列安全条件：

（1）经现场指挥允许；

（2）危险区以外有人监护；

（3）执行任务完毕迅速离开；

（4）操作人员未离开危险区时，不应变更作业内容。

酸化压裂、泡沫压裂和使用易燃易爆物质压裂，应将地面流程内的残液用替置液注入井内。

压裂施工完毕，应按设计要求关井，使用清水清洗泵头内腔，拆卸压裂管线。

冬季施工前应根据实际需要对压裂设备、井口装置等进行预热，中途停泵后，再次启泵前应对压裂设备、地面流程、井口装置进行加热；施工结束，应对所有设备、管汇、配件进行防冻处理。

压裂施工后应按设计要求关井，扩散压力，观察压力变化。按设计要求排液和进行后续施工。

designed.

Before checking the exit spray potential and ejecta, the test of toxic and hazardous gases should be carried out, and the construction personnel should be located in the upwind place. When the ventilation conditions are poor or there is no wind, the location of higher ground should be chosen.

Measuring operation worker should have safety precautions when going to the head of the tank.

### 6.2.1.5 Test Oil (Gas)

The oil testing operation shall comply with the provisions of Chapter 6 of SY/T 5981—2012.

The wellhead, casing and process were pressure tested and passed as required.

The following safety requirements shall be met when pumping:

(1) A person at the wellhead is responsible for marking and observing the drawdown, and no one should stand near the wellhead or wire rope or cross the wire rope;

(2) When the wire rope is twisted or jumps the groove, first use the card to tighten the wire rope, and then relax the rope, and wait until it no longer slides down before you can use tools to lift it;

(3) Pumping spray boxes, rope caps, weighted rods and pumping heads are intact and reliable.

Air should not be used as a medium to perform operations such as air lifting and gasified water well washing.

### 6.2.1.6 Well Operations Containing Toxic and Hazardous Gases

When working in complex stratigraphic areas (e.g., high-pressure gas formations, areas that may contain hydrogen sulfide, etc.),

查看出口喷势和喷出物前，应进行有毒有害气体的检测，施工人员应位于上风处。通风条件较差或无风时，应选择地势较高的位置。

计量液位的人员到罐口应有安全防护措施。

### 6.2.1.5 试油（气）

试油作业应符合 SY/T 5981—2012 中第 6 章的规定。

井口、套管和流程按规定试压合格。

抽汲时应符合下列安全要求：

（1）井口有专人负责做好抽汲记号并观察，井口、钢丝绳附近不应站人，不应跨越钢丝绳；

（2）钢丝绳打扭或跳槽时，先用卡子卡紧钢丝绳后，再放松绳索，待不再下滑时方可用工具解除；

（3）抽汲防喷盒、绳帽、加重杆及抽子头完好可靠。

不应使用空气作为介质实施气举、气化水洗井等作业。

### 6.2.1.6 含有毒有害气体井作业

在地层复杂区域作业（如高压气层、可能含硫化氢区域等），操作人员应经专业知识培训、考核合格，持证上岗。

operators should be trained in specialized knowledge, qualified by examination, and licensed before work.

For the construction of wells containing or likely to contain toxic and hazardous gases, qualified personal protective equipment and corresponding gas monitors should be provided. The monitoring and personal safety protection of hydrogen sulfide should be in accordance with the provisions of Chapter 4 and Chapter 5 in SY/T 6277—2017.

#### 6.2.1.7 Operate Under Pressure

Prior to construction, the pressurized operating device should be prepared in accordance with the design requirements. The blowout preventer set should be pressure tested and qualified according to design requirements.

The outer diameter and length of the steel body of the plugger should be measured before the plugger is lowered into the well, and the components should be checked intact, and the plugger specifications should be matched with the downhole pipe column.

The body of the blowout preventer should be free of corrosion, cracks, bends, and the threads shall be intact.

After completing the blocking, the oil pipe gate should be opened and emptied, and at least 2h should be observed to confirm that no overflow is qualified for blocking.

Plug valves should be installed and closed on the tubing suspensions before installing devices that operate under pressure.

During the pressure test, personnel should stand in a safe area 10m away from the pressurized part, and should not be close to the observation.

Lifting and lowering of pipe columns and

在含有或可能含有有毒有害气体井施工，应配备合格的个人防护用具和相应气体监测仪。对硫化氢的监测和人身安全防护应符合 SY/T 6277—2017 中第 4 章、第 5 章的规定。

#### 6.2.1.7　带压作业

防喷器组应按设计要求试压合格。

堵塞器下井前应测量堵塞器钢体外径和长度，检查各部件是否完好，堵塞器规格应与井下管柱相匹配。

防喷管本体应无腐蚀、裂纹、无弯曲现象，螺纹完好。

完成封堵后，应打开油管阀门放空，至少观察 2h，确认无溢流为封堵合格。

安装带压作业的装置前，应在油管悬挂器上安装旋塞阀并关闭。

在试压过程中人员应站至承压部位 10m 以外的安全区域，不得靠近观察。

起下管柱及带压作业时符合相关规定。

working under pressure should comply with the relevant regulations.

When installing or dismantling, a person should be assigned to direct the installation.

#### 6.2.1.8 Stratigraphic Testing

The stratigraphic testing operation shall comply with SY/T 483 and SY/T 5486.

Before construction, blowout prevention devices should be installed, and well control acceptance should be carried out according to the design requirements.

Equipment such as test vinches and instrument carts are in good working order, with reliable enclosure insulation.

The test tools should be connected in the sequence required by the design.

When connecting the test tool at the wellhead, a person should be assigned to observe the tool above and prevent it from backing up above when connecting tools.

After the control well head is connected to the test tube, it should be connected to another movable line, and the lower end of the line should be tied tightly with a rope on the tube column to prevent it from being thrown out and injuring people when it is rotated and seated.

Safety cords should be added to each live fitting connection of the movable line.

No one is allowed to stand within 10m of the wellhead and pipeline when pumping.

During the test, a person should be assigned to observe the wellhead to prevent blowouts.

Tools should be maintained in a timely manner, and tool surfaces should be free of oil and burrs.

#### 6.2.1.9 Fire and Explosion Proof

The equipping and management of fire-

安装、拆除时，应安排专人指挥。

#### 6.2.1.8 地层测试

地层测试作业应符合 SY/T 483 和 SY/T 5486 的规定。

施工前应按设计要求安装防喷装置并进行井控验收。

测试绞车和仪器车等设备性能良好，外壳绝缘可靠。

测试工具应按设计要求顺序连接。

井口连接测试工具时应设专人观察上方工具，严防连接工具时上方倒扣。

控制头连接到测试管后，应再连上一根活动管线，管线下端应用绳索在管柱上绑紧，以防止旋转坐封时，甩出伤人。

活动管线每一个活接头连接处应加装保险绳。

打压时，井口及管线周围 10m 内不得站人。

测试过程中，应设专人观察井口，防止井喷。

工具应及时进行保养，工具表面应无油污、毛刺。

#### 6.2.1.9 防火防爆

井场消防器材的配备与管理应符合 SY/T

fighting equipment in the well site shall be in accordance with the provisions in SY/T 5225-2019.

Combustible gas detectors should be present on site. Combustible gas detectors should be calibrated and maintained regularly.

On-site heating should be done with no open flame appliances.

Smoke and fire should be prohibited in the well site.

Industrial fires at the well site should be subject to fire procedures.

Exhaust pipes of vehicles entering the well site should be equipped with flame arrestors.

### 6.2.2　Security Management

There should be a HSE organization, and the construction team should have a trained and qualified full-time (part-time) HSE supervisor.

The rules and regulations, job responsibilities and job operating procedures should be complete and effective.

HSE meetings, training, drills, etc. should be organized regularly and recorded in detail.

All technical data and HSE records should be complete and accurate.

Strengthen daily safety inspections and promptly rectify any potential accident hazards.

5225—2019 中的相关规定。

现场应有可燃性气体检测仪。可燃性气体检测仪应定期校验和维护。

现场取暖应采用无明火器具。

井场内应禁止烟火。

井场工业动火应办理动火手续。

进入井场车辆排气管应装有阻火器。

### 6.2.2　安全管理

应有 HSE 组织机构，施工队应设经培训合格的专（兼）职 HSE 监督员。

各项规章制度、岗位职责和岗位操作规程应齐全有效。

应定期组织 HSE 会议、培训、演练等，并详细记录。

施工现场各项技术资料、HSE 记录报表应齐全准确。

加强日常安全检查，对存在的事故隐患及时整改。

## 6.3 Special Safety Signs for Oil and Gas Production (SY/T 6355-2017) (Part)

Table A.1  No Entry Sign

| Serial number | Symbol | Title | Description |
| --- | --- | --- | --- |
| A.1 |  | Prohibition of mixing | Designed to avoid a series of chemical reaction accidents caused by mixing of chemicals. Suitable for chemical storage, use of the warehouse and the corresponding workplace |
| A.2 |  | Don't mess with the valves | Prohibition of tampering with the valve is designed to avoid non-operators to open and close the process pipeline valves. Applicable to oil and gas plants, stations and pipelines and other process pipeline valves |
| A.3 |  | Prohibition of tampering with firefighting equipment | Designed to prevent unauthorized diversion of firefighting equipment. Suitable for sites where firefighting equipment is placed |
| A.4 |  | Prohibition of gasoline rubbing | Designed to prevent fires from occurring when scrubbing objects with flammable liquids such as gasoline when exposed to open flames, high temperatures and static electricity. It is suitable for posts or places where flammable and explosive production areas and mechanical equipment are concentrated |
| A.5 |  | Prohibition of drunkenness on duty | It is designed to prevent "working after drinking", which brings serious safety hazards to normal production operations and even leads to accidents. Suitable for oil and gas development and construction areas |

continued table

| Serial number | Symbol | Title | Description |
|---|---|---|---|
| A.6 | | Prohibition of mixing | Designed to avoid a series of chemical reaction accidents caused by mixing of chemicals. Suitable for chemical storage, use of the warehouse and the corresponding workplace |
| A.7 | | No bumping | It is designed to avoid serious hazards such as deformation, leakage and explosion in flammable and explosive places or containers with flammable and explosive toxic and hazardous media due to impact. It is suitable for the transportation, storage and use of the above workplaces and related containers |
| A.8 | | Prohibition of single-buckle lifting | Designed to prevent accidents caused by imbalance of the lifting piece due to lifting with a single rope. Suitable for lifting and hoisting workplaces |
| A.9 | | Prohibition of passing under the boom | Designed to prevent injuries caused by falling loads or slipping of the boom during lifting and hoisting operations. Suitable for lifting and hoisting operations |
| A.10 | | Prohibition of unauthorized removal | Designed to prevent mechanical injuries caused by unauthorized removal of safety guards during the operation of equipments and facilities. Suitable for workplaces with safety guards during the operation of equipments and facilities |
| A.11 | | Prohibition of fire escape | Designed to prevent serious consequences due to unauthorized occupation or obstruction of fire escape routes. Suitable for fixed production workplaces |

continued table

| Serial number | Symbol | Title | Description |
|---|---|---|---|
| A.12 | | Prohibition of operation with disease | It is designed to prevents "equipments and facilities working with disease" from bringing serious safety hazards to normal production operations and even leading to accidents. Applicable to oil and gas development and other workplaces and construction zones |
| A.13 | | Prohibition of indiscriminate connection of equipments | Designed to prevent accidents such as overloading, electrocution and even fire due to private connection of equipments in the workplace. Suitable for any fixed or mobile power distribution facilities |
| A.14 | | Prohibition of under pressure work | Designed to prevent production sites from causing hazards to personnel or equipment during operation due to pressurized devices or pipelines. Suitable for process system maintenance, remodeling and other workplaces |
| A.15 | | No exposure to sunlight | Designed to prevent containers of flammable and explosive media from being exposed to sunlight, which may cause danger due to the increase of internal pressure. Suitable for outdoor storage of flammable and explosive, toxic and hazardous media containers |
| A.16 | | No closing of gate | To prevent closing the gate at the construction site, due to unauthorized. Suitable for oil and gas exploration, development and other workplaces |
| A.17 | | No cell phones | Designed to prevent fire accidents caused by static electricity at oil and gas well construction sites. Suitable for oil and gas exploration, development and other workplaces |

continued table

| Serial number | Symbol | Title | Description |
|---|---|---|---|
| A.18 | | No fireworks | Designed to prevent fire accidents due to the use of fire at oil and gas well construction sites. Suitable for oil and gas exploration, development and other workplaces |
| A.19 | | No Fires | Designed to prevent fire accidents due to carrying fire at the construction site of oil and gas wells. Suitable for oil and gas exploration, development and other workplaces |
| A.20 | | No smoking | There are places with A, B, C fire hazardous substances and public places where smoking is prohibited, etc., such as carpentry workshops paint workshops, asphalt workshops, textile factories, printing and dyeing factories |
| A.21 | | Prohibit the use of water to extinguish fires | Production, storage and transportation, the use of water is not allowed to extinguish the substance of the place, such as transformer room, acetylene station, chemical medicine storehouse, a variety of oil depots and so on |
| A.22 | | No flammable materials | Workplaces with open-flame equipments or high temperature, such as fire zones, all kinds of welding, cutting, forging, pouring workshops and other places |
| A.23 | | Prohibition of forklifts and other in-plant motorized vehicles | Places where forklifts and other in-plant motorized vehicles are prohibited |

continued table

| Serial number | Symbol | Title | Description |
|---|---|---|---|
| A.24 | | No passengers | Facilities where occupants are vulnerable to injury, such as outdoor transportation cradles, externally operated freight elevator frames, etc |
| A.25 | | Do not approach | Hazardous areas where proximity is not permitted, such as high-voltage test areas, high-voltage lines, and the vicinity of power transmission and transformation equipments |
| A.26 | | No admittance | Entrances to places prone to accidents or injuries to personnel, such as high-voltage equipment rooms, various sources of pollution, etc |
| A.27 | | Prohibition of promotion | Installations or equipments that can be easily tipped over, for instance, station screen doors |
| A.28 | | Prohibition on staying | Places of immediate danger to personnel, such as crushing sites, dangerous intersections, bridge crossings, etc. |
| A.29 | | Prohibition on traffic | Hazardous work areas, such as lifting, blasting sites road construction sites, etc. |

第六章　油气井工程相关的规范与标准
Chapter 6　Specifications and Standards Related to Oil and Gas Well Engineering

continued table

| Serial number | Symbol | Title | Description |
|---|---|---|---|
| A.30 | | Do not cross | Hazardous areas where crossing is prohibited, e.g., dedicated transportation lanes, belt conveyors and other operating lines, ditches, cans, pits, and so on |
| A.31 | | No jumping down. | Dangerous places where jumping is not allowed, such as deep ditches, deep pools, station platforms and tankers, storage tanks, cellars, etc. where toxic substances and easy to produce asphyxiating gases have been loaded |
| A.32 | | No stacking | Firefighting equipment storage, fire escapes and main workshop corridors, etc. |
| A.33 | | No starting | Proximity of suspended equipment, such as equipments repair and replacement parts |
| A.34 | | No Leaning | Locations or parts that cannot be relied upon, like train doors station screen doors, elevator car doors, etc. |
| A.35 | | No sitting | High temperature, corrosive, collapsing, falling, overturning fragile, easily caused by personnel injury equipments and facilities surface |

229

continued table

| Serial number | Symbol | Title | Description |
|---|---|---|---|
| A.36 | | Prohibition of collapse | High temperature, corrosive, collapsing, falling, overturning fragile, easily caused by personnel injury equipments and facilities surface |
| A.37 | | No touching | Near equipment or objects that are prohibited to be touched, such as exposed electrically charged objects, hot objects, toxic and corrosive objects, etc. |
| A.38 | | No reaching | Devices or places where body parts can be easily entrapped, like open motors, crushers, etc. |
| A.39 | | Not for Drinking | Switching places where drinking water is prohibited, like recycled water, industrial water, polluted water, etc. |
| A.40 | | No Throwing | Throwing easy to hurt the location, such as high work site deep ditch (pit), etc. |
| A.41 | | No gloves | Workplaces where gloves can cause hand injuries such as near rotating machining equipments |

第六章　油气井工程相关的规范与标准
Chapter 6　Specifications and Standards Related to Oil and Gas Well Engineering

continued table

| Serial number | Symbol | Title | Description |
|---|---|---|---|
| A.42 | | Chemical fiber clothing is prohibited | Workplaces with static sparks that can cause disasters or hot substances, such as smelting, welding and places with flammable and explosive substances |
| A.43 | | No spiked shoes | Workplaces where static sparks can cause disasters or risk of electric shock, such as workshops with flammable and explosive gases or dusts and electrified workplaces |
| A.44 | | No metal objects or watches | Microwave and electromagnetic fields susceptible to interference from metallic objects, such as magnetic resonance chambers, etc. |
| A.45 | | Prohibition of the carriage of flammable and explosive substances in consignments | Places or means of transportation, such as trains, airplanes, subways, etc., where flammable, explosive and other dangerous goods can not be carried and con signed |
| A.46 | | Prohibition of the carriage of toxic substances and hazardous liquids in consignments | Places or means of transportation, such as trains, airplanes, subways, etc., where toxic substances and hazardous liquids can not be carried in the consignment |

231

## 6.3 《石油天然气生产专用安全标志》(SY/T 6355—2017)(部分)

表 A.1 禁止标志

| 编号 | 图形标志 | 名称 | 说明 |
|---|---|---|---|
| A.1 | | 禁止混放 | 为避免因化学品混放,引起一系列化学反应事故而设计。适用于化学品储存、使用的库房及相应工作场所 |
| A.2 | | 禁止乱动阀门 | 禁止乱动阀门为避免非操作人员开、关工艺管线阀门而设计。适用于油气厂、站及管道等工艺管线阀门 |
| A.3 | | 禁止乱动消防器材 | 为防止擅自挪用消防器材而设计。适用于放置消防器材的场地 |
| A.4 | | 禁止用汽油擦物 | 为防止用汽油等易燃液体擦洗物件时遇明火、高温、静电发生火灾而设计。适用于易燃易爆生产区域和机械设备较集中的岗位或场所 |
| A.5 | | 禁止酒后上岗 | 为防止因"酒后上岗作业",给正常的生产作业带来严重的安全隐患,甚至导致事故发生而设计。适用于油气开发等工作场所及施工区域 |
| A.6 | | 禁止混放 | 为避免因化学品混放,引起一系列化学反应事故而设计。适用于化学品储存、使用的库房及相应工作场所 |

续表

| 编号 | 图形标志 | 名称 | 说明 |
| --- | --- | --- | --- |
| A.7 | | 禁止撞击 | 为避免易燃易爆场所或盛装易燃易爆有毒有害介质的容器，因撞击产生变形泄漏、爆炸等严重危害而设计。适用于上述工作场所及相关容器的运输、储存及使用 |
| A.8 | | 禁止单扣吊装 | 为防止用单绳吊装造成吊件失衡引发事故而设计。适用于起重吊装作业场所 |
| A.9 | | 禁止吊臂下过人 | 为防止在起重吊装作业过程中，因吊装物坠落或吊臂滑落导致物体打击伤害而设计。适用于起重吊装作业现场 |
| A.10 | | 禁止擅自拆除 | 为防止在设备设施运行过程中，因未经许可擅自拆除安全防护设施造成的机械伤害而设计。适用于设备设施运行中具有安全防护设施的工作场所 |
| A.11 | | 禁止占用消防通道 | 为防止因消防通道被违规占用或阻挡造成严重后果而设计。适用于固定生产作业场所 |
| A.12 | | 禁止带"病"运行 | 为了防止因"设备设施带'病'作业"给正常的生产作业带来严重的安全隐患，甚至导致事故发生而设计。适用于油气开发等工作场所及施工区域 |

续表

| 编号 | 图形标志 | 名称 | 说明 |
|---|---|---|---|
| A.13 | | 禁止乱接设备 | 为防止因作业场所私搭乱接设备,导致过载、触电甚至火灾等事故而设计。适用于任何固定式或移动式配电设施 |
| A.14 | | 禁止带压作业 | 为防止生产场所因装置或管线带压,在作业过程中给人员或设备造成危害而设计。适用于工艺系统维修、改造等工作场所 |
| A.15 | | 禁止曝晒 | 为防止盛装易燃易爆介质的容器在阳光曝晒下,因内压增高,引发危险而设计。适用于室外盛放易燃易爆、有毒有害介质容器的存放及使用场所 |
| A.16 | | 禁止合闸 | 为防止在施工现场,因未经许可擅自合闸。适用于油气勘探、开发等工作场所 |
| A.17 | | 禁打手机 | 为防止在油气井施工现场,因静电产生火灾事故而设计。适用于油气勘探、开发等工作场所 |
| A.18 | | 禁止烟火 | 为防止在油气井施工现场,因使用火种发生火灾事故而设计。适用于油气勘探、开发等工作场所 |

续表

| 编号 | 图形标志 | 名称 | 说明 |
|---|---|---|---|
| A.19 | | 禁止带火种 | 为防止在油气井施工现场，因携带火种发生火灾事故而设计。适用于油气勘探、开发等工作场所 |
| A.20 | | 禁止吸烟 | 有甲、乙、丙类火灾危险物质的场所和禁止吸烟的公共场所等，如木工车间油漆车间、沥青车间、纺织厂、印染厂等 |
| A.21 | | 禁止用水灭火 | 生产、储运、使用中有不准用水灭火的物质的场所，如变压器室、乙炔站、化工药品库、各种油库等 |
| A.22 | | 禁止放置易燃物 | 具有明火设备或高温的作业场所，如动火区，各种焊接、切割、锻造、浇注车间等场所 |
| A.23 | | 禁止叉车和其他厂内机动车辆通行 | 禁止叉车和其他厂内机动车辆通行的场所 |
| A.24 | | 禁止乘人 | 乘人时易造成伤害的设施，如室外运输吊篮、外操作载货电梯框架等 |

续表

| 编号 | 图形标志 | 名称 | 说明 |
|---|---|---|---|
| A.25 | | 禁止靠近 | 不允许靠近的危险区域，如高压试验区、高压线、输变电设备的附近 |
| A.26 | | 禁止入内 | 易造成事故或对人员有伤害的场所如高压设备室、各种污染源等入口处 |
| A.27 | | 禁止推动 | 易于倾倒的装置或设备，如车站屏蔽门等 |
| A.28 | | 禁止停留 | 对人员具有直接危害的场所，如粉碎场地、危险路口、桥口等处 |
| A.29 | | 禁止通行 | 有危险的作业区，如起重、爆破现场道路施工工地等 |
| A.30 | | 禁止跨越 | 禁止跨越的危险地段，如专用的运输通道、带式输送机和其他作业流水线，作业现场的沟、坎、坑等 |

续表

| 编号 | 图形标志 | 名称 | 说明 |
|---|---|---|---|
| A.31 | | 禁止跳下 | 不允许跳下的危险地点，如深沟、深池、车站站台及盛装过有毒物质、易产生窒息气体的槽车、贮罐、地窖等处 |
| A.32 | | 禁止堆放 | 消防器材存放处，消防通道及车间主通道等 |
| A.33 | | 禁止启动 | 暂停使用的设备附近，如设备检修更换零件等 |
| A.34 | | 禁止倚靠 | 不能依靠的地点或部位，如列车车门车站屏蔽门、电梯轿门等 |
| A.35 | | 禁止坐卧 | 高温、腐蚀性、塌陷、坠落、翻转易损等易造成人员伤害的设备设施表面 |
| A.36 | | 禁止蹬踏 | 高温、腐蚀性、塌陷、坠落、翻转易损等易造成人员伤害的设备设施表面 |

续表

| 编号 | 图形标志 | 名称 | 说明 |
|---|---|---|---|
| A.37 | | 禁止触摸 | 禁止触摸的设备或物体附近，如裸露的带电体，炽热物体，具有毒性、腐蚀性物体等处 |
| A.38 | | 禁止伸入 | 易于夹住身体部位的装置或场所，如有开口的传动机、破碎机等 |
| A.39 | | 禁止饮用 | 禁止饮用水的开关处，如循环水、工业用水、污染水等 |
| A.40 | | 禁止抛物 | 抛物易伤人的地点，如高处作业现场深沟（坑）等 |
| A.41 | | 禁止戴手套 | 戴手套易造成手部伤害的作业地点如旋转的机械加工设备附近 |
| A.42 | | 禁止穿化纤服装 | 有静电火花会导致灾害或有炽热物质的作业场所，如冶炼、焊接及有易燃易爆物质的场所等 |

第六章 油气井工程相关的规范与标准
Chapter 6 Specifications and Standards Related to Oil and Gas Well Engineering

续表

| 编号 | 图形标志 | 名称 | 说明 |
|---|---|---|---|
| A.43 | | 禁止穿带钉鞋 | 有静电火花会导致灾害或有触电危险的作业场所，如有易燃易爆气体或粉尘的车间及带电作业场所 |
| A.44 | | 禁止携带金属物或手表 | 易受到金属物品干扰的微波和电磁场所，如磁共振室等 |
| A.45 | | 禁止携带托运易燃及易爆物品 | 不能携带和托运易燃、易爆物品及其他危险品的场所或交通工具，如火车、飞机、地铁等 |
| A.46 | | 禁止携带托运有毒物品及有害液体 | 不能携带托运有毒物品及有害液体的场所或交通工具，如火车、飞机、地铁等 |

Table B.1  Warning Sign

| Serial number | Symbol | Title | Description |
| --- | --- | --- | --- |
| B.1 |  | Mind the thunder and lightning | Designed to avoid the injury to the operator caused by lightning. Suitable for oil and gas development and other workplaces and construction zone |
| B.2 |  | Mind the tipping | It is designed to avoid the impact on the safety of personnel or equipment caused by the dumping of buildings, structures, equipments and facilities during production operations. Suitable for oil and gas field construction workplaces |
| B.3 |  | Mind the slippery slope | Designed to avoid the impact of landslides and mudslides on the safety of personnel or equipment during construction work near mountains and ditches. Suitable for construction sites such as mountains and ditches |
| B.4 |  | Mind the over pressure | Designed to remind operators to pay attention to the working pressure and prevent accidents from overpressure. Suitable for boilers, vessels and pipelines used under pressure |
| B.5 |  | Mind the iron filings | Designed to alert the human body to strike injuries caused by flying iron chips during machining operations. Suitable for workplaces in machining operations |
| B.6 |  | Mind the high-pressure pipeline | Designed to alert people in high-pressure pipeline areas to avoid accidents caused by deformation, breakage or rupture of pipelines due to operations. Suitable for any workplace where high pressure pipelines are used or where the work point is located in a high pressure pipeline area |

第六章　油气井工程相关的规范与标准
Chapter 6　Specifications and Standards Related to Oil and Gas Well Engineering

continued table

| Serial number | Symbol | Title | Description |
|---|---|---|---|
| B.7 | | Mind the spillover | Designed to prevent spillage from loading and unloading of oil products and oil field chemicals. Suitable for all places where oil products and oil field chemicals are loaded and unloaded |
| B.8 | | Mind the blowout | Designed to remind construction workers that the well control device should be operated reliably and safely to avoid blowout accidents, it is suitable for drilling, downhole work and other construction workplaces |
| B.9 | | Mind the mess | Designed to prevent accidents caused by tangled ropes and equipment malfunctions in installations and equipment with rollers, this product is suitable for use in production sites where installations and equipments with rollers are used |
| B.10 | | Mind the leak | Designed to prevent the leakage of medium in pipelines under abnormal conditions from causing injury or impact on personnel, equipment or working environment. Applicable to workplaces where pipelines are used to transmit media in oil and gas field development and production |
| B.11 | | Mind the mechanical injuries | Designed to alert to injuries caused to the human body during machining operations. Suitable for workplaces with machining operations |
| B.12 | | Mind the steam and hot water | Suitable for oil and gas exploration and development boiler room use workplace |

241

continued table

| Serial number | Symbol | Title | Description |
|---|---|---|---|
| B.13 | | Mind the corrosion | Suitable for oil and gas exploration, development of drilling fluid drugs and other dangerous chemicals used in the workplace |
| B.14 | | Mind the electric shock | Suitable for oil and gas exploration, development of power distribution room, power distribution cabinet and other use of the workplace |
| B.15 | | Mind the poison | Production, storage and transportation of highly toxic products and toxic substances (substances specified in item 1 of category 6 in GB 12268—2005) and the places where they are used |
| B.16 | | Mind the cable | Locations with exposed cables or construction at cables under the ground |
| B.17 | | Mind the auto-start | Equipment equipped with an automatic starting device |
| B.18 | | Mind the roof | Workplaces with a risk of overtopping, such as mine tunnels |

continued table

| Serial number | Symbol | Title | Description |
|---|---|---|---|
| B.19 | | Mind the potholes | Workplaces with potholes that are prone to injury, such as pre-drilled holes in members and over deep pits of various kinds |
| B.20 | | Mind the falling objects | Locations susceptible to falling hazards, such as underneath elevated work cubic crossings, etc. |
| B.21 | | Mind the head | There are places to generate encounters |
| B.22 | | Mind the crush | There are devices, equipment or places that generate extrusion, such as automatic doors, elevator doors and station screen doors |
| B.23 | | Mind the burns | Workplaces with heat sources prone to injury, such as smelting, forging, casting and heat-treatment shops |
| B.24 | | Mind your hands | Workplaces that are prone to hand injuries, such as glasswork, woodworking and machine shops |

continued table

| Serial number | Symbol | Title | Description |
|---|---|---|---|
| B.25 | | Mind the pinch | There are devices, equipment or places that generate extrusion, such as automatic doors, elevator doors, train car doors, etc. |
| B.26 | | Mind your feet | Workplaces that are prone to foot injuries, such as foundries, carpentry shops, construction sites and places with sharp-edged loose materials |
| B.27 | | Mind the arc light | Various welding workplaces where eye injuries are caused by arc light |
| B.28 | | Mind the hot surfaces | Places with burnt surfaces |
| B.29 | | Mind the hypothermia | Locations prone to frostbite, such as cold storage, gasifier surfaces, and locations where liquefied gases are present |
| B.30 | | Mind the forklift | Sites with forklift access |

第六章 油气井工程相关的规范与标准
Chapter 6 Specifications and Standards Related to Oil and Gas Well Engineering

continued table

| Serial number | Symbol | Title | Description |
|---|---|---|---|
| B.31 | | Mind the vehicle | Inside the factory vehicles, people mixed walking road, the corner of the road, level intersections; vehicles in and out of a large number of factory garages and other outlets |
| B.32 | | Mind the fall | Workplaces prone to fall accidents, such as scaffolding, elevated platforms, deep trenches (pools and troughs) on the ground, building construction and elevated workplaces |
| B.33 | | Mind the obstacles | There are obstacles on the ground, tripping over injury-prone locations |
| B.34 | | Mind the fall | Easy fall locations, such as stairs, steps, etc. |
| B.35 | | Mind the slip | The ground has slipping and falling places that can cause injury, such as oil, ice, water and other substances on the ground and slippery slopes |
| B.36 | | Mind the water | Places or areas where drowning is likely to occur after falling into water such as urban rivers, fire ponds, etc. |

continued table

| Serial number | Symbol | Title | Description |
|---|---|---|---|
| B.37 | | Mind the crevices | Installations, equipment or places with gaps, such as automatic doors, elevator doors, trains, etc. |

表 B.1 警告标志

| 编号 | 图形标志 | 名称 | 说明 |
|---|---|---|---|
| B.1 | | 当心雷电 | 为避免雷电对作业人员造成伤害而设计。适用于油气开发等工作场所及施工区城 |
| B.2 | | 当心倾倒 | 为避免在生产作业过程中，因建筑物、构筑物及设备、设施倾倒，给人员或设备安全造成影响而设计。适用于油气田建设施工作业场所 |
| B.3 | | 当心滑坡 | 为避免在山体、沟渠附近施工作业因滑坡、泥石流给人员或设备安全造成影响而设计。适用于山体、沟渠等施工作业场所 |
| B.4 | | 当心超压 | 为提醒操作人员注意工作压力，防止超压出现意外而设计。适用于带压使用的锅炉、容器及管线等 |

续表

| 编号 | 图形标志 | 名称 | 说明 |
|---|---|---|---|
| B.5 | | 当心铁屑伤人 | 为提醒在机械加工作业中，因铁屑飞溅对人体造成的打击伤害而设计。适用于机械加工作业的工作场所 |
| B.6 | | 当心高压管线 | 为提醒人们在高压管线区，避免因作业致使管线出现变形、破损或断裂造成事故而设计。适用于任何使用高压管线作业或作业点位于高压管线区的工作场所 |
| B.7 | | 当心外溢 | 为防止油品和油田化学剂装卸外溢而设计。适用于各油品、油田化学剂装卸的场所 |
| B.8 | | 当心井喷 | 为提醒施工作业人员注意井控装置应可靠并安全操作，避免井喷事故发生而设计适用于钻井、井下作业等施工作业场所 |
| B.9 | | 当心缠乱 | 为防止带有滚筒的装置、设备，因绳索缠乱而引发设备故障导致事故发生而设计适用于使用带有滚筒的装置、设备的生产作业场所 |
| B.10 | | 当心泄漏 | 为防止管道内介质在异常情况下泄漏对人员、设备或工作环境造成伤害或影响而设计。适用于油气田开发生产中使用管道传输介质的工作场所 |

续表

| 编号 | 图形标志 | 名称 | 说明 |
|------|----------|------|------|
| B.11 | | 当心机械伤人 | 为提醒在机械加工作业中，对人体造成的伤害而设计。适用于机械加工作业的工作场所 |
| B.12 | | 当心蒸汽和热水 | 适用于油气勘探、开发及锅炉房使用工作场所 |
| B.13 | | 当心腐蚀 | 适用于油气勘探、开发及钻井液药品等危险化学品使用工作场所 |
| B.14 | | 当心触电 | 适用于油气勘探、开发及配电房、配电柜等使用工作场所 |
| B.15 | | 当心中毒 | 剧毒品及有毒物质(GB 12268—2005 中第 6 类第 1 项所规定的物质)的生产储运及使用地点 |
| B.16 | | 当心电缆 | 有暴露的电缆或地面下有电缆处施工的地点 |

续表

| 编号 | 图形标志 | 名称 | 说明 |
|---|---|---|---|
| B.17 | | 当心自动启动 | 配有自动启动装置的设备 |
| B.18 | | 当心冒顶 | 具有冒顶危险的作业场所，如矿井隧道等 |
| B.19 | | 当心坑洞 | 具有坑洞易造成伤害的作业地点，如构件的预留孔洞及各种深坑的上方等 |
| B.20 | | 当心落物 | 易发生落物危险的地点，如高处作业立体交叉作业的下方等 |
| B.21 | | 当心碰头 | 有产生碰头的场所 |
| B.22 | | 当心挤压 | 有产生挤压的装置、设备或场所，如自动门、电梯门、车站屏蔽门等 |

续表

| 编号 | 图形标志 | 名称 | 说明 |
|---|---|---|---|
| B.23 | | 当心烫伤 | 具有热源易造成伤害的作业地点,如冶炼、锻造、铸造、热处理车间等 |
| B.24 | | 当心伤手 | 易造成手部伤害的作业地点,如玻璃制品、木制加工、机械加工的车间等 |
| B.25 | | 当心夹手 | 有产生挤压的装置、设备或场所,如自动门、电梯门、列车车门等 |
| B.26 | | 当心扎脚 | 易造成脚部伤害的作业地点,如铸造车间、木工车间、施工工地及有尖角散料等处 |
| B.27 | | 当心弧光 | 由于弧光造成眼部伤害的各种焊接作业场所 |
| B.28 | | 当心高温表面 | 有灼烫物体表面的场所 |

续表

| 编号 | 图形标志 | 名称 | 说明 |
| --- | --- | --- | --- |
| B.29 | | 当心低温 | 易于导致冻伤的场所，如冷库、气化器表面、存在液化气体的场所等 |
| B.30 | | 当心叉车 | 有叉车通行的场所 |
| B.31 | | 当心车辆 | 厂内车、人混合行走的路段，道路的拐角处，平交路口；车辆出入较多的厂房车库等出入口 |
| B.32 | | 当心坠落 | 易发生坠落事故的作业地点，如脚手架、高处平台、地面的深沟（池、槽）、建筑施工、高处作业场所等 |
| B.33 | | 当心障碍物 | 地面有障碍物，绊倒易造成伤害的地点 |
| B.34 | | 当心跌落 | 易于跌落的地点，如楼梯、台阶等 |

续表

| 编号 | 图形标志 | 名称 | 说明 |
|---|---|---|---|
| B.35 | | 当心滑倒 | 地面有易造成伤害的滑跌地点，如地面有油、冰、水等物质及滑坡处 |
| B.36 | | 当心落水 | 落水后有可能产生淹溺的场所或部位如城市河流、消防水池等 |
| B.37 | | 当心缝隙 | 有缝隙的装置、设备或场所，如自动门、电梯门、列车等 |

Table C.1　Order Sign

| Serial number | Symbol | Title | Description |
|---|---|---|---|
| C.1 | | Must wear protective gear | Designed for operators to wear personal labor protection equipments according to the specified requirements. Applicable to the relevant production workplaces |
| C.2 | | Explosion-proof equipment must be used | Designed to prevent explosions caused by static electricity generated by the use of tools in flammable and explosive places. Suitable for flammable and explosive places |

continued Table

| Serial number | Symbol | Title | Show |
|---|---|---|---|
| C.3 | | Shrouds must be installed | Designed to prevent the rotating parts of mechanical equipment from causing harm to people or objects during operation. Suitable for workplaces with transmission and rotating equipments |
| C.4 | | Fire hats must be worn | Designed to prevent motor vehicles from entering flammable and explosive areas and causing fire accidents due to sparks emitted from the exhaust pipe. Suitable for flammable and explosive workplaces |
| C.5 | | Must eliminate static electricity | Designed to eliminate static electricity from the human body before entering flammable and explosive places. Suitable for use in flammable and explosive workplaces |
| C.6 | | Safe voltage must be used | Designed to avoid injury to personnel due to unauthorized use of electricity in confined spaces, humid environments and other specific workplaces |
| C.7 | | Must be tested | Designed to avoid the impact of toxic, hazardous, flammable and explosive media on the safety of personnel or equipments in the premises. Suitable for workplaces where toxic, hazardous, flammable and explosive media gather |
| C.8 | | It has to be partitioned | Designed to avoid hazards to production operations caused by media in installations, vessels or pipelines |

continued Table

| Serial number | Symbol | Title | Show |
|---|---|---|---|
| C.9 | | Must be shut down for maintenance | Designed to avoid injury to personnel during maintenance or overhaul of equipments during operation |
| C.10 | | Dedicated lanes must be used | Designed to avoid injuries caused to personnel during production operations because they do not walk along the prescribed pathways |
| C.11 | | Must ignite before turning on the gas | Designed to prevent accidents caused by reversal of the ignition program in gas installations and other equipments. Suitable for gas installations |
| C.12 | | Must be ventilated | Designed to prevent the gathering of toxic and hazardous, flammable and explosive gases and dusts in the workplace, which may cause serious hazards to the safety of personnel and equipments |
| C.13 | | Must wear protective gloves | Designed to avoid injury to personnel who do not use protective equipment according to the regulations in the process of production and operation due to the use of chemicals. Suitable for oil and gas exploration, development drilling fluid drugs and other dangerous chemicals used in the workplaces |
| C.14 | | Dust masks must be worn | Designed to avoid injury to personnel who do not use protective equipment according to the regulations in the process of production and operation due to the use of chemicals. Suitable for oil and gas exploration, development drilling fluid drugs and other dangerous chemicals used in the workplaces |

continued Table

| Serial number | Symbol | Title | Show |
|---|---|---|---|
| C.15 | | Must wear ear protection | Designed to avoid injury to personnel who do not use protective equipment according to the regulations in the process of production and operation due to the use of chemicals. Suitable for oil and gas exploration, development drilling fluid drugs and other dangerous chemicals used in the workplaces |
| C.16 | | Must wear goggles | Designed to avoid injury to personnel who do not use protective equipment according to the regulations in the process of production and operation due to the use of chemicals. Suitable for oil and gas exploration, development drilling fluid drugs and other dangerous chemicals used in the workplaces |
| C.17 | | Protective clothing must be worn | Workplaces with radiation, microwaves, high temperatures and others requiring protective clothing |
| C.18 | | Must wear protective gloves | Workplaces that are prone to hand injuries, such as locations with corrosive, contaminated, burning, freezing and electrocution hazards |
| C.19 | | Protective shoes must be worn | Workplaces that are prone to foot injuries, such as workplaces with corrosion, burns, electric shock, smashing (stabbing) and other hazards |
| C.20 | | Seat belts must be worn | Designed for the use of safety belts by operators of elevated operations in accordance with the specified requirements. Applicable to the relevant production workplaces |

continued Table

| Serial number | Symbol | Title | Show |
| --- | --- | --- | --- |
| C.21 | | Hold the handrail when going up and down the ladder | Suitable for use in relevant production workplaces |
| C.22 | | Must keep a safe distance | Suitable for workshops and construction sites where electrical welding is used |
| C.23 | | Lifting must be directed | Suitable for lifting work sites |
| C.24 | | Must wash hands | After relieving toxic and hazardous materials operations |
| C.25 | | Must be locked | Locations such as highly toxic and hazardous materials warehouses |
| C.26 | | Must be grounded | Lightning protection and anti-static places |

第六章　油气井工程相关的规范与标准
Chapter 6　Specifications and Standards Related to Oil and Gas Well Engineering

continued Table

| Serial number | Symbol | Title | Show |
|---|---|---|---|
| C.27 | | Must be unplugged | In case of equipment maintenance, malfunction, prolonged deactivation and unattended state |

表 C.1　指令标志

| 编号 | 图形标志 | 名称 | 说明 |
|---|---|---|---|
| C.1 | | 必须穿戴防护用品 | 为操作人员按规定要求，穿戴个人劳动防护用品而设计。适用于有关的生产作业场所 |
| C.2 | | 必须用防爆器具 | 为防止在易燃易爆场所因使用工具产生的静电导致爆炸事故而设计。适用于易燃易爆等场所 |
| C.3 | | 必须装设护罩 | 为防止机械设备运转时，转动部位对人或物造成危害而设计。适用于有传动、转动设备的工作场所 |
| C.4 | | 必须戴防火帽 | 为防止机动车辆进入易燃易爆区，因排气管喷出火星引起火灾事故而设计。适用于易燃易爆工作场所 |

257

续表

| 编号 | 图形标志 | 名称 | 说明 |
|---|---|---|---|
| C.5 | | 必须消除静电 | 为进入易燃易爆场所前消除人体静电而设计。适用于易燃易爆工作场所 |
| C.6 | | 必须使用安全电压 | 为避免在密闭空间，潮湿环境等特定工作场所内违规用电，给人员造成伤害而设计 |
| C.7 | | 必须检测 | 为避免场所内有毒有害、易燃易爆介质对人员或设备安全造成影响而设计。适用于有毒有害、易燃易爆介质聚集的工作场所 |
| C.8 | | 必须隔断 | 为避免装置、容器或管线内介质对生产作业造成危害而设计 |
| C.9 | | 必须停机检修 | 为避免在设备运行过程中对其维护或检修给人员造成伤害而设计 |
| C.10 | | 必须使用专用通道 | 为避免生产作业过程中，因人员不按规定通道行走，对其造成伤害而设计 |

续表

| 编号 | 图形标志 | 名称 | 说明 |
|---|---|---|---|
| C.11 | | 必须先点火后开气 | 为防止燃气装置等设备，因点火程序颠倒引发事故而设计。适用于燃气装置 |
| C.12 | | 必须通风 | 为防止工作场所内有毒有害，易燃易爆气体及粉尘聚集，对人员及设备安全造成严重危害而设计 |
| C.13 | | 必须戴防护手套 | 为避免生产作业过程中，因使用化学药品，人员不按规定使用防护用品，对其造成伤害而设计。适用于油气勘探、开发及钻井液药品等危险化学品使用工作场所 |
| C.14 | | 必须戴防尘口罩 | 为避免生产作业过程中，因使用化学药品，人员不按规定使用防护用品，对其造成伤害而设计。适用于油气勘探、开发及钻井液药品等危险化学品使用工作场所 |
| C.15 | | 必须戴护耳器 | 为避免生产作业过程中，因使用化学药品，人员不按规定使用防护用品，对其造成伤害而设计。适用于油气勘探、开发及钻井液药品等危险化学品使用工作场所 |
| C.16 | | 必须戴护目镜 | 为避免生产作业过程中，因使用化学药品，人员不按规定使用防护用品，对其造成伤害而设计。适用于油气勘探、开发及钻井液药品等危险化学品使用工作场所 |

续表

| 编号 | 图形标志 | 名称 | 说明 |
| --- | --- | --- | --- |
| C.17 | | 必须穿防护服 | 具有放射、微波、高温及其他需穿防护服的作业场所 |
| C.18 | | 必须戴防护手套 | 手部易受伤害的作业场所，如具有腐蚀、污染、灼烫、冰冻及触电危险的作业等地点 |
| C.19 | | 必须穿防护鞋 | 脚部易受伤害的作业场所，如具有腐蚀、灼烫、触电、砸（刺）伤等危险的作业地点 |
| C.20 | | 必须系安全带 | 为登高作业操作人员按规定要求，使用安全带而设计。适用于有关的生产作业场所 |
| C.21 | | 上下梯子扶好扶手 | 适用于有关的生产作业场所 |
| C.22 | | 必须保持安全距离 | 适用于车间、施工现场动用电气焊作业现场 |

续表

| 编号 | 图形标志 | 名称 | 说明 |
|---|---|---|---|
| C.23 | | 起吊必须有人指挥 | 适用于吊装作业现场 |
| C.24 | | 必须洗手 | 解除有毒有害物质作业后 |
| C.25 | | 必须加锁 | 剧毒品、危险品库房等地点 |
| C.26 | | 必须接地 | 防雷、防静电场所 |
| C.27 | | 必须拔出插头 | 在设备维修、故障、长期停用、无人值守状态下 |

Table D.1 Prompt Sign

| serial number | symbol | title | description |
| --- | --- | --- | --- |
| D.1 | | Emergency exit | Emergency exits for safe evacuation, combined with directional arrows at stairways leading to emergency exits, etc. |
| D.2 | | | |
| D.3 | | Refuge | Hazardous hiding places in railroad bridges, road bridges, mines and tunnels |
| D.4 | | Refuge | Hazardous hiding places in railroad bridges, road bridges, mines and tunnels |
| D.5 | | Access route | Designed according to the need to ensure safe passage of people in production workplaces |

continued table

| serial number | symbol | title | description |
|---|---|---|---|
| D.6 | | Escape routes | Designed in accordance with the emergency response needs of emergency escape and evacuation of personnel in case of emergencies (accidents) |
| D.7 | | Underground pipeline | Designed to meet the safety needs of buried pipelines in oil fields |
| D.8 | | Keep away from danger. | Suitable for oil and gas field exploration and development with sand sinking pits, sewage pits and other construction work sites |
| D.9 | | Emergency shelter | Designed in accordance with the emergency response needs of emergency escape and evacuation of personnel in case of emergencies (accidents) |
| D.10 | | Flammable area | Places where open fires may be used, as designated by the relevant authorities |
| D.11 | | Crushed plate | Must break the surface to gain access to the exit |

表 D.1 提示标志

| 编号 | 图形标志 | 名称 | 说明 |
| --- | --- | --- | --- |
| D.1 | | 紧急出口 | 便于安全疏散的紧急出口处，与方向箭头结合设在通向紧急出口的通道楼梯口等处 |
| D.2 | | | |
| D.3 | | 避险处 | 铁路桥、公路桥、矿井及隧道内躲避危险的地点 |
| D.4 | | 避险处 | 铁路桥、公路桥、矿井及隧道内躲避危险的地点 |
| D.5 | | 通行路线 | 根据生产作业场所内确保人员安全通行需要而设计 |

续表

| 编号 | 图形标志 | 名称 | 说明 |
|---|---|---|---|
| D.6 | | 逃生路线 | 根据突发事件(事故)人员应急逃生、疏散的应急处置需要而设计 |
| D.7 | | 地下管线 | 根据油田埋地管线的安全生产需要而设计 |
| D.8 | | 危险请勿靠近 | 适用于油气田勘探、开发有沉砂坑、污水坑等施工作业现场 |
| D.9 | | 紧急集合地 | 根据突发事件(事故)人员应急逃生、疏散的应急处置需要而设计 |
| D.10 | | 可动火区 | 经有关部门划定的可使用明火的地点 |
| D.11 | | 击碎板面 | 必须击开板面才能获得出口 |

## Post-class exercises

Eliminate venting, minimize flaring and other emissions

Target zero venting and minimal flaring of natural gas during well completion and seek to reduce fugitive and vented greenhouse-gas emissions during the entire productive life of a well. Best practice is to recover and market gas produced during the completion phase of a well, and public authorities need to consider imposing restrictions on venting and flaring and specific requirements for installing equipment to help minimize emissions. Measures in this area will also lower emissions of conventional pollutants, including VOCs. Operators should consider setting targets on emissions as part of their overall strategic policies to win public confidence that they are acting to minimize the environmental impact of their activities, taking into account the financial benefits of commercializing the gas that would otherwise: be vented or fared, The gas industry as a whole, including conventional gas producers and companies operating in the midstream and down-stream, needs to demonstrate that they are just as concerned by methane emissions beyond the production stage, for example in transportation and distribution.

Minimize air pollution from vehicles, drilling rig engines, pump engines and compressors. Pollution from vehicles and equipment is often controlled by existing environmental and fuel-efficiency standards (it is the responsibility of governments to ensure that appropriate standards are in place). Operators and service providers should consider the advantages of deploying the cleanest vehicles and equipment available, for example, electric vehicles and gas-powered rig engines , to reduce both local air and noise pollution.

## 课后练习

消除排放、最小化烧气和其他排放

目标是在井完井期间实现零排放天然气，并尽可能减少整个生产生命周期中井的泄漏和排放温室气体。最佳做法是在完井阶段回收和销售产生的天然气，公共当局需要考虑对排放和烧气施加限制，并制定具体要求以帮助减少排放。在这个领域的措施还将降低常规污染物排放，包括挥发性有机化合物。运营商应考虑将排放目标纳入其整体战略政策，以赢得公众对他们在最小化其活动环境影响方面的行动的信心，考虑到商业化否则会排放或烧掉的天然气所带来的财务利益。整个天然气行业，包括常规天然气生产商和从事中游和下游运营的公司，需要证明他们同样关注生产阶段以外（例如在运输和分配中）的甲烷排放。

最小化车辆、钻机发动机、泵发动机和压缩机的空气污染。车辆和设备的污染通常通过现有的环保和燃油效率标准来控制（确保适当的标准是政府的责任）。运营商和服务提供商应考虑部署最清洁的车辆和设备的优势，例如电动汽车和燃气驱动的钻机发动机，以减少当地空气和噪声污染。

# Chapter 7 Laws and Regulations Related to Oil and Gas Well Engineering

The important laws followed by China's petroleum and petrochemical industry include the *People's Republic of China Work Safety Law* and *the People's Republic of China Environmental Protection* Law. In the field of oil and gas well engineering, *the People's Republic of China Work Safety Law* is particularly important. Section 1 of this chapter extracts parts of Chapters 1、2、3 and 5 of the People's Republic of China Work Safety Law for reference. Section 2 of this chapter excerpts part of the *Unified Regulations on the Rational and Comprehensive Utilization of Mineral Resources* in the laws and regulations of the oil and gas industry in Kazakhstan in the field of oil and gas well engineering; Section 3 of this chapter excerpts some of the laws and regulations of the oil and gas industry in Russia that deal with the field of well engineering in the regulations on oil and gas field development.

# 第七章 油气井工程相关的法律法规

我国石油石化行业遵循的重要法律有《中华人民共和国安全生产法》《中华人民共和国环境保护法》。油气井工程领域,《中华人民共和国安全生产法》尤为重要。本章第一节摘取《中华人民共和国安全生产法》第一、二、三、五章的部分内容以供参阅。本章第二节摘录了哈萨克斯坦油气行业法律法规中《矿产资源合理和综合利用统一规定》中涉及油气井工程领域的部分内容；本章第三节摘录了俄罗斯油气行业法律法规中油气田开发法规中涉及油气井工程领域的部分内容。

## 7.1 Work Safety Law of the People's Republic of China (Excerpt)

### 7.1.1 Chapter 1 General Provisions

Article 1: This Law is formulated so as to strengthen work safety, prevent and reduce production safety accidents, ensure the safety of people's lives and property, and promote sustained and healthy economic and social development.

Article 2: This Law shall apply to the production safety of units engaged in production and business activities within the territory of the People's Republic of China (hereinafter collectively referred to as production and business units); Where relevant laws and administrative regulations have other provisions on fire safety, road traffic safety, railway traffic safety, water traffic safety, civil aviation safety, nuclear and radiation safety, and special equipment safety, those provisions shall apply. Article 3 Work safety adheres to the leadership of the Communist Party of China.

Article 3: Work safety adheres to the leadership of the Communist Party of China.

Work safety should be people-oriented, adhere to the supremacy of the people and the supremacy of life, put the protection of people's lives and safety in the first place, firmly establish the concept of safe development, adhere to the principle of safety first, prevention first, and comprehensive management, and prevent and resolve major safety risks from the source.

## 7.1 《中华人民共和国安全生产法》(节选)

### 7.1.1 第一章 总则

第一条 为了加强安全生产工作，防止和减少生产安全事故，保障人民群众生命和财产安全，促进经济社会持续健康发展，制定本法。

第二条 在中华人民共和国领域内从事生产经营活动的单位（以下统称生产经营单位）的安全生产，适用本法；有关法律、行政法规对消防安全和道路交通安全、铁路交通安全、水上交通安全、民用航空安全以及核与辐射安全、特种设备安全另有规定的，适用其规定。

第三条 安全生产工作坚持中国共产党的领导。

安全生产工作应当以人为本，坚持人民至上、生命至上，把保护人民生命安全摆在首位，树牢安全发展理念，坚持安全第一、预防为主、综合治理的方针，从源头上防范化解重大安全风险。

The implementation of safety production work must manage the safety of the industry, the safety of the business, and the safety of production and operation, strengthen and implement the main responsibility of production and business operation units and government supervision responsibilities, and establish a mechanism for production and business operation units to be responsible, employees to participate, government supervision, industry self-discipline and social supervision.

Article 4: Production and business operation entities must comply with this Law and other laws and regulations related to production safety, strengthen the management of production safety, establish and improve the responsibility system for production safety and the rules and regulations for production safety, increase the investment in production safety funds, materials, technology and personnel, improve production safety conditions, strengthen the standardization and informatization of production safety, build a dual prevention mechanism for hierarchical control of safety risks and investigation and management of hidden dangers, improve the risk prevention and resolution mechanism, improve the level of production safety, and ensure production safety.

### 7.1.2 Chapter 2 Production and Business Operation Units of Safety Production Guarantees

Article 24: Mining, metal smelting, construction and transportation units, as well as units of producing, operating, storing, loading and unloading dangerous goods, shall set up production safety management institutions or

安全生产工作实行管行业必须管安全、管业务必须管安全、管生产经营必须管安全，强化和落实生产经营单位主体责任与政府监管责任，建立生产经营单位负责、职工参与、政府监管、行业自律和社会监督的机制。

第四条　生产经营单位必须遵守本法和其他有关安全生产的法律、法规，加强安全生产管理，建立健全全员安全生产责任制和安全生产规章制度，加大对安全生产资金、物资、技术、人员的投入保障力度，改善安全生产条件，加强安全生产标准化、信息化建设，构建安全风险分级管控和隐患排查治理双重预防机制，健全风险防范化解机制，提高安全生产水平，确保安全生产。

### 7.1.2 第二章 生产经营单位的安全生产保障

第二十四条　矿山、金属冶炼、建筑施工、运输单位和危险物品的生产、经营、储存、装卸单位，应当设置安全生产管理机构或者配备专职安全生产管理人员。

assign full-time production safety management personnel.

Article 27: The main person in charge of the production and business operation entity and the safety production management personnel must have the safety production knowledge and management ability corresponding to the production and business activities engaged in by the unit.

The main responsible persons and safety production management personnel of the units of production, operation, storage, loading and unloading dangerous goods, as well as the mining, metal smelting, construction and transportation units, should be qualified by the competent department responsible for the supervision and management of production safety and their safety production knowledge and management ability. There is no charge for the assessment.

Units engaged in the production, storage, loading and unloading of dangerous goods, as well as mines and metal smelting units, shall have registered safety engineers engaged in safety production management. Encourage other production and business operation units to employ registered safety engineers to engage in safety production management. Registered safety engineers should be managed according to professional classification, and the specific measures should be formulated by the human resources and social security department of the State Council and the emergency management department of the State Council in conjunction with the relevant departments of the State Council.

Article 28: Production and business operation entities shall carry out education and training on production safety for employees,

第二十七条　生产经营单位的主要负责人和安全生产管理人员必须具备与本单位所从事的生产经营活动相应的安全生产知识和管理能力。

危险物品的生产、经营、储存、装卸单位以及矿山、金属冶炼、建筑施工、运输单位的主要负责人和安全生产管理人员，应当由主管的负有安全生产监督管理职责的部门对其安全生产知识和管理能力考核合格。考核不得收费。

危险物品的生产、储存、装卸单位以及矿山、金属冶炼单位应当有注册安全工程师从事安全生产管理工作。鼓励其他生产经营单位聘用注册安全工程师从事安全生产管理工作。注册安全工程师按专业分类管理，具体办法由国务院人力资源和社会保障部门、国务院应急管理部门会同国务院有关部门制定。

第二十八条　生产经营单位应当对从业人员进行安全生产教育和培训，保证从业人员具备必要的安全生产知识，熟悉有关的安全生产规章制度和安全操作规程，掌握本岗位的安全操作技能，了解事故应急处理措施，知悉自身在安全生产方面的权利和义务。未经安全生产教育和培训合格的从业人员，不得上岗作业。

ensure that employees shall have the necessary knowledge of production safety, and familiar with the relevant production safety rules and regulations and safety operation procedures, master the safety operation skills of their positions, understand the emergency response measures for accidents, and know their rights and obligations in production safety. Employees who have not been educated and trained in safety production shall not be allowed to work.

If a production and business operation entity uses dispatched workers, it shall include the dispatched workers in the unified management of its employees, and educate and train the dispatched workers on post safety operation procedures and safe operation skills. The labor dispatch unit shall provide the dispatched workers with necessary education and training on work safety.

If a production and business operation entity accepts students from secondary vocational schools and institutions of higher learning for internships, it shall conduct corresponding safety production education and training for the intern students and provide necessary labor protection articles. The school shall assist the production and business operation units to carry out safety production education and training for intern students.

Production and business operation entities shall establish safety production education and training files, truthfully record the time, content, participants and assessment results of safety production education and training.

Article 29: Production and business operation entities that adopt new processes, new technologies, new materials or use new equipment must understand and master their

生产经营单位使用被派遣劳动者的，应当将被派遣劳动者纳入本单位从业人员统一管理，对被派遣劳动者进行岗位安全操作规程和安全操作技能的教育和培训。劳务派遣单位应当对被派遣劳动者进行必要的安全生产教育和培训。

生产经营单位接收中等职业学校、高等学校学生实习的，应当对实习学生进行相应的安全生产教育和培训，提供必要的劳动防护用品。学校应当协助生产经营单位对实习学生进行安全生产教育和培训。

生产经营单位应当建立安全生产教育和培训档案，如实记录安全生产教育和培训的时间、内容、参加人员以及考核结果等情况。

第二十九条　生产经营单位采用新工艺、新技术、新材料或者使用新设备，必须了解、掌握其安全技术特性，采取有效的安全防护措施，并对从业人员进行专门的安全生产教育和培训。

safety technical characteristics, take effective safety protection measures, and conduct special safety production education and training for employees.

Article 30: The special operations personnel of the production and business operation units must undergo special safety operation training and obtain the corresponding qualifications in accordance with the relevant provisions of the State before they can take up their posts.

The scope of special operations personnel is to be determined by the emergency management department of the State Council in conjunction with the relevant departments of the State Council.

Article 31: The safety facilities of the new, reconstructed and expanded projects (hereinafter referred to as construction projects) of production and business operation entities must be designed, constructed, and put into production and use at the same time as the main project. Investment in safety facilities shall be included in the estimated budget of the construction project.

Article 32: Mine and metal smelting construction projects and construction projects used for the production, storage, loading and unloading of dangerous goods should be subject to safety assessment in accordance with the relevant provisions of the State.

Article 33: The designer and design unit of the safety facilities of the construction project should be responsible for the design of the safety facilities. The design of safety facilities for mines, metal smelting construction projects and construction projects for the production, storage, loading and unloading of dangerous goods should be submitted to the relevant

第三十条　生产经营单位的特种作业人员必须按照国家有关规定经专门的安全作业培训，取得相应资格，方可上岗作业。

特种作业人员的范围由国务院应急管理部门会同国务院有关部门确定。

第三十一条　生产经营单位新建、改建、扩建工程项目（以下统称建设项目）的安全设施，必须与主体工程同时设计、同时施工、同时投入生产和使用。安全设施投资应当纳入建设项目概算。

第三十二条　矿山、金属冶炼建设项目和用于生产、储存、装卸危险物品的建设项目，应当按照国家有关规定进行安全评价。

第三十三条　建设项目安全设施的设计人、设计单位应当对安全设施设计负责。矿山、金属冶炼建设项目和用于生产、储存、装卸危险物品的建设项目的安全设施设计应当按照国家有关规定报经有关部门审查，审查部门及其负责审查的人员对审查结果负责。

departments for review in accordance with the relevant provisions of the State, and the review department and its personnel responsible for the review should be responsible for the results of the review.

Article 34: The construction units of mines, metal smelting construction projects and construction projects used for the production, storage, loading and unloading of dangerous goods shall design and construct the safety facilities in accordance with the approved regulations, and should be responsible for the engineering quality of the safety facilities.

Before the construction project of mine and metal smelting and the construction project for the production, storage, loading and unloading of dangerous goods are completed and put into production or use, the construction unit should be responsible for organizing the acceptance of the safety facilities; After passing the acceptance, it can be put into production and use. The departments responsible for the supervision and management of production safety should strengthen the supervision and verification of the acceptance activities and acceptance results of the construction unit.

Article 35: Production and business operation entities shall set up obvious safety warning signs in production and business operation sites and related facilities and equipment with relatively large risk factors.

Article 36: The design, manufacture, installation, use, testing, maintenance, modification and scrapping of safety equipments shall conform to national standards or industry standards.

Production and business operation units must carry out regular maintenance and maintenance of safety equipment, and

第三十四条 矿山、金属冶炼建设项目和用于生产、储存、装卸危险物品的建设项目的施工单位必须按照批准的安全设施设计施工，并对安全设施的工程质量负责。

矿山、金属冶炼建设项目和用于生产、储存、装卸危险物品的建设项目竣工投入生产或者使用前，应当由建设单位负责组织对安全设施进行验收；验收合格后，方可投入生产和使用。负有安全生产监督管理职责的部门应当加强对建设单位验收活动和验收结果的监督核查。

第三十五条 生产经营单位应当在有较大危险因素的生产经营场所和有关设施、设备上，设置明显的安全警示标志。

第三十六条 安全设备的设计、制造、安装、使用、检测、维修、改造和报废，应当符合国家标准或者行业标准。

生产经营单位必须对安全设备进行经常性维护、保养，并定期检测，保证正常运转。维护、保养、检测应当做好记录，并由有关人员签字。

生产经营单位不得关闭、破坏直接关系生产安全的监控、报警、防护、救生设备、设施，或者篡改、隐瞒、销毁其相关数据、信息。

regularly inspect to ensure normal operation. Maintenance, maintenance, and testing should be recorded and signed by relevant personnel.

If the production and business operation units in the catering and other industries use gas, they shall install combustible gas alarm devices and ensure their normal use.

Article 37: The containers and means of transport of dangerous goods used by production and business operation entities, as well as special equipments for offshore oil exploitation and special equipments for underground mines that involve personal safety and are more dangerous, must be produced by professional production units in accordance with the relevant provisions of the State, and should be tested and inspected by a testing and inspection agency with professional qualifications, and shall obtain a safety use certificate or safety mark before they can be put into use. Testing and inspection institutions are responsible for the results of testing and inspection.

Article 38: The State implements a system of elimination of processes and equipments that seriously endanger production safety, and the specific catalogue should be formulated and published by the emergency management department of the State Council in conjunction with the relevant departments of the State Council. Where laws and administrative regulations have other provisions on the formulation of catalogs, apply those provisions.

The people's governments of provinces, autonomous regions, and municipalities directly under the Central Government, on the basis of the actual conditions of their respective regions, formulate and publish specific catalogs, and eliminate processes and equipment that endanger production safety

餐饮等行业的生产经营单位使用燃气的，应当安装可燃气体报警装置，并保障其正常使用。

第三十七条　生产经营单位使用的危险物品的容器、运输工具，以及涉及人身安全、危险性较大的海洋石油开采特种设备和矿山井下特种设备，必须按照国家有关规定，由专业生产单位生产，并经具有专业资质的检测、检验机构检测、检验合格，取得安全使用证或者安全标志，方可投入使用。检测、检验机构对检测、检验结果负责。

第三十八条　国家对严重危及生产安全的工艺、设备实行淘汰制度，具体目录由国务院应急管理部门会同国务院有关部门制定并公布。法律、行政法规对目录的制定另有规定的，适用其规定。

省、自治区、直辖市人民政府可以根据本地区实际情况制定并公布具体目录，对前款规定以外的危及生产安全的工艺、设备予以淘汰。

other than those provided for in the preceding paragraph.

Production and business operation entities should not use the process and equipments that should be eliminated and endanger production safety.

Article 41: Production and business operation entities shall establish a hierarchical management and control system for safety risks, and take corresponding control measures in accordance with the classification of safety risks.

Production and business operation entities shall establish, improve and implement a system for the investigation and management of hidden dangers of production safety accidents, and take technical and management measures to discover and eliminate potential accidents in a timely manner. The investigation and management of potential accidents should be truthfully recorded, and the employees should be notified through the staff meeting or the workers'congress, information bulletin boards, etc. Among them, the investigation and management of major accidents shall be reported to the department responsible for the supervision and management of production safety and the staff meeting or the workers'congress in a timely manner.

The departments responsible for the supervision and management of production safety of local people's governments at or above the county level shall incorporate major accident hazards into the relevant information system, establish and improve the supervision system for the management of major accident hazards, and urge production and business operation units to eliminate major accident hazards.

生产经营单位不得使用应当淘汰的危及生产安全的工艺、设备。

第四十一条　生产经营单位应当建立安全风险分级管控制度，按照安全风险分级采取相应的管控措施。

生产经营单位应当建立健全并落实生产安全事故隐患排查治理制度，采取技术、管理措施，及时发现并消除事故隐患。事故隐患排查治理情况应当如实记录，并通过职工大会或者职工代表大会、信息公示栏等方式向从业人员通报。其中，重大事故隐患排查治理情况应当及时向负有安全生产监督管理职责的部门和职工大会或者职工代表大会报告。

县级以上地方各级人民政府负有安全生产监督管理职责的部门应当将重大事故隐患纳入相关信息系统，建立健全重大事故隐患治理督办制度，督促生产经营单位消除重大事故隐患。

Article 42: Workshops, shops and warehouses that produce, operate, store and use dangerous goods should not be in the same building as the staff dormitory, and shall keep a safe distance from the staff dormitory.

Production and business premises and staff dormitories should be equipped with exits and evacuation passages that meet the requirements of emergency evacuation, are clearly marked and kept unblocked. It is forbidden to occupy, lock or block the exits and evacuation channels of production and business premises or employee dormitories.

Article 43: Production and business operation entities carrying out blasting, hoisting, ignition, temporary electricity and other dangerous operations stipulated by the emergency management department of the State Council in conjunction with the relevant departments of the State Council shall arrange special personnel to carry out on-site safety management to ensure compliance with operating procedures and the implementation of safety measures.

Article 44: Production and business operation entities shall educate and supervise employees to strictly implement the rules and regulations on production safety and safety operation procedures of their own units; And truthfully inform employees of the risk factors, preventive measures and accident emergency measures existing in the workplace.

Production and business operation entities shall pay attention to the physical and psychological conditions and behavioral habits of employees, strengthen psychological counseling and spiritual comfort for employees, strictly implement job safety production responsibilities, and prevent accidents caused by abnormal behavior of employees.

第四十二条　生产、经营、储存、使用危险物品的车间、商店、仓库不得与员工宿舍在同一座建筑物内，并应当与员工宿舍保持安全距离。

生产经营场所和员工宿舍应当设有符合紧急疏散要求、标志明显、保持畅通的出口、疏散通道。禁止占用、锁闭、封堵生产经营场所或者员工宿舍的出口、疏散通道。

第四十三条　生产经营单位进行爆破、吊装、动火、临时用电以及国务院应急管理部门会同国务院有关部门规定的其他危险作业，应当安排专门人员进行现场安全管理，确保操作规程的遵守和安全措施的落实。

第四十四条　生产经营单位应当教育和督促从业人员严格执行本单位的安全生产规章制度和安全操作规程；并向从业人员如实告知作业场所和工作岗位存在的危险因素、防范措施以及事故应急措施。

生产经营单位应当关注从业人员的身体、心理状况和行为习惯，加强对从业人员的心理疏导、精神慰藉，严格落实岗位安全生产责任，防范从业人员行为异常导致事故发生。

Article 45: Production and business operation entities must provide employees with labor protection articles that meet national or industry standards, and supervise and educate employees to wear and use them in accordance with the rules of use.

Article 46: The production safety management personnel of the production and business operation entity, shall in accordance with the production and operation characteristics of the units, conduct regular inspections on the production safety status; Safety issues discovered during the inspection shall be dealt with immediately; and where it cannot be handled, it should be promptly reported to the relevant responsible person of the unit, and the relevant responsible person shall promptly handle it. The circumstances of the inspection and disposition should be truthfully recorded in the case file.

Production and business operation units of production safety management personnel found in the inspection of major accidents, in accordance with the provisions of the preceding paragraph to the relevant responsible person of the unit report, the relevant person in charge of the timely processing, production safety management personnel may be in charge of the responsibility for the supervision and management of production safety of the department report, received the report of the department shall be dealt with in a timely manner in accordance with the law.

Article 47: Production and business operation entities shall arrange funds for the provision of labor protection articles and safety production training.

Article 48: Where two or more production and business operation entities carry out production and business activities in the same operation areas and may endanger the

第四十五条 生产经营单位必须为从业人员提供符合国家标准或者行业标准的劳动防护用品，并监督、教育从业人员按照使用规则佩戴、使用。

第四十六条 生产经营单位的安全生产管理人员应当根据本单位的生产经营特点，对安全生产状况进行经常性检查；对检查中发现的安全问题，应当立即处理；不能处理的，应当及时报告本单位有关负责人，有关负责人应当及时处理。检查及处理情况应当如实记录在案。

生产经营单位的安全生产管理人员在检查中发现重大事故隐患，依照前款规定向本单位有关负责人报告，有关负责人不及时处理的，安全生产管理人员可以向主管的负有安全生产监督管理职责的部门报告，接到报告的部门应当依法及时处理。

第四十七条 生产经营单位应当安排用于配备劳动防护用品、进行安全生产培训的经费。

第四十八条 两个以上生产经营单位在同一作业区域内进行生产经营活动，可能危及对方生产安全的，应当签订安全生产管理协议，明确各自的安全生产管理职责和应当采取的安全措施，并指定专职安全生产管理人员进行安全检查与协调。

production safety of the other party, they shall sign a production safety management agreement, clarify their respective production safety management responsibilities and safety measures to be taken, and designate full-time production safety management personnel to conduct safety inspections and coordination.

Article 49: Production and business operation entities shall not contract or lease production and business operation projects, sites and equipments to units or individuals that do not have the conditions for safe production or the corresponding qualifications.

If a production or business operation project or site is contracted or leased to other units, the production and business operation entity shall sign a special safety production management agreement with the contracting unit or the lessee unit, or stipulate their respective safety production management responsibilities in the contract or lease contract; The production and business operation entity shall coordinate and manage the safety production work of the contractor and the tenant in a unified manner, conduct regular safety inspections, and if safety problems are found, they shall promptly supervise and urge rectification.

The construction units of mines, metal smelting construction projects and construction projects used for the production, storage, loading and unloading of dangerous goods shall strengthen the safety management of the construction projects, and shall not resell, lease, lend, attach or otherwise illegally transfer construction qualifications, and shall not subcontract all the construction projects contracted by them to a third party or subcontract all the construction projects

第四十九条　生产经营单位不得将生产经营项目、场所、设备发包或者出租给不具备安全生产条件或者相应资质的单位或者个人。

生产经营项目、场所发包或者出租给其他单位的，生产经营单位应当与承包单位、承租单位签订专门的安全生产管理协议，或者在承包合同、租赁合同中约定各自的安全生产管理职责；生产经营单位对承包单位、承租单位的安全生产工作统一协调、管理，定期进行安全检查，发现安全问题的，应当及时督促整改。

矿山、金属冶炼建设项目和用于生产、储存、装卸危险物品的建设项目施工单位应当加强对施工项目的安全管理，不得倒卖、出租、出借、挂靠或者以其他形式非法转让施工资质，不得将其承包的全部建设工程转包给第三人或者将其承包的全部建设工程分解以后以分包的名义分别转包给第三人，不得将工程分包给不具备相应资质条件的单位。

contracted by them to a third party in the name of subcontracting after dismantling, and shall not subcontract the project to a unit that does not have the corresponding qualifications.

Article 50: When a production safety accident occurs in a production and business operation entity, the main person in charge of the unit shall immediately organize a rescue and shall not leave his post without permission during the investigation and handling of the accident.

Article 51: Production and business operation entities must participate in work-related injury insurance in accordance with the law and pay insurance premiums for employees. The State encourages production and business operation units to purchase production safety liability insurance; Production and business operation units in high-risk industries and fields stipulated by the state shall purchase production safety liability insurance. The specific scope and implementation measures shall be formulated by the emergency management department of the State Council in conjunction with the financial department of the State Council, the insurance regulatory authority of the State Council and the competent departments of relevant industries.

## 7.1.3　Chapter 3 Rights and Obligations of Employees in Work Safety

Article 52: The labor contract concluded between a production and business operation entity and its employees shall specify matters related to ensuring the labor safety of employees and preventing occupational hazards, as well as matters concerning the handling of work-related injury insurance for

第五十条　生产经营单位发生生产安全事故时，单位的主要负责人应当立即组织抢救，并不得在事故调查处理期间擅离职守。

第五十一条　生产经营单位必须依法参加工伤保险，为从业人员缴纳保险费。国家鼓励生产经营单位投保安全生产责任保险；属于国家规定的高危行业、领域的生产经营单位，应当投保安全生产责任保险。具体范围和实施办法由国务院应急管理部门会同国务院财政部门、国务院保险监督管理机构和相关行业主管部门制定。

## 7.1.3　第三章　从业人员的安全生产权利义务

第五十二条　生产经营单位与从业人员订立的劳动合同，应当载明有关保障从业人员劳动安全、防止职业危害的事项，以及依法为从业人员办理工伤保险的事项。生产经营单位不得以任何形式与从业人员订立协议，免除或者减轻其对从业人员因生产安全事故伤亡依法应承担的责任。

employees in accordance with law. Production and business operation entities shall not enter into agreements with employees in any form to exempt or reduce their liability for injuries or injuries caused by production safety accidents in accordance with the law.

Article 53: Employees of production and business operation entities have the right to understand the risk factors, preventive measures and emergency measures existing in their workplaces and workplaces, and have the right to make suggestions on the work safety of their units.

Article 54: Employees have the right to criticize, report or accuse problems existing in their unit's work safety; It has the right to refuse illegal commands and forced risky operations.

Production and business operation entities shall not reduce their wages, benefits and other benefits or terminate the labor contract concluded with employees because they criticize, report, accuse or refuse to command in violation of regulations or force risky operations to work in violation of regulations.

Article 55: When employees discover an emergency that directly threatens their personal safety, they have the right to stop operations or evacuate the workplace after taking possible emergency measures.

Production and business operation entities shall not reduce their wages, benefits and other benefits or terminate the labor contract concluded with employees because they stop working or take emergency evacuation measures in the emergency situation in the preceding paragraph.

Article 56: After a production safety accident occurs in a production and business

第五十三条　生产经营单位的从业人员有权了解其作业场所和工作岗位存在的危险因素、防范措施及事故应急措施，有权对本单位的安全生产工作提出建议。

第五十四条　从业人员有权对本单位安全生产工作中存在的问题提出批评、检举、控告；有权拒绝违章指挥和强令冒险作业。

生产经营单位不得因从业人员对本单位安全生产工作提出批评、检举、控告或者拒绝违章指挥、强令冒险作业而降低其工资、福利等待遇或者解除与其订立的劳动合同。

第五十五条　从业人员发现直接危及人身安全的紧急情况时，有权停止作业或者在采取可能的应急措施后撤离作业场所。

生产经营单位不得因从业人员在前款紧急情况下停止作业或者采取紧急撤离措施而降低其工资、福利等待遇或者解除与其订立的劳动合同。

第五十六条　生产经营单位发生生产安全事故后，应当及时采取措施救治有关人员。

# 第七章 油气井工程相关的法律法规
## Chapter 7 Laws and Regulations Related to Oil and Gas Well Engineering

operation entity, it shall take timely measures and treat the relevant personnel.

Employees who have suffered damage due to production safety accidents have the right to claim compensation in accordance with the relevant civil laws if they still have the right to compensation in addition to work-related injury insurance in accordance with the law.

Article 57: In the course of work, employees shall strictly implement their post safety responsibilities, abide by their unit's production safety rules and regulations and operating procedures, obey management, and correctly wear and use labor protection equipments.

Article 58: Employees shall receive education and training on work safety, master the knowledge of work safety required for their own work, improve their skills in work safety, and enhance their ability to prevent accidents and deal with emergencies.

Article 59: Employees who discover potential accidents or other unsafe factors shall immediately report to the on-site safety production management personnel or the person in charge of the unit; Personnel who receive the report shall promptly handle it.

Article 61: Where a production or business operation entity uses dispatched workers, the dispatched workers shall enjoy the rights of employees as provided for in this Law and shall perform the obligations of employees as provided for in this Law.

## 7.1.4 Chapter 5 Emergency Rescue and Investigation and Handling of Production Safety Accidents

Article 79: The State strengthens the capacity for emergency response to production

因生产安全事故受到损害的从业人员，除依法享有工伤保险外，依照有关民事法律尚有获得赔偿的权利的，有权提出赔偿要求。

第五十七条　从业人员在作业过程中，应当严格落实岗位安全责任，遵守本单位的安全生产规章制度和操作规程，服从管理，正确佩戴和使用劳动防护用品。

第五十八条　从业人员应当接受安全生产教育和培训，掌握本职工作所需的安全生产知识，提高安全生产技能，增强事故预防和应急处理能力。

第五十九条　从业人员发现事故隐患或者其他不安全因素，应当立即向现场安全生产管理人员或者本单位负责人报告；接到报告的人员应当及时予以处理。

第六十一条　生产经营单位使用被派遣劳动者的，被派遣劳动者享有本法规定的从业人员的权利，并应当履行本法规定的从业人员的义务。

## 7.1.4 第五章 生产安全事故的应急救援与调查处理

第七十九条　国家加强生产安全事故应急能力建设，在重点行业、领域建立应

safety accidents, establishes emergency rescue bases and emergency rescue teams in key industries and fields, and has the State production safety emergency rescue institutions coordinate and command them in a unified manner; Encourage production and business operation units and other social forces to establish emergency rescue teams, equipped with corresponding emergency rescue equipment and materials, and improve the professional level of emergency rescue.

The emergency management department of the State Council shall take the lead in establishing a unified national emergency rescue information system for production safety accidents, and the relevant departments of transportation, housing and urban-rural construction, water conservancy, civil aviation and other relevant departments of the State Council and local people's governments at or above the county level shall establish and improve the emergency rescue information system for production safety accidents in relevant industries, fields and regions, so as to achieve interconnection and information sharing, and improve the precision and intelligence level of supervision through the implementation of online safety information collection, safety supervision and monitoring and early warning.

Article 80: The local people's governments at or above the county level shall organize the relevant departments to formulate emergency rescue plans for production safety accidents within their respective administrative areas and establish an emergency rescue system.

Township people's governments and neighborhood offices, as well as development zones, industrial parks, port areas, scenic spots,

急救援基地和应急救援队伍，并由国家安全生产应急救援机构统一协调指挥；鼓励生产经营单位和其他社会力量建立应急救援队伍，配备相应的应急救援装备和物资，提高应急救援的专业化水平。

国务院应急管理部门牵头建立全国统一的生产安全事故应急救援信息系统，国务院交通运输、住房和城乡建设、水利、民航等有关部门和县级以上地方人民政府建立健全相关行业、领域、地区的生产安全事故应急救援信息系统，实现互联互通、信息共享，通过推行网上安全信息采集、安全监管和监测预警，提升监管的精准化、智能化水平。

第八十条　县级以上地方各级人民政府应当组织有关部门制定本行政区域内生产安全事故应急救援预案，建立应急救援体系。

乡镇人民政府和街道办事处，以及开发区、工业园区、港区、风景区等应当制定相应的生产安全事故应急救援预案，协助人民政府有关部门或者按照授权依法履行生产安全事故应急救援工作职责。

etc., shall formulate corresponding emergency rescue plans for production safety accidents, and assist the relevant departments of the people's governments or perform their duties for emergency rescue work of production safety accidents in accordance with the law as authorized.

Article 81: Production and business operation entities shall formulate emergency rescue plans for production safety accidents of their own units, which should be linked with the emergency rescue plans for production safety accidents organized and formulated by the local people's governments at or above the county level where they are located and organize regular drills.

Article 82: Units producing, trading, and storing dangerous goods, as well as mining, metal smelting, urban rail transit operations, and construction units, shall establish emergency rescue organizations; Where the scale of production and operation is small, an emergency rescue organization may not be established, but part-time emergency rescue personnel should be designated.

Units engaged in the production, operation, storage and transportation of dangerous goods, as well as mining, metal smelting, urban rail transit operation and construction units, shall be equipped with necessary emergency rescue equipment, equipment and materials, and carry out regular maintenance and maintenance to ensure normal operation.

Article 83: After a production safety accident occurs in a production and business operation entity, the relevant personnel at the scene of the accident shall immediately report to the person in charge of the unit.

After receiving the accident report, the

第八十一条 生产经营单位应当制定本单位生产安全事故应急救援预案,与所在地县级以上地方人民政府组织制定的生产安全事故应急救援预案相衔接,并定期组织演练。

第八十二条 危险物品的生产、经营、储存单位以及矿山、金属冶炼、城市轨道交通运营、建筑施工单位应当建立应急救援组织;生产经营规模较小的,可以不建立应急救援组织,但应当指定兼职的应急救援人员。

危险物品的生产、经营、储存、运输单位以及矿山、金属冶炼、城市轨道交通运营、建筑施工单位应当配备必要的应急救援器材、设备和物资,并进行经常性维护、保养,保证正常运转。

第八十三条 生产经营单位发生生产安全事故后,事故现场有关人员应当立即报告本单位负责人。

单位负责人接到事故报告后,应当迅速采取有效措施,组织抢救,防止事故扩大,减少人员伤亡和财产损失,并按照国家有关规定立即如实报告当地负有安全生产监督管理职责的部门,不得隐瞒不报、谎报或者迟报,不得故意破坏事故现场、毁灭有关证据。

person in charge of the unit shall promptly take effective measures to organize rescue, prevent the accident from expanding, reduce casualties and property losses, and immediately and truthfully report to the local department responsible for the supervision and management of production safety in accordance with the relevant provisions of the state.

Article 84: After receiving the accident report, the department responsible for the supervision and management of production safety shall immediately report the accident in accordance with the relevant provisions of the State. The departments responsible for the supervision and management of production safety and the relevant local people's governments shall not conceal or fail to report, make false reports or delay in reporting the circumstances of the accident.

Article 85: After receiving the report of a production safety accident, the responsible person of the relevant local people's government and the department responsible for the supervision and management of production safety shall immediately rush to the scene of the accident and organize the rescue of the accident in accordance with the requirements of the emergency rescue plans for the production safety accident.

Departments and units involved in accident rescue shall obey unified command, strengthen coordination and linkage, adopt effective emergency rescue measures, and take vigilance, evacuation and other measures according to the needs of accident rescue, to prevent the expansion of accidents and the occurrence of secondary disasters, reduce casualties and property losses.

第八十四条　负有安全生产监督管理职责的部门接到事故报告后，应当立即按照国家有关规定上报事故情况。负有安全生产监督管理职责的部门和有关地方人民政府对事故情况不得隐瞒不报、谎报或者迟报。

第八十五条　有关地方人民政府和负有安全生产监督管理职责的部门的负责人接到生产安全事故报告后，应当按照生产安全事故应急救援预案的要求立即赶到事故现场，组织事故抢救。

参与事故抢救的部门和单位应当服从统一指挥，加强协同联动，采取有效的应急救援措施，并根据事故救援的需要采取警戒、疏散等措施，防止事故扩大和次生灾害的发生，减少人员伤亡和财产损失。

Necessary measures shall be taken in the process of accident rescue to avoid or reduce the harm caused to the environment. All units and individuals should support and cooperate with the rescue of the accident and provide all convenient conditions.

## 7.2 Unified Regulations on the Rational and Comprehensive Utilization of Mineral Resources in Kazakhstan (Excerpt)

### 7.2.1 Chapter 13 Well Structure and Drilling and Completion

165.All drilling operations should be carried out in accordance with the drilling technical design plans. The drilling technology design plans should be prepared by the design unit with relevant operation qualifications.

166.Prepare drilling technical design plans in accordance with the requirements stipulated in the drilling standard technical documents approved by the authorized authorities in the oil and gas field. Without an approved technical design, drilling operations are not allowed.

167.Drilling operations are not allowed without an approved technical design scheme.

168.The drilling design is based on the following provisions:

事故抢救过程中应当采取必要措施，避免或者减少对环境造成的危害。任何单位和个人都应当支持、配合事故抢救，并提供一切便利条件。

## 7.2 哈萨克斯坦矿产资源合理和综合利用统一规定（节选）

### 7.2.1 第13章 井身结构和钻完井

165. 所有的钻井作业按照钻井技术设计方案实施。钻井技术设计方案由具有相关作业资质的设计单位编写。

166. 按照油气领域的授权机关批准的钻井标准技术文件规定的要求，编写钻井技术设计方案。没有批准的技术设计方案，不允许钻井作业。

167. 没有批准的技术设计方案，不允许钻井作业。

168. 钻井设计依据以下规定：

(1) Implement drilling in accordance with the group or individual drilling technology design scheme;

(2) The technical design scheme is the main document for adjusting the drilling process;

(3) In the design plans, give the good production layer and cementing that will be drilled and ensure the reliability of the well, and fulfill all the requirements of the basic design documents and oil and gas exploration and development design documents;

(4) In the drilling design, comply with the current standard documents on various types of work and environmental protection. Prepare technical design documents according to the drilling plan design task book, which is prepared by the mineral resource user according to the basic design documents and oil and gas exploration and development design documents;

(5) The comprehensiveness and reliability of the initial design data should be the responsibility of the mineral resource utilizer, and the quality of the design scheme should be jointly responsible for the mineral resource user and the design unit;

(6) Drilling is carried out in accordance with the contract between the drilling unit and the mineral resource utilizer, or in the case of holding the relevant license, the mineral resource user can carry out the drilling on its own;

(7) According to the requirements of the authorized authorities in the field of environmental protection, oil and gas and other state agencies within their jurisdiction, the technical design documents shall be modified to improve the quality and safety of operations;

（1）按照成组的或单个的钻井技术设计方案，实施钻井；

（2）技术设计方案是调整钻井过程的主要文件；

（3）在设计方案中给出会钻遇的好产层、固井并保证井的可靠性，履行基础设计文件和油气勘探开发设计文件的所有要求；

（4）在钻井设计时，遵守现行各类工作和环境保护方面的标准性文件。依据钻井方案设计任务书编写技术设计文件，钻井方案设计任务书是矿产资源利用者根据基础设计文件和油气勘探开发设计文件编写的；

（5）初始设计资料的全面性和可靠性由矿产资源利用者负责，设计方案的质量由矿产资源利用者和设计单位共同负责；

（6）根据钻井单位和矿产资源利用者之间的承包合同实施钻井，或者在持有相关许可证的情况下，矿产资源利用者可以自行实施；

（7）根据环境保护领域、油气领域的授权机关及其权限范围内其他国家机关的要求，对技术设计文件进行修改，以提高作业质量和安全性；

(8) In the case that the numerical error between the wellbore depth and casing length and the technical design plan is within ±250m (within ±300m for directional wells and horizontal wells), it is allowed to modify the technical design scheme without the consent of the design unit;

(9) supervise the implementation of the technical design scheme, which is implemented by the customer and the design unit that prepares the technical design documents;

(10) The implementation of the drilling design plans are responsible for the mineral resource user and the drilling unit.

169.The drilling technology design plan should be approved by the mineral resource utilizer.

170.All drilling operations should be carried out in accordance with the requirements of the approved drilling technical design scheme.

171.The drilling method and the drill pipe, drill bit, drilling system, drilling fluid type and formula corresponding to it shall meet the requirements of the approved drilling technology design plans.

172.In the drilling technology design plans, it is necessary to stipulate and demonstrate the way of producing the layer under different formation pressures.

173.Drilling technical issues should be listed in advance in the basic design documents and oil and gas exploration and development design documents, and should be studied in detail in the drilling technology design plans.

174.The amount of dangerous drilling fluid is determined in the technology design plans.

175.The drilling circulation system of

（8）在井筒深度和套管长度与技术设计方案规定的数值误差在 ±250m 内（定向井和水平井为 ±300m 内）的情况下，允许不经设计单位同意，对技术设计方案进行修改；

（9）对技术设计方案的执行情况进行监督，由客户和编写技术设计文件的设计单位实施；

（10）钻井设计方案实施由矿产资源利用者和钻井单位负责。

169.钻井技术设计方案由矿产资源利用者批准。

170.所有钻井作业按照批准的钻井技术设计方案的要求实施。

171.钻井方式及与其相符的钻杆、钻头、钻井制度、钻井液类型和配方，应符合批准的钻井技术设计方案的要求。

172.在钻井技术设计方案中需规定和论证不同地层压力生产层的方式。

173.钻井技术问题，在基础设计文件和油气勘探开发设计文件中提前列举，并在钻井技术设计方案中详细研究。

174.危险钻井液量在技术设计方案中确定。

175.高油气比和地层压力异常高的油井、气井和凝析气井钻井循环系统应保证使用专业设备对钻井液进行连续脱气。

oil wells, gas wells and condensate wells with high oil-gas ratio and abnormally high formation pressure should ensure the continuous degassing of drilling fluid using professional equipments.

176.The drilling characteristics of oil and gas fields with high sulfur content, formation salt content, abnormally high formation pressure and high temperature and in offshore oilfields should be studied separately in the drilling technology design plans in accordance with the basic design documents and oil and gas exploration and development design documents.

177.The well casing structure is a complete set of casing that meets the requirements of diameter and length, and the space outside the pipe is fixed with cement, including the equipments near the well zone of the production layer and the wellhead equipment.

178.The well casing structure ensures the reliability, manufacturability and safety of drilling and well production, including:

(1) The optimal production of the diameter of the production casing and the bottom hole structure can maximize the use of the production capacity of the development layer in the well production process;

(2) the possibility of optimizing the production mode and mode of the well with effective equipments under the conditions of adopting the designed methods of maintaining the formation pressure or using the natural energy of the mineral deposit;

(3) Carry out safe, accident-free and complex-free operations at all stages of drilling and well production;

(4) obtain the required mineral geological information of the drill profile;

176.高含硫、地层含盐、地层压力异常高和高温的油气田及在海上油田的钻井特点，按照基础设计文件和油气勘探开发设计文件，单独在钻井技术设计方案中研究。

177.井身结构是符合直径和长度要求的一整套套管，管外空间注入水泥固定，包括产层近井地带设备和井口设备。

178.井身结构保证钻井和井生产的可靠性、工艺性和安全性，其中包括：

（1）生产套管直径和井底结构的最优化生产，可以在井生产过程中最大可能地利用开发层系的产能；

（2）在采用设计的地层压力保持方法或使用矿藏天然能量的条件下，采用有效设备优化井的生产方式和生产模式的可能性；

（3）在钻井和井生产的各个阶段，进行安全、无事故和无复杂情况作业；

（4）获得所需的钻开剖面矿产地质信息；

(5) The protection of mineral resources mainly depends on the solidity and durability of the well support, the tightness of the casing and its blocking annulus, and the isolation between the fluid-bearing horizon and the isolation of the fluid-bearing horizon from the ground;

(6) The specifications and sizes of casing and wellbore should be standardized as much as possible;

(7) Ensure the working conditions of underground maintenance and production during well production;

(8) Possibility of installing throttle valves, packers and other devices.

179.The well structure that is planned to be produced by gas lift method should meet the requirements for the gas well structure.

180.The structure of the injection well for hot water, steam and gas injection should be demonstrated in advance in the basic design documents and the oil and gas exploration and development design documents and the detailed technical design plan for drilling.

181.The structure of the evaluation wells of oil and gas fields which production capacity has been proven shall meet the requirements that can be used during production.

182.The well structure and well profile should be listed in advance in the basic design documents and the oil and gas exploration and development design documents, and should be studied and demonstrated in detail in the drilling technology design plan.

183.According to the purpose and use of the well, the specific drilling geology-technical possibilities, the ground conditions and the presence or absence of protection areas, etc., the wellbore profile is designed when drilling.

（5）保护矿产资源，主要是依靠井支架的坚固性和耐用性、套管及其封锁环空的密封性，实现含流体层位之间的隔离及含流体层位与地面的隔离；

（6）套管和井筒规格尺寸尽量标准化；

（7）在井生产时，保证井下维修和生产的工作条件；

（8）安装节流阀、封隔器和其他装置的可能性。

179.计划采用气举方式生产的井结构，应满足对气井结构提出的要求。

180.注热水、注蒸汽和注气的注入井结构，在基础设计文件和油气勘探开发设计文件及钻井详细技术设计方案中提前论证。

181.产能已探明的油气田的评价井结构，应符合在生产时能够使用的要求。

182.井结构和井剖面，在基础设计文件和油气勘探开发设计文件中提前列举，并在钻井技术设计方案中详细研究和论证。

183.根据井的目的用途、具体的钻井地质—技术可能性、地面条件和有无保护区等情况，在钻井时设计井筒剖面。

184.Adopt vertical profile, directional oblique profile, and the wellbore of the production interval section is the horizontal section.

185.According to the purpose and use of the well, the specific drilling geology and technology conditions, design the directional inclined wellbore profile.

186.The selection of directional inclined wellbore profile type, drilling tool combination, drilling system parameters, wellbore deepening rate and other combination measures shall ensure that:

(1)Based on the existing drilling operation technology and process level, the well is drilled to the design depth without any complications;

(2) Drilling is qualified, and the time and capital consumption is minimized;

(3) Within the allowable deviation rate, the vertical displacement of the bottom hole design in a given direction is reached;

(4) The number of bending of the wellbore is the least, and the bending radius does not exceed the allowable value;

(5) Drilling tool combination and casing and downhole equipments and tools can pass freely in the process of production and downhole maintenance;

(6) Prevent casing direction, channeling, tools and geophysical equipments from tightening and jamming.

187.The borehole profile of horizontal wells in the production layer should be demonstrated when designing the test production plans or using horizontal wells for industrial development of oil and gas fields.

188.In accordance with the overall drilling technology design scheme to ensure the completion of all geophysical studies,

184.采用垂直剖面、定向斜剖面，产层段井筒为水平段。

185.根据井的目的用途、具体的钻井地质及技术条件，设计定向斜井筒剖面。

186.选择定向斜井筒剖面类型、钻具组合、钻井制度参数、井筒加深速率和其他组合措施，应保证：

（1）基于现有的钻井作业技术和工艺水平，钻井至设计深度，且不出现任何复杂情况；

（2）钻井合格，且时间和资金消耗最低；

（3）在容许的偏差率范围内，达到给定方向的井底设计垂向位移；

（4）井筒折曲数量最少，且弯曲半径不超过容许值；

（5）钻具组合和套管及在生产和井下维修过程中井下设备用具能够自由通过；

（6）防止套管走向、窜槽、工具和地球物理设备拉紧和卡塞。

187.产层水平井井身剖面，应在设计试采方案或利用水平井对油气田进行工业开发时论证。

188.按照保证完成所有地球物理研究的整体的钻井技术设计方案，采用多分支井、定向井和定向水平井钻井方式。

the drilling methods of multi-lateral wells, directional wells and directional horizontal wells are adopted.

189. When drilling, surface facilities and wellhead equipments should be closely integrated with the drilling conditions under specific geological and technical conditions.

190. The choice of the type of drilling equipments are based on the large hook load of the drilling rig, which depends on the weight of the suspended drill pipe or the weight of the heaviest casing and its pipe section; The hook load shall exceed at least 40% of the weight of the heaviest suspended drill string.

191. Drilling is prohibited without mechanical purification of drilling fluids.

192. After the surface casing or intermediate casing is lowered, and before the casing continues to be lowered below, it is possible to drill the gas layer, condensate gas layer, oil layer or water layer, then the wellhead should be installed with a blowout prevention device.

193. In order to complete the subsequent process operations, the selection of blowout prevention devices, manifolds (throttle pipelines and kill lines), hydraulic control stations, throttle tables and flare devices depends on the specific mineral geological conditions of the following technical operations:

(1) In the case of the drill pipe has been lowered and the drilling pipe has not been lowered, seal the wellhead;

(2) According to the process adopted, flush the fluid in the well;

(3) After the lower BOP is closed, hang the drill string on the ram of the lower BOP;

(4) scraping drilling fluid;

189. 在钻井时，地面设施和井口设备应与具体地质技术条件下的钻井条件紧密结合。

190. 钻井设备类型的选择基于钻机的大钩载荷，该载荷取决于悬挂钻杆重量或最重套管及其管段的重量；大钩载荷应至少超过悬挂最重钻柱的重量40%。

191. 没有对钻井液进行机械净化，则禁止进行钻井。

192. 下入表层套管或中间套管后，在其下面继续下入套管之前，有可能钻遇气层、凝析气层、油层或水层，则井口应安装防喷装置。

193. 为了完成后续工艺作业，选择防喷装置、管汇（节流管线和压井管线）、水力控制站、节流台和火炬装置，取决于以下技术操作的具体矿产地质条件：

（1）在已下放钻杆及未下放钻杆情况下，密封井口；

（2）按照采用的工艺，冲洗井中流体；

（3）下部防喷器关闭后，在下部防喷器闸板上悬挂钻柱；

（4）刮钻井液；

(5) Observe the condition of the well when killing the well;

(6) Lift the drill string up and down to prevent it from getting stuck;

(7) In the case of sealing the wellhead, lower or lift part or all of the drilling pipes.

194. When drilling abnormal high-pressure gas layers, oil layers and water layers, as well as hydrogen sulfide-containing formations (content exceeding 3.5%), at least 3 BOP preventers are installed at the wellhead, one of which is a universal blowout preventer.

195. When drilling formations with abnormal high pressure and hydrogen sulfide content exceeding 3.5%, at least 4 BOP preventers should be installed, including a BOP with shear ram and a universal BOP.

196. During the drilling process, the natural conditions of the production zone near the well zone should be protected to the greatest extent possible.

197. According to the characteristics of geology-geophysical structure, formation reservoir and permeability, combined with the research objectives and methods in the drilling process, the types and parameters of drilling fluid used in drilling formations are demonstrated in the drilling technology design scheme. The use of drilling fluids shall ensure maximum protection of the reservoir's natural permeability and oil saturation, as well as the ability to carry out the required comprehensive geophysical studies.

198. The quality control of drilling in the production layer should be the responsibility of the technical department and the geological department of the mineral resource user and its contractor.

199. In order to maintain the natural permeability of pore and pore-fracture reservoirs,

（5）在压井时观察井的状况；

（6）上下提动钻柱，防止其遇卡；

（7）在密封井口情况下，下放或提升部分或全部钻杆。

194. 当钻开异常高压气层、油层和水层，以及含硫化氢地层时（含量超过3.5%），在井口至少安装3个防喷器，其中有一个万能防喷器。

195. 在钻遇异常高压及硫化氢含量超过3.5%的地层时，应至少安装4个防喷器，其中有一个带剪切闸板的防喷器和一个万能防喷器。

196. 在钻井过程中钻开产层，应保证最大可能地保护产层近井地带的自然状况。

197. 根据地质—地球物理构造的特点、地层储集性和渗透性，结合钻井过程中的研究目的和方法，在钻井技术设计方案中论证钻井地层用的钻井液的类型和参数。钻井液的使用应能保证最大限度地保护储层的天然渗透率和含油饱和度，以及可以进行所需的综合地球物理研究。

198. 产层钻开的质量监控，由矿产资源利用者及其承包商的技术部和地质部负责。

199. 在进行套管水泥胶结作业时，为了保持孔隙型和孔隙—裂缝型储层的天然渗透率，采用固井液尽可能不渗透，且总矿化度与用于钻揭这些层位时采用的钻井液的矿化度相近。

the cementing fluid is as impermeable as possible, and the total salinity is similar to that of the drilling fluid used to drill these horizons.

200.In oil and gas fields containing hydrogen sulfide, carbon dioxide and other corrosive compounds, use corrosion-resistant casing and cementing cement.

201.Casing cementing quality and formation sealing quality are controlled by professional geophysical research. If there is no cement outside the pipe for more than 50% of the design return height, the cement repair operation outside the pipes must be carried out.

202.Integrated geophysical studies shall ensure:

(1) Control and record the actual diameter and wall thickness of the casing;

(2) Control and record the actual position of the original process equipments of the inlet string;

(3) obtain information on the distribution of cement outside the pipes;

(4) Discover the possible channels and gaps between cement and casing, between cement and rock, and discover channeling;

(5) Find out if there are gases and liquids in the space outside the pipes.

203.By testing the tightness of the well structure, the cementation degree of the casing is detected.

204.By perforating the cemented casing, installing a filter in the case of the casing not cementing, or by retaining the open-hole bottom well, ensure the communication between the production layer and the wellbore.

205.Perforation drilling is the most commonly used method.

206.Before the production of casing perforation, according to the drilling technology

200. 在含有硫化氢、二氧化碳和其他腐蚀性化合物的油气田，采用抗腐蚀的套管和固井水泥。

201. 套管固井质量和地层封隔质量，由专业的地球物理研究控制。设计返高的50%以上管外没有水泥的情况下，必须进行管外水泥的修复作业。

202. 综合地球物理研究应保证：

（1）控制和记录套管的实际直径和壁厚；

（2）控制和记录下入管柱工艺装备原件的实际位置；

（3）获得有关管外水泥分布的资料；

（4）发现水泥和套管之间、水泥和岩石之间可能出现的通道和间隙，发现窜流；

（5）查明管外空间内是否有气体和液体。

203. 通过测试井身结构的密封性，检测套管的水泥环胶结程度。

204. 通过对水泥胶结的套管进行射孔，在套管不进行水泥胶结的情况下安装过滤器，或通过保留裸眼井底，保证产层和井筒连通。

205. 射孔钻开产层是最常用的方式。

206. 在生产套管射孔之前，根据钻井技术设计方案和已批准的工艺流程图，井口安装射孔阀或防喷装置，并注满钻井液（液体），钻井液中固相含量最低，其密度可防止出现油气显示，且能保证最大限度地保护储层的天然渗透率和含油饱和度。

design plans and the approved process flow chart, the wellhead is installed with a perforation valve or blowout prevention device, and the well is filled with drilling fluid (liquid), the solid content in the drilling fluid is the lowest, and its density can prevent the appearance of oil and gas display, and can ensure the maximum protection of the natural permeability and oil saturation of the reservoir.

207.After obtaining the actual logging data, before entering the casing, the geological department of the customer unit designs the formation drilling method and perforating well section.

208.Combined with the geology-site characteristics of the stratum, the perforation method, type and density are selected according to the scope and conditions adopted by the perforation method, and the casing and cement can not be damaged twice.

209.Before loading the ejection gun to go down the well, enter the well gauge with a deep manometer in order to check the passability of the equipments and measures the pressure in the perforating belt casing.

210.Observe the liquid level at the wellhead during perforation, and do not allow the liquid level to drop.

211.Before installing the Christmas tree at the wellhead, the pressure test should be carried out according to the rated pressure, and after installation, the pressure test should be carried out according to the pressure test value equal to the production casing.

212.After the production casing has been decentralized, tests are carried out on pre-exploration and evaluation wells (downhole strata) with the aim of obtaining information about the characteristics of the liquid flow,

207. 获得实际测井资料后，在下入套管之前，由客户单位的地质部门设计地层钻开方式和射孔井段。

208. 结合层系的地质—现场特点，根据射孔方法采用的范围和条件，选择射孔方式、类型和密度，且不能造成套管和水泥发生二生损坏。

209. 装弹射孔枪下井之前，下入带有深层压力计的通井规，以便检查设备的通过性并测量射孔带套管中的压力。

210. 在射孔时观察井口液位，不允许液位下降。

211. 在井口安装采油树之前，应按照额定压力进行试压，安装后应按照等于生产套管的试压值进行试压。

212. 在下放生产套管之后，对预探井和（或）评价井（井内地层）进行测试，目的是获得有关液流特点的信息，该测试是钻井过程的主要组成部分。

which is the main component of the drilling process.

213.The testing of production wells for the purpose of obtaining industrial oil and gas flows, which is a major component of the drilling process.

214.Only if the technical conditions are met and the technology equipments and materials specified in the drilling technology design plans can the test work of the well be started.

215.Wells are tested according to a standard scheme or individually with the aim of determining the hydrodynamic characteristics of the formation and the optimal production mode.

216.Well testing work shall ensure:

(1) Remove the well washing fluid in the formation near the well zone to the greatest extent;

(2) protect the strata in the near-well zone;

(3) Prevent the intrusion of bottom water and air top gas;

(4) Thermohydrodynamic studies to determine the quantitative and qualitative characteristics of formations and geophysical parameters;

(5) Prevent uncontrollable oil, gas and water display and open injection;

(6) Prevent the deformation of the production casing;

(7) Protection of mineral resources and the environment.

217.Drill and expose well tests in formations under complex geological conditions (abnormally high formation pressure, containing hydrogen sulfide and other acid gases, high temperatures, high oil-gas ratios) and prepare separate plans.

213. 对采油井进行测试，目的是获得工业油气流，该测试是钻井过程的主要组成部分。

214. 只有符合工艺条件并具备钻井技术设计方案规定的技术设备和物资，方可开始井的测试工作。

215. 按照标准方案或单独方案，对井进行测试，目的是确定地层的流体动力特点和最佳生产模式。

216. 井测试工作应保证：

（1）最大限度地除去地层近井地带的洗井液；

（2）保护近井地带的地层；

（3）防止底水和气顶气侵入；

（4）热力流体动力研究，测定地层的数量和质量特点及地球物理参数；

（5）防止出现不可控的油气水显示和敞喷；

（6）防止生产套管变形；

（7）保护矿产资源和环境。

217. 在复杂地质条件下（异常高地层压力、含有硫化氢和其他酸性气体、高温、高油气比）钻揭地层的井测试，编写单独的计划。

218. In the process of well testing, comprehensive temperature and pressure studies and hydrodynamic studies are carried out, as well as formation liquid sampling and research, and the water content of the produced liquid is determined.

219. A study of formation liquid samples should be carried out for each completed well (after testing). The comprehensive study shall determine all parameters, including saturation pressure, gas content, volume coefficient, formation fluid viscosity and density, the composition of the separated gas and degassed crude oil, and draw a diagram of the relationship between the main parameters of formation crude oil and the pressure (stage degassing or staged degassing).

220. A well is considered to have been tested if the formation productivity has been determined by the operation and the flow that characterizes the formation has been obtained. If the test results of the wells tested and completed in accordance with the provisions and requirements of the technology design plans are not satisfactory, the cause should be found and the follow-up operation plan should be approved.

221. If necessary, the productivity of the well is restored through repeated perforation of the formation or treatments of the near-well zone, and the treatment method, process and parameters of the near-well zone are selected according to the geological-physical properties of the reservoir.

222. According to the technology design documents for oil and gas exploration and exploitation, the user of mineral resources shall select the production mode, select and install downhole equipment, and carry out

218. 在井测试过程中，实施综合温压研究和流体动力研究，以及地层液体采样和研究，测定产液的含水率。

219. 应对每口完钻井进行地层液体样品研究（测试后）。综合研究应测定全部参数，其中包括饱和压力、含气量、体积系数、地层流体黏度和密度、分离出的气体和脱气原油的成分，绘制地层原油主要参数与压力的关系图（级次脱气或分级脱气）。

220. 如果通过作业已测定地层产能，并获得表征该层系的液流，则该井视为已测试。如果按照技术设计方案的规定和要求测试的井及完钻井的测试结果不理想，则应查找原因并批准后续作业计划。

221. 必要时，通过对地层重复射孔或对近井地带进行处理，恢复井的产能，近井地带处理方式、工艺和参数是根据油藏的地质—物理性质选择的。

222. 根据油气勘探和开采技术设计文件，矿产资源利用者选择生产方式，挑选和安装井下设备，实施提高勘探井、评价井或采油井产能及达到注入井设计吸收能力的后续工作。

follow-up work to improve the productivity of exploration wells, evaluation wells or oil production wells and achieve the design absorption capacity of injection wells.

223.Drilling is deemed to be completed after all the work specified in the drilling technology design plans and the well test plans has been completed.

224.The procedure for transferring the well from the contractor to the employer after the completion of drilling should be determined by the drilling contract signed between the two.

225.After the completion of drilling, the drilling units shall submit the following follow-up documents and work handover orders to the employers (mineral resources utilizer):

(1) Well location recording;

(2) Drilling design scheme (standard geological and technical task list);

(3) Drilling start and end records;

(4) Casing wellhead elevation measurement record;

(5) All geophysical research data and geophysical research conclusions;

(6) Casing design, parameters, diameter, wall thickness, steel grade and other necessary characteristics of non-metallic string;

(7) Casing cementing records, cementing design, laboratory analysis and density determination results of cement slurry quality during cementing, wellhead cement slurry output or cement slurry return data (cement tester records), pipe measurement, string records, drilling fluid density data in the well before cementing;

(8) Tightness test records of all casings;

(9) Test work plan for each layer;

223. 完成钻井技术设计方案和井测试计划规定的所有工作后，钻井视为结束。

224. 井完钻后由承包方转交给发包方的程序，由两者之间签署的钻井承包合同确定。

225. 完钻后的各井，钻井单位向发包方（矿产资源利用者）提交下列后续文件及工作交接单：

（1）井位记录；

（2）钻井设计方案（标准地质技术任务单）；

（3）钻井开始和结束记录；

（4）套管井口海拔测量记录；

（5）所有地球物理研究资料及地球物理研究结论；

（6）套管设计，参数，直径，壁厚，钢级及非金属管柱的其他必要特征；

（7）套管固井记录，固井设计，固井过程中水泥浆质量实验室分析和密度测定结果，井口水泥浆输出量或水泥浆返高数据（水泥测定器记录），管具测量、管柱记录，固井前井内钻井液密度数据；

（8）所有套管的密封性测试记录；

（9）各层系测试工作计划；

(10) Casing perforation record, indicating perforation layer, perforation method and perforation quality;

(11) Test records of each layer, including research data attachments (production, pressure, production capacity, oil, gas and water analysis);

(12) The size and type of tubing, indicating the installation depth of the equipments and the lower valve (hole);

(13) Geological report, including description of the entire drilling and well testing process;

(14) Core description;

(15) Drilling certificate, including drilling process, oil and gas display and well structure data;

(16) Casing tension recording;

(17) Wellhead equipment records;

(18) Geological documents of each well are transferred to the record;

(19) Block reclamation records.

226. Prepare standard technology documents according to these provisions and the drilling technology design plans, and formulate the drilling technology design plans based on the successful experience of mineral resource utilization.

227. The modification and supplement of the exploration operation design plans, the trial production design plans or the deposit development design plans involve the parameters of the drilling (well construction) technology design plans prepared according to the modification and supplement, the modification and supplement must be incorporated into the corresponding drilling (well construction) technical design document.

228. Mining of mineral resources is prohibited without the confirmation of the mineral resources user and the approval of

（10）套管射孔记录，注明射孔层段、射孔方式和射孔质量；

（11）各层系测试记录，含研究数据附件（产量、压力、产能、油（气、水）分析）；

（12）油管尺寸和类型，注明设备、下放阀门（孔）安装深度；

（13）地质报告，含整个钻井和井测试过程描述；

（14）岩心描述；

（15）钻井合格证，含钻井过程、油气显示和井身结构数据；

（16）套管张力记录；

（17）井口设备记录；

（18）各井的地质文件转交记录；

（19）区块复垦记录。

226. 根据本规定和钻井技术设计方案编写标准技术文件，依据矿产资源利用的成功经验，制定钻井技术设计方案。

227. 对勘探作业设计方案、试生产设计方案或矿床开发设计方案的修改和（或）补充，涉及依据修改和（或）补充编写的钻井（建井）技术设计方案的参数，则必须将修改和（或）补充纳入相应的钻井（建井）技术设计文件中。

228. 没有矿产资源利用者确认的并获得规定评审钻井（建井）技术设计方案，禁止进行矿产资源开采作业。

the technical design plan for drilling (well construction).

229.During the period of oil and gas exploration and development, drilling should be carried out according to the type of well.

230.According to the solutions approved by the basic design documents and the oil and gas exploration and development design documents, the well classification can be modified in the process of oil and gas exploration and industrial exploitation.

231.For some reason, the production of wells lacks economic viability, and the production wells are temporarily converted into mothballed wells in accordance with the provisions on oil and gas exploration and development and the sealing and scrapping of wells during uranium exploitation approved by the authorized authorities in the field of oil and gas and uranium exploitation.

232.All wells that have completed their mission and whose use for other purposes are unreasonable or impossible shall be scrapped in accordance with the provisions on the sealing and decommissioning of wells for oil and gas exploration and development and uranium exploitation approved by the authorized authorities in the field of oil and gas and uranium exploitation.

### 7.2.2 Chapter 14 Drilling Uses

233.During the period of oil and gas exploration and exploitation, the categories of wells shall be classified according to their use.

234.For the purpose of exploration and discovery of oil and gas reservoirs, exploratory wells are designed in well-drilled blocks and structural areas with prospects for geo-geophysical research.

229. 在油气勘探和开发期，按照井别进行钻井。

230. 根据基础设计文件和油气勘探开发设计文件批准的解决方案，可以在油气勘探和工业开采过程中修改井别。

231. 基于某种原因，井生产缺乏经济可行性，按照油气和铀开采领域的授权机关批准的油气勘探开发及铀开采时井封存和报废规定，临时由生产井转为封存井。

232. 完成使命的所有井，并且用作其他目的也不合理或不可能，按照油气和铀开采领域的授权机关批准的油气勘探开发及铀开采时井封存和报废规定，应进行报废。

### 7.2.2 第14章 钻井用途

233. 在油气勘探和开采期内，按照井的用途对井的类别进行划分。

234. 为了勘探和发现油气藏，在已经钻井区块和具备地质—地球物理研究前景的构造区进行设计的井为探井。

235.In order to conduct geological research and delineate the scope of oil and gas reservoirs, obtain the original data required for calculating oil and gas reserves and designing exploration plans, the wells designed in the discovered oil and gas areas are exploration and evaluation wells.

236.Production evaluation wells carry out the task of studying the characteristics of reservoir seepage and carry out the field research work of determining the parameters of oil and gas reservoirs.

237.Wells designed for test production (advanced production wells) and industrial exploitation in oil and gas fields are production wells:

(1) Development wells (oil and gas wells) :to build the development and recovery systems for oil, gas, condensate and water;

(2) Injection wells: maintain reservoir energy by injecting water, gas (gas-water mixture) or other displacement agents to maintain formation pressure, inject gas or secondary associated components separated from the reservoir for temporary hold, and extract oil and gas during taking.

238.Design and drill monitoring wells in order to systematically supervise the changes in the fluid interface (oil-water interface, oil-gas interface, gas-water interface) and other parameters in the process of reservoir development (including the oil, gas and water saturation of the formation).

239.In order to supervise the changes in formation pressure and temperature, pressure monitoring wells are designed.

240.With the refinement of the stratigraphic structure, in order to achieve the designed oil and gas recovery, the stratigraphic backup wells are

235. 为了对油气藏进行地质研究和圈定范围，获得计算油气储量和设计勘探方案所需的原始资料，在已发现油气区域设计的井为勘探评价井。

236. 生产评价井执行储层渗流特征研究任务，执行油气藏参数测定的现场研究工作。

237. 在油气田进行试采（超前生产井）和工业开采时设计的井为生产井：

（1）开发井（油井和气井）：为了建立石油、天然气、凝析油和水的开发和采收系统；

（2）注入井：保持油藏能量，通过注入水、气体（气水混合物）或其他驱替剂，以保持地层压力，注入气体或从油藏中分离出的二级伴生组分，以进行临时保持，以及在采取期间开采油气。

238. 为了对流体界面（油水界面、油气界面、气水界面）变化情况及油藏开发过程中其他参数（其中包括地层的含油气水饱和度）的变化情况实施系统性监督，设计和钻探监测井。

239. 为了对地层压力和温度变化情况实施监督，设计压力监测井。

240. 随着层系结构的精细化，为了实现设计的油气采收率，在钻探过程中部署层系备用井。

deployed during the drilling process.

### 7.2.3 Chapter 16 Requirements for the Development of Injection-Production Wells

264.Depending on the production capacity and the degree of flooding, the production wells are produced by self-injection or mechanical exploitation. Mechanical mining methods include different pumps and gas lifts.

265.The user of mineral resources shall include all wells drilled in the area of the mineral resource block in the balance sheet and monitor them in accordance with this law.

266.The self-injection method is to lift the well production fluid from the bottom of the well to the surface with the help of formation energy in the early stage of reservoir development (waterless period).

267.With the natural flooding of the well, the average density of the produced liquid increases, and the free gas component in the well product decreases, resulting in a decrease in production, which will cause the well to stop self-blowing, even if the formation pressure remains at the original formation pressure level.

268.Due to the decline in production, the use of self-spraying production can not obtain economic value, and the well is converted to a more suitable mechanical production method.

269.In accordance with the natural climatic conditions, the following specialized pumping equipment is used in the establishment of the mining system and the development of ore deposits:

(1) rod pumps for deep wells;

(2) Electric centrifugal pump.

270.When the production conditions of the well tend to be complex (pumping out

### 7.2.3 第16章 注采井开发要求

264. 根据产能和水淹程度，采油井采用自喷或机械开采方式生产。机械开采方式包括各种泵和气举方式。

265. 矿产资源利用者应将矿产资源区块区域内已完钻的所有井纳入平衡表，并按照本法对其进行监测。

266. 自喷方式是在油藏开发初期（无水期）借助地层能量将井产液从井底提升至地面。

267. 随着井的自然水淹，采出液体的平均密度增大，井产物中的游离气成分降低，导致产量下降，即使地层压力保持在原始地层压力水平，也会导致井停止自喷。

268. 由于产量下降，利用自喷方式生产不能获得经济价值，将井转为更为合适的机械生产方式。

269. 根据自然气候条件，在组建开采体系和矿床开发时，设备支行和维修系统采用以下专业泵设备：

（1）深井有杆泵；

（2）电动离心泵。

270. 当井的生产条件趋于复杂（抽出高黏度液体，采出产物中机械杂质含量升高，井深处动液面低），必须使用专业的泵设备：

high-viscosity liquids, increasing the content of mechanical impurities in the produced products, and low dynamic liquid level in the deep wells), it is necessary to use professional pumping equipments:

(1) electric screw pumps;

(2) Electric diaphragm pumps;

(3) Hydraulic piston pumps.

271.When the liquid is high viscosity and the content of mechanical impurities in the produced products is high, the user of mineral resources should formulate the optimal operation mode of the pump device under complex conditions and select the required protective devices.

272.When the well adopts the gas lift mode for production, the following main gas lift production modes are used according to its characteristics, natural gas resources and the availability of gas injection equipment in the well and on the surface:

(1) Compressed gas lift;

(2) No compressed air lift;

(3) gas lift in the well;

(4) continuous gas lift;

(5) Periodic gas lift.

273.The level and speed of liquid production in the production layer, the bottomhole and wellhead pressure of the oil production well, the limit pressure of self-injection, the conversion of the well group to mechanical mining, and the selection of mechanical mining methods are demonstrated in the deposit development and design documents, and are implemented by the mineral resource users in accordance with the geological-technical measures plan.

274.Tubing is required to implement a production method whose material, size and

（1）电动螺杆泵；

（2）电动隔膜泵；

（3）液压活塞泵。

271.当液体高黏度且采出产物中机械杂质含量高的井生产时，矿产资源利用者应制定复杂条件下泵装置最佳运行模式，并选择所需的防护装置。

272.井采用气举方式生产时，根据其特点、天然气资源量及井内和地面注气设备的可用性，使用以下主要气举生产模式：

（1）压缩气举；

（2）无压缩气举；

（3）井内气举；

（4）连续气举；

（5）周期性气举。

273.生产层系采液水平和速度，采油井井底和井口压力，自喷极限压力，井组转为机械开采，以及机械开采方式的选择，在矿床开发设计文件中论证，由矿产资源利用者按照地质—技术措施计划实施。

274.必须具备油管才能实施的生产方式，其材料、尺寸和下入深度取决于采出液体的性质、井下温压条件、生产方式，且按照批准的方法和建议确定。

depth of entry depend on the nature of the produced liquid, downhole temperature and pressure conditions, production method, and are determined in accordance with approved methods and recommendations.

275.The selection of the type, size and depth of the downhole equipment in the well production mode is completed by the mineral resource utilizer.

276.When selecting production facilities for oil production wells, it is necessary to ensure:

(1) the well is reliable and accident-free;

(2) a given standard for liquid recovery;

(3) High efficiency of equipments and maintenance period;

(4) Lowest cost compared to other methods;

(5) The development process and the operation mode of the wells can be supervised and regulated.

277.In the production of self-injection wells, in order to realize the best utilization of formation energy, prolong the self-injection period and ensure the operation mode (no wave), a downhole equipment scheme is provided:

(1) the lower part of the string is installed with a tubing, a packer or a special funnel that seals the space outside the pipe to capture most of the gas separated from the crude oil and send it into the string;

(2) Install a packer-throttle valve to seal the space outside the pipe and cut off (block) the oil-gas mixed flow of the tubing string in case of emergency;

(3) Install the bottom hole nozzle to ensure that the operation mode of the well can be adjusted and the energy generated when the gas separated from the crude oil rises to the

275. 井生产方式中井下设备的类型尺寸和下入深度的选择，由矿产资源利用者完成。

276. 在选择采油井生产设施时需保证：

（1）井可靠、无事故工作；

（2）给定的液体采出标准；

（3）设备高效率和检修期；

（4）与其他方式相比，费用最低；

（5）可以对开发过程和井的运行模式进行监督和调控。

277. 在井自喷生产时，为了实现地层能量的最佳利用，延长自喷期并保障的运行模式（无波动），提供一种井下设备方案：

（1）管柱下部安装油管、密封管外空间的封隔器或专门的漏斗，捕捉原油中分离出的大部分气体并将其送入管柱；

（2）安装封隔器—节流阀，密封管外空间并在紧急情况下截断（堵住）油管管柱的油气混合物流；

（3）安装井底接管嘴，保障可调整井的运行模式和最大限度地利用从原油中分离出的气体上升至地面时产生的能量；

surface is maximized;

(4) Install one (or several) downhole valve chambers to prevent air lift valves. The gas lift valve ensure that the gas enters the tubing string in the outer space of the gas self-injection production, or ensures the operation of the well produced by the gas lift mode after the self-injection is completed, if the mine development design documents have this provision.

278.The use of natural gas and associated gas as a working agent is only permitted with the help of compressed gas lift if the gas used is fully and efficiently utilized. At the same time, the well structure should meet the requirements for gas wells.

279.In the case of pumps, professional guards (gas separators, gas anchors, sand anchors, etc.) must be used to prevent gas, sand and mechanical impurities from entering the pump units.

280.In the production of high-temperature mineral deposits, the downhole equipment is selected by combining the operation possibilities of the downhole equipments under the conditions of high temperature and high content of corrosive components (carbon dioxide, hydrogen sulfide, etc.).

281.Only with downhole and surface equipments can two or more strata be operated separately at the same time using a single well, ensuring separate measurement of produced products and field study of each stratum.

282.The commissioning, production sequence and duration of the injection well are specified in the basic design documents.

283.Injection wells located within the oil-bearing zone can initially be used as production wells and connected to the oil collection pipe.

（4）安装一个（或几个）井下阀室，用来防止气举阀。气举阀保障自喷方式生产时气体自管外空间进入油管管柱，或保障自喷结束后气举方式生产的井运行，如果矿藏开发设计文件有此规定。

278.使用天然气和（或）伴生气作为工作剂，只有在使用气体全部有效利用的情况下，才允许借助无压缩气举进行生产。同时，井身结构应满足对气井提出的要求。

279.在采用泵进行生产时，为了防止气、砂和机械杂质进入泵装置，必须使用专业的防护装置（气体分离器、气锚和砂锚等）。

280.在高温度矿藏生产时，结合井下设备在高温、腐蚀成分（二氧化碳、硫化氢等）含量高条件下的运行可能性，以选择井下设备。

281.只有采用井下和地面设备，才允许利用一口井同时对两个或多个层系分别操作，保证分别计量采出物并对各层系进行现场研究。

282.注入井的投产、生产顺序和期限，在基础设计文件中规定。

283.位于含油区范围的注入井，最初可以作为采油井使用，并将其接入集油管。

284.Oil wells with uncontrollable gas channeling along the formation or along the space outside the pipe are not allowed to produce.

285.Self-blowing wells that pass through the space between pipes (outside the pipes) or wells that are converted to mechanical mining are not allowed to be produced.

286.Wells whose oil-gas ratio exceeds the design value specified in the basic design documents or oil and gas exploration and development design documents are not allowed to be produced.

287.When the pressure is lower than the allowable value of the bottom hole pressure, the liquid is not allowed to be extracted, unless otherwise specified in the design documents.

## 7.2.4 Chapter 17 Well Work System Management and Development

288.The number of production and injection wells, the sequence of commissioning and the optimal working system are specified in the basic design documents, and the oil and gas exploration and development design documents, depend on the indicators employed, including the level, velocity and dynamics of formation oil and gas and liquid extraction and the level, velocity and dynamics of displacement agent injection.

289.Combined with the main development indicators adopted, as well as the various restrictions and suggestions stipulated in the basic design documents and oil and gas exploration and development design documents or development analysis reports, the mineral resource user shall stipulate the liquid production standard of each oil production well and the displacement agent injection amount (absorption capacity) of each

284. 沿地层或沿管外空间发生不可控气窜的油井，不允许生产。

285. 通过管间（管外）空间的自喷井或转为机械开采的井，不允许生产。

286. 油气比超过基础设计文件或油气勘探开发设计文件规定设计值的井，不允许生产。

287. 当压力低于井底压力容许值时，不允许提液，除非设计文件另有规定。

## 7.2.4 第17章 井的工作制度管理和制定

288. 采油井和注入井的数量、投产顺序及最佳工作制度，由基础设计文件和油气勘探开发设计文件规定，取决于采用的指标，包括地层油气和液体开采水平、速度和动态及驱替剂注入水平、速度和动态。

289. 结合采用的主要开发指标，以及基础设计文件和油气勘探开发设计文件或开发分析报告规定的各种限制和建议，由矿产资源利用者规定每口采油井的采液标准和每口注入井的驱替剂注入量（吸收能力），并形成井的工艺运行制度。

injection well, and form a process operation system for the well.

290. In order to supervise the production of wells, and to count the amount of liquid produced and the implementation of geological-technical measures, the user of mineral resources shall prepare and keep the documents of the oilfield in the form of electronic and paper versions throughout the operation cycle of mineral resources exploitation, including the oilfield trial production:

(1) Make daily measurement of oil, water, natural gas and condensate gas produced on the well;

(2) Make daily and monthly reports of oil, water, natural gas and condensate gas produced in oilfields.

291. According to the stability of the stratigraphic mining conditions, the oil production well process operation system is prepared by the mineral resource utilizer, and is prepared once a month or quarter; Prepare and approve the geological-technical measures plan at the same time as the production well process operation mode to ensure that the liquid volume of the well and the production layer is reasonably recovered.

292. In the process operation system of oil production wells, the following main parameters are clarified according to the production mode:

(1) liquid production, oil production, condensate production, gas production, water content and oil-gas ratio;

(2) Bottom and wellhead pressure, or the position of the moving fluid level in the well;

(3) Nozzle diameter, tubing diameter and depth (self-injection well);

(4) Plunger diameter, stroke, stroke, pump

290. 为了对井的生产情况进行监督，并统计产液量和实施的地质—技术措施，矿产资源利用者应在整个矿产资源开采作业周期内，以电子版和纸质版的形式编制并保存油田相关资料文件，其中包括油田试采时：

（1）按照规定格式，做好井上采出石油、水、天然气和凝析气计量日报；

（2）按照规定格式，油田采出石油、水、天然气和凝析气计量日报和月报。

291. 根据层系开采条件的稳定情况，采油井工艺运行制度由矿产资源利用者编写，月或季度编写一次；与采油井工艺运行模式同时编写并批准地质—技术措施计划，保障井和生产层系的液量采出合理。

292. 在采油井工艺运行制度中，根据生产方式，明确以下主要参数：

（1）产液量、产油量、凝析气产量、产气量、含水率和油气比；

（2）井底和井口压力，或井内动液面位置；

（3）油嘴直径、油管直径和下入深度（自喷井）；

（4）柱塞直径、冲次、冲程、泵规格尺寸和下入深度（采用泵生产时）；

size and depth (when using pump production);

(5) Gas unit flow and working pressure, starting valve and working valve installation depth (gas lift production);

(6) Packers, air anchors, metering valves, bottom-hole nozzle types and depths, etc.;

(7) Perforated well section and open hole well section.

293. The user of mineral resources shall supervise the implementation of the process operation system specified in the oil production well.

294. If there is no equipment to measure production and conduct well research alone or in groups, production of oil production wells is not allowed according to the basic design documents and oil and gas exploration and development design documents.

295. In order to supervise the system of operation of wells, the measuring tools used in the process shall comply with the Law of the Republic of Kazakhstan on the guarantee of measuring units.

296. In order to monitor the production of the well, a comprehensive study is required to determine the following main parameters: formation fluid composition, viscosity, density, saturation pressure, gas content, volume factor, composition of separated gas and degassed crude oil, as well as formation fluid gas content, volume coefficient and density vs. pressure. The geological production department of the enterprise shall formulate a sampling plan for oil/condensate deep samples based on the commissioning of new wells and their uniform distribution in the reservoir range. Deep sampling and study of formation fluids should be carried out for each production horizon, and the number of wells studied per year should

（5）气体单位流量和工作压力，启动阀和工作阀安装深度（气举生产）；

（6）封隔器、气锚、计量阀、井底接管嘴类型和下入深度等；

（7）射孔井段、裸眼井段。

293. 由矿产资源利用者对采油井规定的工艺运行制度执行情况进行监督。

294. 如果没有设备单独或成组测定产量和进行井研究，根据基础设计文件和油气勘探开发设计文件，不允许采油井生产。

295. 为了监督井的运行制度，在工艺流程中使用的测量工具，应符合哈萨克斯坦共和国测量单位保障法。

296. 为了监督井的生产状况，需要进行综合研究来测定以下主要参数：地层流体成分组成，黏度、密度、饱和压力、含气量、体积系数、分离出的气体和脱气原油的成分，以及地层流体含气量、体积系数和密度与压力的关系。由企业地质生产处结合新井投产及其在油气藏范围的均匀分布情况制订石油/凝析气深层样品取样计划表。应对每个生产层系进行地层流体深层取样和研究，每年研究的井数量应不少于总采油井数的5%。

not be less than 5% of the total number of producing wells.

297.For oil and gas reservoirs developed on the offshore continental shelf, the sampling period and research workload are determined in combination with all possible risk factors (high pressure, high temperature, hydrogen sulfide content).

298.The data of the well operation system should be preserved, analyzed and summarized. The user of mineral resources shall effectively supervise and analyze the implementation of the determined process mode, find out the reasons for violating the process mode, and implement measures to improve the efficiency of wells and production equipments.

299.The users of mineral resources summarize the operation mode and analysis results of each development layer, each well area and each production mode well, and reflect them in the annual report.

300.The mineral resource utilizer manages the technology documents of each injection well, which reflects all the indicators of well production, as well as the implementation of geological-technical measures and effects, and checks the reliability and tightness of the wellhead equipment and production casing.

301.By analyzing the pressure recovery curve, using a deep flowmeter, resistivity measuring instrument, electric thermometer, radiological isotope, etc., and using the packer on the tube to test the casing in sections, the sealing of the casing and whether the injection well has extra-tube circulation were determined.

302.The technical status of the production wells and downhole equipments should be ensured:

297.针对在海上大陆架开发的油气藏，结合所有可能存在的危险因素（高压、高温、硫化氢的含量），确定取样周期和研究工作量。

298.井运行制度的资料应保存、分析并总结。矿产资源利用者对确定的工艺模式执行情况进行有效监督和分析，查找违背工艺模式的原因，实施提高井和生产设备工作效率的措施。

299.矿产资源利用者总结各开发层系、各井区、各生产方式井的运行模式和分析结果，并在年度报告中加以体现。

300.矿产资源利用者管理每口注入井的技术文件，技术文件体现井生产的所有指标，以及实施的地质—技术措施及效果，检查井口设备和生产套管的可靠性和密封性。

301.通过分析压力恢复曲线，利用深层流量计、电阻率测量仪、电动温度计、放射学同位素进行研究，借助管上封隔器对套管进行分段试压等方式，确定套管密封性及注入井是否有管外循环。

302.采油井和井下设备的技术状态应保证：

(1) the implementation of well production in accordance with the approved process mode within a certain period of time;

(2) Supervise the parameters of the well operation mode (measure the wellhead and extra-tube space pressure, liquid production and gas production, output moisture content, working pressure and gas unit flow, pump inlet pressure and its production capacity, wellhead sampling);

(3) the implementation of hydrodynamic studies with the purpose of supervising the condition of wells and downhole equipments, determining the nature dynamics of formations and produced products, and supervising and adjusting the development process;

(4) implement measures to prevent complications during well production;

(5) Implement the formation wellbore parts and the work areas near the well.

303. If the new well is not equipped with technical equipments related to individual metering and single-well dynamic studies, it will not allowed to be put into production.

304. Production wells that are not equipped with the following devices are prohibited from production: wellhead pressure gauge and bottom hole pressure gauge, device for wellhead sampling and wellhead temperature determination, and reinforcing pads and lubricants for lowering downhole instruments (pressure gauge, thermometer, flow meter, sampler). Except for the following cases:

(1) Wells produced by gas lift mode need to install wellhead exhaust fitting devices, and need to install pressure gauges, flow meters and other equipments used to measure and regulate pressure and working gas flow;

（1）按照一定时期内批准的工艺模式实施井生产；

（2）监督井运行模式参数（测量井口和管外空间压力、产液量和产气量、产出物含水率、工作压力和气体单位流量、泵入口压力及其生产能力、井口采样）；

（3）实施流体动力研究，目的是监督井和井下设备的状况，确定地层和采出物的性质动态，监督和调整开发过程；

（4）实施预防和防止井生产时出现复杂情况的措施；

（5）实施地层井筒部分和近井地带工作。

303. 新井未安装用于单独计量和单井动态研究相关的技术设备，则不允许投产。

304. 没有安装以下装置的采油井禁止进行生产：井口压力计和井底压力计，用来进行井口取样和测定井口温度的装置及用来下入井下仪器（压力计、温度计、流量计、取样器）的加强垫和润滑剂。以下情况除外：

（1）采用气举方式生产的井，需安装井口排气配件装置，另外需要安装压力计、流量计和其他用来测量调节压力和工作气流量的设备；

(2) The wellhead of the rod pump production well needs to be installed to complete the following operations: calibration of the well power meter, measurement of the liquid level in the well with an echo sounder or wavelength meter, and taking gas samples from the oil jacket space to determine the cause of the pressure between the pipes;

(3) A management station should be set up at the wellhead of the electric centrifugal pump production well, which can monitor and change the operation mode of the device. A special remote-controlled mechanical device is installed underground to ensure the measurement of pump inlet pressure and temperature;

(4) The water injection wells produced by downhole and surface equipments are used to monitor the water absorption capacity, water injection pressure and water flooding of the injection wells.

305.The interaction between injection and production wells, the movement path of injection reagents along the formation, and the pressure changes at different locations in the formation were studied by interfering with well testing, geophysical methods, adding indicators to the injected water and observing the occurrence of indicators in the well output.

306.In accordance with the approved integrated geophysical and hydrodynamic study plan, the periodicity and workload of the downhole study work are determined by the mineral resource utilizer in conjunction with the requirements of the basic design documents, oil and gas exploration and production design documents.

307.When the operation mode of the production well is destroyed, immediate

（2）有杆泵生产井的井口，需要安装用来完成以下作业的装置：井功率计校准，用回声测深仪或波长表测量井内液位，从油套空间取气样用以确定产生管间压力的原因；

（3）电动离心泵生产井的井口应设置管理站，可以监测和改变装置的运行模式。井下安装专门的可遥控机械装置，保障泵入口压力和温度的测量；

（4）利用井下和地面设备进行生产的注水井，对注入井的吸水能力、注水压力、水驱波及情况进行监测。

305.通过干扰试井、地球物理方法、向注入水中添加指示剂并观察指示剂在油井产出物中的出现情况，研究注水井和采油井之间的相互影响、注入试剂沿地层的移动路径，以及地层不同位置的压力变化。

306.根据批准的综合地球物理研究和流体动力学研究计划，由矿产资源利用者结合基础设计文件、油气勘探和开采设计文件的要求，确定井下研究工作的周期性和工作量。

307.当采油井的运行模式被破坏时，立即采取措施查明引起不同开发阶段井的实际工作参数与设计值产生偏差的原因（井下形成砂堵，井底气窜或水窜、石蜡、盐水合物、腐蚀性产出物沉淀等），并采取处理措施。

measures should be taken to find out the reasons for the deviation between the actual working parameters of the well and the design value at different development stages (sand plugging formation downhole, gas channeling or water channeling at the bottom of the well, paraffin, salt, hydrate, corrosive output precipitation, etc.), and take treatment measures.

308.Wells with severe sand production need to take measures to reinforce the area near the well. Depending on the situation, the reinforcement methods (installation of screens, cement sealing, gum, polymer treatment, etc.) are selected.

309.According to the cause of the complex situation of gas channeling or water channeling at the bottom of the well, it can be eliminated by changing the process mode of the well or implementing plugging measures.

310.The treatment methods and means of other complex situations (salt, paraffin, hydrate precipitation, string/equipment erosion/corrosive wear) are selected according to the specific conditions and effects.

311.The characteristics and severity of the complications (decreased suction capacity of the well, uneven suction profile, and destruction of the tightness of the casing and cement sheath) depend on the mode of operation of the injection well and the degree to which the injection reagent parameters, characteristics, and design values of the injection well are consistent with each other.

312.When injecting gas (air) into the formation, the structure of the injection well should meet the design requirements of the gas injection well.

313.When injecting different heat carriers (hot water, steam) into the formation, special

308. 出砂严重的井需要采取措施，加固近井地带。根据具体情况，选择加固方法（安装筛管，水泥封固，胶质、聚合物处理等）。

309. 根据井底气窜或水窜复杂情况出现的原因，可以通过改变井的工艺模式或实施封堵措施加以消除。

310. 其他复杂情况（盐、石蜡、水合物沉淀、管柱/设备侵蚀/腐蚀性磨损）的处理办法和手段，根据具体条件和效果，进行选择。

311. 注入井生产时，复杂情况的特点和严重程度（井的吸入能力下降、吸入剖面不均衡、套管和水泥环的密封性被破坏）取决于注入井的工作模式，以及注入井注入试剂参数、特点和设计值的吻合程度。

312. 在向地层注气（空气）时，注入井的结构应满足注气井的设计要求。

313. 在向地层注入不同热载体（热水、蒸汽）时，需要规定专门的措施降低套管—水泥环体系的热张力，尤其是在井工作制度不稳定的情况下。

measures need to be prescribed to reduce the thermal tension of the casing-cement sheath system, especially when the well working system is unstable.

314.In order to improve the productivity and injection capacity of the well, improve the hydrodynamic connection between the well and the formation, balance the liquid production profile and the injection profile, and speed up the progress of well testing and production, the mineral resource users plan and implement various treatment methods for the near-well zone and the near-well areas of the formation (various acidizing treatment, hydraulic fracturing, vibration deplugging, thermal washing, hydrodynamic action methods and a combination of various methods).

315.According to the comprehensive research results of the formation near the well zone, the rock and liquid composition, as well as the systematic summary and study of the results of the implementation of different treatment methods of the well and the study strata by the workover contractor or service unit, the mineral resource users select the specific treatment methods.

316.In daily (downhole) workover including the following work:

(1) Replace all or part of the downhole equipment due to wear or accidental failure of the downhole equipment;

(2) Clean up all kinds of sediments (sand, paraffin, salt and corrosive outputs, etc.) at the bottom of the well.

The overhaul of the well includes the following work:

(1) the maintenance of the well‐plugging work;

314.为了提高井的产能和注入能力，改善井与地层的流体动力学联系，平衡产液剖面和注入剖面，加快井测试和投产进度，由矿产资源利用者计划并实施各种近井地带和地层近井部分的处理办法（各种酸化处理、水力压裂、震动解堵、热洗、流体动力学作用方法及各种方法的组合）。

315.根据地层近井地带状况、岩石和液体成分的综合研究结果，以及修井承包单位或服务单位对井和研究层系的不同处理办法实施结果的系统总结和研究，由矿产资源利用者选择具体的处理办法。

316.在日常（井下）修井包括以下工作：

（1）因井下设备磨损或意外故障，全部或部分更换井下设备；

（2）清理井壁和井底的各种沉淀物（砂、石蜡、盐和腐蚀性产出物等）。

井的大修包括以下工作：

（1）井的维修—封堵工作；

(2) Layer replacement or lamination;

(3) change the type of well;

(4) Deal with accidents that occur during well production or maintenance (extraction tubing, electric centrifugal pump, deep well rod pump and cleaning of well wall, etc.);

(5) Well maintenance with packer-throttle valve and sidetracking equipment for the implementation of two formation separation;

(6) Injection well maintenance: adjust the suction profile, limit the injection water from entering other formations, repair the casing, restore its integrity and tightness, etc.;

(7) Hole filling and underground blasting;

(8) Well sealing and scrapping.

318.In order to avoid damaging the tightness of the oil and gas reservoir, hydraulic fracturing is prohibited for layered oil and gas reservoirs with small caprock thickness.

319.In order to strengthen oil recovery, in wells with intact technology, near-well zone and formation near-well part treatment operations should be carried out, including formation hydraulic fracturing, radial drilling of formation, use of liquid flow direction modulation and flooding technology, acoustic plugging, thermochemical action, electrical action, wave treatment, chemical treatment and formation maintenance-plugging operations, with the purpose of preventing water from entering the production well from entering the oil production well through the high permeability part of the formation.

320.The treatment work of the near-well zone and the near-well part of the formation does not belong to the overhaul of the well or the daily (downhole) maintenance.

321.Wellhead and wellbore equipments should be installed to adjust the density of the

（2）换层或合层；

（3）改变井的类型；

（4）处理在井生产或维修过程中出现的事故（提取油管、电动离心泵、深井有杆泵和清理井壁等）；

（5）安装有实施两个地层分采的封隔器—节流阀及侧钻设备的井维修；

（6）注入井维修：调整吸入剖面，限制注入水进入其他地层，修复套管，恢复其完整性和密封性等；

（7）补孔和井下爆破；

（8）井封存和报废。

318.为了避免破坏油气藏的密封性，禁止对盖层厚度小的层状油气藏进行水力压裂。

319.为了强化采油，在工艺完好的井中，可以进行近井地带和地层近井部分处理作业，包括地层水力压裂、径向钻开地层、采用液流转向调驱技术、声波解堵、热化学作用、电作用、波处理、化学处理及地层维修—封堵作业，目的是防止水通过地层高渗透部位从注入井窜入采油井。

320.近井地带和地层近井部分处理工作不属于井大修和（或）日常（井下）维修。

321.应安装井口和井筒设备，调整工作液的密度，以防止发生油气溢出。

working fluid to prevent oil and gas overflow.

322.For workover in accordance with the work plan approved by the management of the mineral resources users (units), the design documents need to contain the name of the unit, the name of the oil and gas reservoir, the exploration or development stage and design scheme of the oil and gas reservoir, the well number and geographical coordinates, the design and actual well depth, the date of drilling and completion, the actual well structure, the planned measures and construction period, the contract number, and the reason for construction.

323.The content of the workover work, the maintenance cycle of the equipments and the wells, and the technologies and economic effects of the implementation of the work are kept by the mineral resource user, and the storage period is the entire production cycle of the layer.

### 7.2.5 Chapter 18 Oil & Gas Field Development Monitoring

324.In order to evaluate the effectiveness of the adoption of the development system and to obtain the information necessary to develop measures to improve the development systems, it is necessary to monitor the development process of the production strata of the oil and gas field.

325.Mineral resource users shall organize the monitoring of mineral resources status and the supervision of oil and gas field development.

326.In the development of oil fields and oil and gas fields, the necessary comprehensive field studies including:

(1) the use of downhole pressure gauges

322.按照矿产资源利用者（单位）管理层批准的工作计划进行修井，设计书中需要包含单位名称、油气藏名称、油气藏所处的勘探或开发阶段和设计方案、井号及地理坐标、设计和实际井深、开钻和完钻日期、实际井身结构、计划的措施及施工期、合同编号、施工原因简单论述。

323.由矿产资源利用者保存修井工作的内容、采用设备和井的检修周期及实施工作的技术—经济效果的信息，保存期为层系的整个生产周期。

### 7.2.5 第18章 油气田开发监测

324.为了评价采用开发系统的效果，获得制定完善开发系统措施所需的信息，需要对油气田生产层系的开发过程实施监测。

325.矿产资源利用者应组织进行矿产资源状况监测和油气田开发监督。

326.在开发油田和油气田时，必要的综合现场研究包括：

（1）利用井下压力计和其他方式，测定整个层系和包含多套开发层系内地层压力和井底压力；

and other methods to determine the entire strata and the internal strata and bottomhole pressures including multiple sets of strata;

(2) Determination of surface oil and gas production and liquid production by means of a single or mobile measuring device with rope ladder and metering container, or at a collection station with the help of an automatic grouping device;

(3) Downhole flow measurement tools or flowmeters (PLT) are used to measure the production of multiple strata of wells;

(4) Measure the ratio of production gas to oil in each production layer;

(5) Determine the moisture content of the well produced fluid according to the liquid sample obtained on the discharge pipes or measuring devices;

(6) Use the wellhead pressure gauge to measure the injection pressure of each injection well, use the flow meter of the meter or cluster pumping station to measure the injection amount of working agent in each well, and measure the suction capacity of a single layer containing multiple sets of development layers;

(7) Deep flow meter or other means;

(8) Hydrodynamic studies of production and injection wells in stable mode and unstable mode;

(9) Draw the current cumulative recovery of oil, gas and liquid, isobar diagram;

(10) Mine geophysical studies to determine the original and current oil, gas and water saturation of the formation and the technical state of the well;

(11) Downhole and surface sampling of well outputs and laboratory studies;

(12) Monitor whether the injected working agent meets the requirements of the

（2）利用单独的或带绳梯和计量容器的移动式测量装置，测定地面油气产量和产液量，或借助自动成组装置在收集站测定；

（3）利用井下流量测量工具或流量计（PLT），测量开发多个层系的井单地产量；

（4）测量各个生产层系的生产气油比；

（5）根据在排出管线或测量装置上获得的液样，测定井产液的含水率；

（6）利用井口压力计，测量各注入井的注入压力，利用计量器或丛式泵站的流量计，测量各井工作剂注入量，测量包含多套开发层系的单层的吸入能力；

（7）深层流量计或其他方式；

（8）稳定模式和不稳定模式下采油井和注入井的流体动力学研究；

（9）绘制油、气和液体当前累计采出量平面图及等压线图；

（10）测定地层原始和当前含油、气、水饱和度及井的技术状态的矿场地球物理研究；

（11）进行井产出物的井下和地面取样，并开展实验室研究；

（12）监测注入的工作剂是否符合设计的要求。

design.

327.Wells that have not completed an integrated mine study alone are not allowed to be put into production.

328.In addition to the list of systematic measurement plans, special studies are carried out according to a separate plan to monitor the temperature of the development layer and the state of injection of working agents, inject tracers to evaluate formation performance, study the possibility of formation wax formation, observe sulfate reduction, and observe inter-well disturbances. If validated by laboratory studies, including core studies, individual water injection quality standards can be developed.

329.The research on the development and monitoring of the production layer should be carried out by the mineral resource user himself or by the professional units with the relevant operation license hired by the mineral resource user in accordance with the monthly plan prepared by the mineral resource user.

330.During the whole period of reservoir development, the first-hand information of the development and monitoring of the production strata should be kept by the mineral resource utilizer.

331.Combined with the geological-physical conditions of the production strata and the proposed development systems, the characteristics and periodicity of the comprehensive measurement methods are demonstrated in the development and design documents of the production strata.

332.The workload and periodicity of reservoir monitoring studies at each stage of development are individually formulated for each production horizon.

333.The comprehensive study of the development and monitoring of production

327.尚未单独完成综合矿场研究的井，不允许投产。

328.除了系统测量计划清单，按照单独的计划实施专项研究，以监测开发层系温度和工作剂注入状态，注入示踪剂评价地层性能，研究地层结蜡可能性，观察硫酸盐还原作用，井间干扰情况。如果经实验室研究论证（包括岩心研究），则可以制定单独的注水质量标准。

329.生产层系开发监测研究，按照矿产资源利用者编写的月度计划，由矿产资源利用者自行实施，或由其聘请的有相关作业许可证的专业单位实施。

330.在整个油气藏开发期内，由矿产资源利用者保管生产层系开发监测的第一手资料。

331.结合生产层系的地质—物理条件及建议的开发系统，在生产层系开发设计文件中论证综合测量方法的特点及周期性。

332.为每个生产层系单独制定各开发阶段油气藏监测研究的工作量和周期性。

333.生产层系开发监测综合研究规定系统性的（周期性的）和单一性的（单次的）测量。

layers prescribes systematic (periodic) and single (single) measurements.

334.When conducting systematic studies, the periodicity of each study is recommended as follows:

(1) Formation pressure measurement:in the first three development phases, once a quarter;in the last stage of development, once every six months.

(2) Bottomhole pressure measurement in production wells and injection wells, at least once a quarter.

(3) Weekly measurement of production well production and injection well suction capacity.

(4) Measure the water cut of the well every week.

335.When the formation pressure and bottomhole pressure are higher than the saturation pressure, the gas-oil ratio is measured once a year; when the bottomhole pressure drops below the saturation pressure, the gas-oil ratio is measured quarterly; when the formation pressure is equal or lower than the saturation pressure, the gas-oil ratio is measured monthly.

336.For new wells and old wells, before and after the implementation of a certain process or technical measure (near-well zone treatment, hydraulic fracturing and plugging operations, etc.), a comprehensive measurement should be carried out, and the period of comprehensive measurement after that should be indicated.

337.As needed, the fluid dynamics studies of the pressure recovery (liquid level recovery) method and the stable drainage method were carried out on each well after the well was put into production.

338.Measure the amount of suspended

334. 在进行系统性研究时，各项研究的周期建议如下：

（1）地层压力测量：在前三个开发阶段，每季度一次；在最后一个开发阶段，每半年一次。

（2）在产采油井和注入井井底压力测量，每季度至少一次。

（3）每周测量采油井产量和注入井吸入能力。

（4）每周测量井的含水率。

335. 当地层压力和井底压力高于饱和压力时，每年测定一次气油比；当井底压力降至饱和压力以下时，每季度测定气油比；当地层压力与饱和压力持平或低于饱和压力时，每月测定气油比。

336. 对于新井及老井在某项工艺措施或技术措施（近井地带处理、水力压裂和封堵作业等）实施前后，应进行一次列举的综合测量，并注明之后综合测量的周期。

337. 根据需要，在井投产后对每口井进行压力恢复（液面恢复）法和稳定排液法流体动力学研究。

338. 根据需要，测量注入水的悬浮颗粒、油品和其他杂质的含量。

particles, oils, and other impurities injected into the water as required.

339. A single (single) determination can complete a full set of comprehensive studies or partial studies at the same time, and can be carried out in each newly completed well and before and after the implementation of a certain process or technical measure (near-well zone treatment, overhaul and replacement of equipments, etc.) in the old well.

340. A single measurement consists of measuring the oil and gas saturation of the formation, and the study workload has increased significantly since the well was flooded. The hydrodynamic study of detecting the interaction between wells and formations and measuring the reservoir profile is also a single measurement.

341. Development monitoring should carried out in production wells, injection wells, observation wells and pressure monitoring wells in use; The number and location of monitoring wells are determined by the industrial development design scheme.

342. In the development of gas fields and condensate gas fields, the monitoring of production strata including:

(1) systematic and supervised measurement and determination of formation pressure, downhole pressure and wellhead pressure;

(2) the liquid level of the pressure monitoring well;

(3) the position of the gas-water interface (when there is an oil ring, it is the oil-gas interface and the oil-water interface);

(4) Determination of the production and chemical composition of natural gas, condensate and water (petroleum).

343. All of the above studies are carried

339. 单一（单次）测定可以同时完成全套综合研究或部分研究，可在每口新完钻井中及老井某项工艺或技术措施（近井地带处理、大修和更换设备等）实施前后进行。

340. 单次测定包括测定地层的含油气水饱和度，并研究工作量自井水淹后明显增加。检测井和地层间相互影响及测量油气藏剖面的流体动力学研究，也属于单次测定。

341. 在采油井、注入井及在使用的观察井和压力监测井中，实施开发监测；监测井的数量和位置由工业开发设计方案确定。

342. 气田和凝析气田开发时，生产层系监测包括：

（1）地层压力、井底压力和井口压力的系统性和监督性测量和测定；

（2）压力监测井的液面；

（3）气水界面的位置（有油环时则为油气界面和油水界面）；

（4）测定天然气、凝析油和水（石油）的产量和化学成分。

343. 在井测试时、停产或封存期结束后投产之前，进行上述所有研究。

out at the time of well testing, before production put into production after the shutdown or mothballing period has ended.

344. According to the research results, determine and regularly update:

(1) reservoir operation system and temperature conditions;

(2) original and current oil and gas reserves;

(3) pressure distribution of oil and gas reservoirs and production strata;

(4) the interaction between different regions of the reservoir;

(5) the strength and characteristics of water (oil) propulsion in each block;

(6) gas-producing intervals with yield evaluation;

(7) The utilization of reserves in the development process;

(8) Identify possible out-of-pipe channeling.

345. Periodic hydrostatic measurements were performed on all wells. At the beginning of development, at least once a quarter; and gradually adjust the frequency to once a year at the end of development.

346. In reservoirs with a large number of wells and a long pressure recovery period (more than five days), the measurement period can be adjusted.

347. When heterogeneous reservoirs are treated, the formation pressure in different areas of the reservoir will decrease unevenly. Therefore, in the area where the pressure drop is greatest, a static pressure measurement should be performed on a single well group and production of the well should be stopped at the same time.

348. Periodic measurement of the static

344. 根据研究结果，确定和定期更新：
（1）油气藏运行制度和温度状况；
（2）原始和当前油气储量；
（3）油气藏、生产层系压力分布；
（4）油气藏各区域间的相互影响；
（5）各个区块水（油）推进强度和特点；
（6）带有产量评价的产气层段；
（7）开发过程储量动用情况；
（8）查明可能存在的管外窜流。

345. 对所有井进行定期静压测量。在开发初期，每季度至少一次；在开发末期将频率逐渐调整为每年一次。

346. 在井数多和压力恢复期长（超过五天）的油气藏，测量周期可进行调整。

347. 对非均质储层进行处理时，油气藏不同区域的地层压力将不均衡下降。因此，在压降最大的区域，应当对单个井组进行静压测量，同时停止井的生产。

348. 定期测量井口的静压，需要结合压力恢复曲线。根据生产层位的特点与地层压力恢复时间的关系，确定测量周期。

pressure at the wellhead needs to be combined with a pressure recovery curve. According to the relationships between the characteristics of the production horizon and the formation pressure recovery time, the measurement period is determined.

349.The formation pressure measurement period of each well is determined by the industrial development and design scheme according to the gas production rate and the demonstration of the formation pressure drop, and the formation pressure need to be selected and calculated to determine it, so as to ensure that the formation pressure drop value shall not exceed three times the value caused by the measurement error during the interval between the two groups of average formation pressure measurements.

350.Development monitoring is carried out in production wells, injection wells, observation wells and pressure monitoring wells in use; The number and location of monitoring wells are determined by the industrial development design scheme.

351.Within the gas-bearing range of the production horizon, the well drilled in the production horizon can be used as an observation well. Observation wells are not in production and are used to measure pressure and observe changes in the gas-water interface (oil-gas interface and oil-water interface). Once the problem is solved, the observation well can be converted into a production well.

352.In the water-bearing range of the production horizon, the well drilled into the production horizon belong to the pressure monitoring well, which used to observe the decline of the liquid level of edge water or bottom water.

349. 各井地层压力测量周期，由工业开发设计方案根据采气速度及论证的地层压力下降情况确定，需要选择地层压力并进行计算加以确定，以保证两组平均地层压力测量间隔期内，地层压力下降值不超过因测量误差引起值的三倍。

350. 在采油井、注入井及在使用的观察井和压力监测井中，实施开发监测；监测井的数量和位置由工业开发设计方案确定。

351. 在生产层位含气范围内，钻遇生产层位的井，可作为观察井。观察井不进行生产，用于测定压力、观察气水界面（油气界面和油水界面）的变化。问题解决后，观察井可转为生产井。

352. 在生产层位含水范围内，钻遇生产层位的井，属于压力监测井，用于观察边水或底水液位下降情况。

353. When determining the number and location of observation wells and pressure monitoring wells, it is necessary to use the drilled exploration wells as much as possible; for small oil and gas reservoirs, only such wells can be used.

354. Observation wells and pressure monitoring wells are measured at least once every 1.5~2 months.

355. For oil and gas reservoirs with high gas content and oil and gas reservoirs with complex structure, it is necessary to have the pressure distribution data in the reservoir plane and longitudinally, that is, the data of each part of the production horizon in the longitudinal direction.

356. For each flooded gas well, a study should be carried out to find out the cause of the flooding.

357. Monitoring the intrusion of formation water into a reservoir during development can be implemented using hydro chemical, mine-geophysical and hydrodynamic methods.

358. An effective water chemistry detection methods require a systematic observation of the variation of representative ion content in the effluent from all production wells. Representative ions of different sedimentary layers and regions are determined experimentally; water samples should be taken on a quarterly basis (for special analysis) and monthly (for full analysis) in wells with original flooding characteristics.

359. Geophysical monitoring of the mine site, implemented through specialized radio-logging, determines the uplift of the gas-water interface in both production and observation wells. The research period should be determined according to specific conditions, and it should be no less than 1~2 times a year.

353. 在确定观察井和压力监测井的数量和位置时，需尽可能使用完钻的探井；对于小型油气藏，只能使用此类井。

354. 观察井和压力监测井每 1.5~2 个月至少进行一次测量。

355. 对于高含气的油气藏，以及构造复杂的油气藏，必须有油藏平面和纵向上的压力分布数据，即生产层位纵向各部位的数据。

356. 针对每口水淹的气井，应进行研究，查明水淹原因。

357. 在开发过程中监督地层水侵入油气藏的情况，可以利用水化学、矿场—地球物理和流体动力学方法实施。

358. 有效的水化学检测方法要求对所有生产井携出水中的代表性离子含量变化情况进行系统性观察。不同沉积层和区域的代表性离子通过实验确定；应按季度取水样（用于特别分析），而在有原始水淹特征的井中，应每月取水样（用于全面分析）。

359. 矿场地球物理监测，通过专门的放射性测井实施，其可确定生产井和观察井中气水界面的抬升。根据具体条件确定研究周期，每年应不少于 1~2 次。

360.Natural gas development accounting should reflect the effective use of gas exploitation, well research, various purge pipeline gas and natural gas losses in the event of an unexpected blowout. These and other potential losses are reflected in the mineral resource utilizer's reserve balance sheet.

361.If a significant loss of natural gas occurs before production begins, in order to assess the loss, formation pressures within all existing wells need to be measured. The results of the evaluation should be included in the balance sheet of reserves, with an explanation of the cause of the loss.

362.Under well operating conditions, studies are conducted twice a year on each well to determine the condensate content, including the determination of the original condensate and the content of the stabilized condensate at the time of cryogenic separation. Based on these studies, the relationships between formation pressure and condensate content are shown in the form of graphs.

363.The main physicochemical properties of the stabilized condensate were determined according to the same period to obtain the curve relationship between the formation pressure and the specific gravity and molecular weight of the condensate.

## 7.2.6　Chapter 19　Oil and Gas Field Development Adjustments

364.The purpose of development and adjustment is to reasonably change the seepage direction and speed of formation fluid, and create favorable conditions for formation oil discharge. Adjustments are implemented throughout the development period of the reservoir.

360. 天然气开发核算应反映有效利用气的开采、井研究和各种吹扫管线用气及意外井喷时的天然气损失。这些损失及其他可能的损失，在矿产资源利用者储量平衡表中都要加以体现。

361. 如果在开始生产之前发生大量的天然气损失，为了评估损失情况，需要测量所有现有井内的地层压力。评价结果应写入储量平衡表，并对损失原因做出解释。

362. 在井工作条件下，每年对每口井进行两次测定凝析油含量的研究，包括在低温分离时测定原始凝析油和稳定凝析油的含量。基于这些研究，用图表形式展示地层压力与凝析油含量的关系。

363. 按照相同的周期，测定稳定凝析油的主要物理化学性质，以获得地层压力与凝析油比重和分子量的曲线关系。

## 7.2.6　第19章　油气田开发调整

364. 开发调整的目的在于合理改变地层流体的渗流方向和速度，为地层泄油创造有利条件。在油气藏的整个开发期内实施调整。

365.By adjusting and improving the development process, the following objectives are achieved:

(1) Ensure that the annual oil and gas production of the development layer specified in the design documents can be realized;

(2) the conditions required to achieve the designed oil and gas recovery;

(3) Optimize the economic indicators, while maximizing the use of completed wells, reducing the cost of displacement reagent injection, and reducing the amount of associated water recovery.

366.The demonstration and selection of development and adjustment methods,and methods depend on the defined objectives and tasks, and the specific geo-physical conditions. Combined with the general principle of development and adjustment, the management idea of the development process of the production layer based on scientific basis, the adjustment methods are selected.

367.The principle of adjustment varies according to different geo-physical conditions. When using water flooding, the following principles can be adopted:

(1) In a single-layer, relatively homogeneous production layer, the oil-bearing boundary or the leading edge of the injected water is uniformly advanced towards the central discharge;

(2) Statistical plane permeability heterogeneity in the single-layer production layer with obvious hyperpermeability bands;

(3) Accelerate the production of large parts of the oil and gas reservoir, inject water to naturally cut the reservoir into multiple low-permeability blocks, and then supplement the development of these blocks;

(4) In a multi-layered system formed by

365. 通过调整和完善开发过程，实现以下目标：

（1）保障设计文件规定的开发层系油气年度产量得以实现；

（2）有利于实现设计油气采收率的所需条件；

（3）优化经济指标，同时最大化使用完钻井，减少驱替试剂注入费用，减少伴生水采出量等。

366. 开发调整方法和方式的论证和选择，取决于确定的目标和任务及具体的地质—物理条件。结合通用的开发调整原则，即以科学依据为基础的生产层系开发过程管理思路，选择调整方式。

367. 调整原则因不同的地质—物理条件而异。在采用水驱时，可以采用以下原则：

（1）在单层、较均质生产层系中，将含油边界或注入水前缘向中心排状均匀推进；

（2）在有明显高渗条带的单层生产层系中，统计平面渗透率非均质性；

（3）加快动用油气藏产能较大部位，注入水将油藏自然切割成多个低渗透率区块，后续对这些区块进行补充开发；

（4）在由渗透性相近的多个地层形成的多层层系中，当含油边界（注入水的前缘）均匀推进时，所有地层均以相同速度推进；

multiple formations with similar permeability, all formations advance at the same speed when the oil-bearing boundaries (the leading edge of the water injected) are uniformly advanced;

(5) Priority should be given to the lower strata, and when the thickness and permeability of the strata increase from bottom to top, it is necessary to avoid the flooded layer in the multi-layered system;

(6) Ensure that the oil-water interface is relatively uniform on the plane of the block reservoir with high oil saturation. This adjustment principle is also applicable to the development of oil and gas reservoirs under other geo-physical conditions.

368.On the basis of the selected principles, the organization develops and improves the work to ensure the achievements of the set goals in the case of small economic losses.

369.Development adjustments based on the current status of the production layer can be implemented through completed drilling without major modifications or adjustments to the development systems.

370.Without modifying the development system, within the framework of the current development systems, the main methods and means of adjusting the development are:

(1) Changing the working system of the injection well, including increasing or limiting the injection volume, and redistributing the injection volume between wells by changing the injection pressure;

(2) Change the working system of oil production wells, including increasing or limiting the amount of liquid produced in each well or well group, changing the oil production

（5）优先动用下部地层，当地层厚度和渗透率自下而上增大时，需避开多层层系中的水淹层；

（6）保证高含油饱和度的块状油藏平面上油水界面相对均匀抬升。该调整原则也适用于其他地质—物理条件的油气藏开发。

368.在选定的原则基础上，组织开发完善工作，保障在经济损失小的情况下实现既定目标。

369.根据生产层系现状进行开发调整，可以通过已完钻井实施，而无须对开发系统做重大修改或调整。

370.不修改开发系统，在现行开发系统框架内，调整开发的主要方法和手段有：

（1）改变注入井的工作制度，其中包括增大或限制注入量，通过改变注入压力在井间重新分配注入量等；

（2）改变采油井的工作制度，其中包括增大或限制各井或各井组采液量，外排井采油改为内排井采油，关闭高含水井和气油比过高的井，提液等；

from external discharge wells to oil production from internal discharge wells, closing high water cut wells and wells with too high gas-oil ratio, and extracting liquid etc.;

(3) Optimize the perforation method and change the perforation layer of the development layer;

(4) Near-well zone treatment, through acidification, surfactant injection, formation fracturing, etc., to improve the hydrodynamic perfection of the well;

(5) Establish various barriers to isolate or restrict the flow of associated water in the well by injecting cement, injecting other substances and chemical reagent solutions;

(6) With the help of segmented testing, chemical reagents and mechanical additives, such as inert gas injection, thickened water, etc., are used to selectively plug the high permeability interlayer, and adjust the liquid production profile and water absorption profile;

(7) Reliable equipments for layered production of production wells and layered water injection of injection wells;

(8) Drill supplementary wells on individual blocks using the backup wells specified in the design documents;

(9) supplement the pressure of the oil production well by drilling a new injection well or utilizing the flooded oil production well to convert to a water injection well;

(10) Carry out spot water injection;

(11) Change the direction of the liquid flow and the periodic water injection.

371.With the consent of the design units that prepares the development design documents, the mineral resources user shall implement the adjustment and improvement of the design and development system.

（3）优化射孔方式，改变开发层系射孔层段；

（4）近井地带处理，通过酸化、注入表面活性剂、地层压裂等提高井的流体动力学完善程度；

（5）通过注水泥、注入其他物质和注入化学试剂溶液，建立各种屏障，隔离或限制井内伴生水流动；

（6）借助分段测试，利用化学试剂和机械添加物，如注入惰性气体、稠化水等，选择性封堵高渗透率夹层，调整产液剖面和吸水剖面；

（7）采用采油井分层生产和注入井分层注水的可靠设备；

（8）利用设计文件中规定的备用井，在个别区块上钻补充井；

（9）通过钻新的注入井或利用已水淹的采油井转为注水井的方式补充采油井压力；

（10）开展点状注水；

（11）改变液流方向和周期注水。

371.经编写开发设计文件的设计单位同意，由矿产资源利用者实施设计开发系统的调整和完善。

372. It is not allowed to selectively drill wells or infill well patterns in the places with the largest production capacity of oil and gas reservoirs and production strata and the thickest oil and gas reservoirs in order to adjust or increase oil production.

373. When the development system improvement measures can not ensure the effective management of the oil recovery process, the development system should be modified in the following ways:

(1) comprehensive or selective (in blocks with poor reservoir parameters) well pattern encryption;

(2) divide the multi-layer system into several layers with small thickness, and drill independent well patterns in each layer;

(3) change the ways of maintaining the energy of the formation or the types of water flooding;

(4) Significant increase in injection pressure.

374. Save the development system modifications to a previously approved design file attachment or in a new design file containing an evaluation of economic effects and process effects.

375. Gas reservoirs and condensate reservoirs (reservoirs and reservoirs) are developed and adjusted with the aim of improving oil recovery. The development and adjustment of gas reservoir and condensate gas layer including the following measures:

(1) By reducing the gas production to reduce the pressure difference, the formation is prevented from sand and water from rushing into the production well;

(2) improve the production capacity of the well through supplementary perforation of

372. 不允许为了调整或增大采油量，在油气藏、生产层系产能最大的地方及厚度最大的油气藏区块选择性地钻井或加密井网。

373. 当实施的开发系统完善措施不能保障有效管理采油过程时，应通过以下方式进行开发系统修改：

（1）全面或选择性（在储层参数变差的区块）进行井网加密；

（2）将多层层系划分为几个厚度小的层系，在每个层系钻独立的井网；

（3）改变保持地层能量的方式或水驱类型；

（4）显著增加注入压力。

374. 将开发系统修改措施保存到之前批准的设计文件附件中，或保存在包含经济效果和工艺效果评价的新的设计文件中。

375. 进行气层和凝析气层（油气藏和油气层）开发调整，目的是提高凝析气采收率。气层和凝析气层开发调整包括以下措施：

（1）通过降低产气量以减小压差，防止地层出砂，水突进到生产井；

（2）通过生产层段补充射孔、近井地带酸化处理、地层水力压裂等，提高井的生产能力；

the production layer, acidification treatment of the near-well zone, and formation hydraulic fracturing;

(3) Maintain the development of formation pressure and improve the degree of gas and condensate recovery by injecting the leading edge of the working agent, changing the working system of the production well and the injection well, and switching the well cycle;

(4) If the observation and injection wells have completed the tasks assigned at the initial stage, the observation wells and injection wells can be converted into production wells to improve the scope of displacement spread.

376. In the development of multi-layered strata, other supplementary measures can be taken:

(1) Statistical permeability differences of production strata through strata mining and stratified injection of reagents (when reliable equipments are available);

(2) Plugging the formation water or injection water section of the oil production well by injection (cement, chemicals, etc.).

377. In the process of gas field and condensate gas field development, with the consent of the design documents preparation unit, the previously undeveloped strata and horizons can be merged into an existing production horizon, and drilled in the process of production well drilling and subsequent exploration drilling.

378. Stratigraphic merging are allowed in the following situations:

(1) The current formation pressure values are similar, and there is no channeling between the main formation and the incorporated formation (horizon);

（3）通过工作剂注入前缘、改变采油井和注入井的工作制度、开关井周期等，保持地层压力开发，提高气体和凝析油采出程度；

（4）如果观察井和注入井已完成初期赋予的任务，则可以将观察井和注入井转为采油井，提高驱替波及范围。

376. 在开发多层层系时，可采取其他补充措施：

（1）通过分层开采和分层注入试剂（有可靠设备时），统计生产层系地层的渗透性差异；

（2）通过注入（水泥、化学药剂等）方式，封堵采油井的产地层水或注入水层段。

377. 在气田和凝析气田开发过程中，经设计文件编写单位同意，可以将之前未开发的地层、层位合并到一个现有的生产层系，在生产井钻井及后续勘探钻井过程中钻开。

378. 以下情况允许层系合并：

（1）当前地层压力值相近，主要地层和并入地层（层位）之间没有窜流；

(2) The production characteristics and properties of the incorporated strata and the fluid composition of the strata are similar to those of the main strata;

(3) Stratigraphic merging does not lead to dilution of beneficial components mined from the main strata;

(4) In the well, the cement outside the pipe is located on top of the incorporated layer, and can form a reliable plugging.

379. Adopt the recommended measures for the adjustment of the development process, use equipment and supervision methods, ensure that the effect evaluation can be carried out, and implement in the process of development supervision.

380. The development adjustment plans and follow-up measures are an integral part of the development analysis and should be considered when making adjustments and supplements to the development and design scheme.

381. The production mode of gas wells and condensate gas wells are determined by the following geological-technical conditions:

(1) Formation pressure value and the working output of the well;

(2) Physico-chemical properties and commodity properties of natural gas;

(3) The production horizon and the physical properties of the overlying rock (abnormally high stratigraphic pressure and abnormally low stratigraphic pressure);

(4) The rmodynamic conditions of well operation;

(5) The conditions for the formation of hydrates in the wellbore and in the on-site gas pipeline network;

(6) The number of strata mined in a

（2）并入地层的生产特点和性质、地层流体成分与主要层系相近；

（3）层系合并不会导致从主要层系开采出来的有益组分的稀释；

（4）在井中，管外水泥位于并入层之上，且能形成可靠封堵。

379.采用开发过程调整的推荐措施，使用设备和监督办法，保障能实施效果评价，并在开发监督过程中落实。

380.开发调整计划及后续措施，是开发分析的组成部分，在对开发设计方案做出调整和补充时应予以考虑。

381.气井和凝析气井生产方式，由以下地质—技术条件确定：

（1）地层压力值和井的工作产量；

（2）天然气的物理—化学性质和商品性质；

（3）生产层位及上覆岩石的物理性质（异常高地层压力和异常低地层压力）；

（4）井工作的热力学条件；

（5）井筒内和现场输气管网中形成水化物的条件；

（6）单井开采的地层层数及生产层位的钻开程度；

single well and the degree of drilling of the production horizon;

(7) Transform the development conditions under formation pressure into surface conditions to meet the conditions for natural gas gathering and transportation to users or natural gas processing plants;

(8) The location of the gas-water interface or oil-water interface and the possible fracture wells.

382.For gas and condensate wells, depending on the location, one of the following process systems should be specified for a certain period of time:

(1) Constant pressure gradient: when it is possible to damage the production zone. Such a system may be replaced by a system of constant pressure-drop operation, but such an alternative must be justified in each particular case;

(2) Constant gas flow rate in the near-wellbore zone: when there is a risk of damaging the production zone and bringing drilling fluid out of the near wellbore zone of the formation;

(3) Constant pressure drop : when there is a risk of coning and tongue flooding;

(4) Constant wellhead pressure: when the well is working without an oil nozzle, or when a certain pressure is maintained before the installation of natural gas pretreatment equipment at the wellsite;

(5) Constant production : When there are no constraints (other than casing capacity), determine the duration of the constant production regime based on the change in production with formation pressure drop.

383.Gas production through production casing shall not allowed when no flow tubing is running in a gas well. The exception is that the

（7）将地层压力下开发条件转变为地面条件，以满足天然气集输并运输至用户或天然气处理厂使用的条件；

（8）存在气水界面或油水界面及可能存在的断裂井的位置。

382.对于气井和凝析气井，根据位置在一定时期内，需指定以下工艺制度中的一种：

（1）恒定压力梯度：在可能破坏生产层时。这种制度可以被恒定压降工作制度所代替，但在每种具体情况下，均需证明这种替代是合理的；

（2）地层近井地带天然气渗流恒定速度：存在破坏生产层，以及带出地层近井地带钻井液的风险时；

（3）恒定压降：当存在锥进和舌进水淹风险时；

（4）井口压力恒定：当不带油嘴的井工作时，或在井场安装天然气预处理设备之前保持一定的压力时；

（5）恒定产量：当没有任何限制条件时（套管通过能力除外），根据产量随地层压力下降的变化情况确定恒定产量制度持续的时间。

383.气井中没有下入自喷油管时，不允许通过生产套管采气。以下情况例外，产层的地层压力不超过生产套管试压压力值，当气体中不存在腐蚀性成分时，为了携带出井筒中的全部冷凝液和地层液体，允许沿管外空间吹扫井筒，但井筒内不能形成砂堵。

formation pressure in the production zone does not exceed the production casing pressure test value. When there is no corrosive component in the gas, the wellbore shall allowed to be blown along the outer space in order to carry out all the condensate and formation fluid in the wellbore, but no sand plug is formed in the wellbore.

384.The diameter of the flow tubing depending on:

(1) The working output of the well;

(2) The allowable pressure drop and temperature of the wellbore content;

(3) To achieve the required flow speed in the flow pipe;

(4) Production casing diameter.

385. In order to remove liquid and mechanical impurities from the bottom of gas and condensate Wells, foaming surfactants, small diameter tubing and hydrodynamic sprays are recommended.

386.The gas tree used in any gas well production methods shall ensure that the downhole equipments and the gas temperature and pressure at the wellhead can be measured during its operation.

387.Underground gas storage can be constructed in depleted gas fields, aquifers and underground salt mines. The operation of underground gas storage do same as that of gas field, and the gas injection and production mode is adopted, and the cushion volume is established in advance.

### 7.2.7 Chapter 20 Protection of Mineral Resources and the Natural Environment During Exploration and Development of Oil and Gas Fields

388.The safeguarding of mineral resources

384. 自喷油管直径取决于：

（1）井的工作产量；

（2）井筒内容许的压降和温度；

（3）达到自喷油管中所需的流动速度；

（4）生产套管直径。

385. 为排除气井和凝析气井井底的液体和机械杂质，建议使用发泡表面活性剂、小直径油管和流体动力学喷洒器。

386. 用于任何气井生产方式的采气树应确保在其工作过程中可以下入井下设备及测定井口的气体温度和压力。

387. 可在枯竭的气田、含水层和地下盐矿，建造地下储气库。地下储气库的运行和气田一样，采用注采气模式，并提前建立垫气量。

### 7.2.7　第20章 油气田勘探、开发过程中矿产资源和自然环境保护

388. 保护矿产资源要求采取一系列措

necessitates a comprehensive set of measures designed to ensure the optimal extraction of oil and gas, promote the rational and integrated utilization of minerals, preserve the energy characteristics above ground, mitigate man-made disasters (such as earthquake, landslide, flooding, soil subsidence), and avert natural calamities during drilling, development, and operational phases. This is particularly crucial due to the cross-flow dynamics among oil, gas, liquids, as well as recycling production waste that may lead to domestic water contamination affecting groundwater sources.

389.Protective measures for mineral resources and environmental conservation are delineated in foundational technology design documents.

390.Users engaged in mineral resource extraction must adhere to established monitoring protocols aimed at environmental protection responsibilities.

391.The development process for oil and gas fields relies on findings from engineering geological assessments, hydrogeological studies, geoecological evaluations, among other investigations; any necessity for supplementary research should be determined by either the mineral resource user or designer in accordance with relevant environmental protection standards documentation.

392.Throughout exploration activities involving drilling for oil or gas, including condensate fields, only environmentally sustainable technologies and chemical products are permitted alongside modern equipment that meets either national standards set forth by Kazakhstan or international benchmarks (provided these do not fall below Kazakhstani regulations), especially when operating under

施，其目的是确保从矿产资源中充分开采石油和天然气，合理、综合性利用矿产，保护土壤上部能量状态特性，以防止发生人为灾难（地震、山体滑坡、水淹、土壤深陷），防止在钻井、开发和运营的过程中，由于油气和液体发生窜流，以及回收生产垃圾和生活用水对地下水造成污染。

389.矿产资源和自然环境的保护措施在基础文件和技术设计文件中规定。

390.矿产资源利用者应遵守保护环境的各项措施并负责对其进行监督。

391.油气田的开发需根据工程地质、水文地质、地球生态和其他研究的结果进行。是否需要开展其他研究，由矿产资源利用者或设计单位，根据保护环境标准文件的要求确定。

392.勘探、钻探和开发油田、油气田、气田、凝析气田过程中，仅允许使用环保的技术和化学产品，以及符合哈萨克斯坦共和国标准或国际标准（如果国际标准不低于哈萨克斯坦标准）高度可靠的现代技术和设备，其中包括能适应高硫化氢含量的地下条件。

conditions characterized by elevated hydrogen sulfide concentrations.

393. Oil and gas field development programmes need to include associated gas processing (recovery) components.

394. The reliability, technicality and safety of the drilling structure to ensure the protection of mineral resources and the natural environment primarily depend on the strength and durability of the well walls, the tightness of the casing and annulus, and the isolation of the fluid bearing layer and permeable rock from the surface.

395. Normally, the external electric power should be used to drive the drilling rig for drilling. If a diesel generator is used to drive a drilling rig, the untreated exhaust gas discharged by the equipment into the atmosphere must meet its technical characteristics and environmental requirements.

396. In selecting the location of the drilling platform, consideration should be given to the natural slope of the terrain, to ensure the flow of wastewater in the direction of the sedimentation tank, to consider factors such as soil cover and lithological composition of the soil, the depth of groundwater (especially fresh water), the existence of protected areas, the latest structural data, seismic hazard, aerospace monitoring, the distance of the drilling rig from drinking water or ponds, etc.

397. Before drilling, the following equipments and sites should be checked for good condition: steam lines, circulation systems, drilling fluid preparation and treatment units, chemical reagent storage warehouses, areas under the drilling rig, fuel and lubricating oil tanks and other objects that may leak toxic liquids.

398. When drilling on fertile land and

393. 油气田开发方案需包括伴生气加工（回收）部分。

394. 钻井结构应具有可靠性、工艺性和安全性，为保护矿产资源和自然环境提供保障，主要取决于井壁的强度和耐久性，套管和环空的密封性，以及将含流体层及渗透性岩石与地表隔离开来。

395. 正常应使用外网电力驱动钻机进行钻井。如果使用柴油发电机驱动钻机钻井，则该设备向大气中排放的未经处理的废气必须符合其技术特征和环境要求。

396. 选择钻井平台地点时，应考虑地形的自然坡度，确保废水向沉淀池方向流动，考虑诸如土壤覆盖范围和土壤的岩性成分，地下水的埋深（尤其是淡水），是否存在保护区，最新的构造数据、地震危险性、航空航天监测、钻机距离饮用水或池塘的距离等因素。

397. 在钻井前，应检查以下设备和场所是否处于良好状态：蒸汽管道、循环系统、钻井液制备和处理单元、化学试剂存储仓库、钻井架下方区域、燃油箱和润滑油箱及其他可能发生有毒液体泄漏的物体。

398. 在肥沃的土地和农田钻井准备安装设备时，应剥离沃土层并将其分开存放，以便随后进行土地复垦。

farmland in preparation for installation of equipments, the fertile soil layer should be stripped and stored separately for subsequent land reclamation.

399.When drilling, damage to vegetation and soil outside the area to be drilled is not permitted.

400.In order to prevent drilling waste from entering the drilling platform and toxic substances from spreading to the natural environment, an engineering system should be in place for the organized collection and storage of waste and the waterproofing of the process site.

401.The discharge of wastes from the use of mineral resources into surface water bodies and mineral resources are not permitted.

402.When drilling aquifers of drinking water sources, the toxicity of the chemical agents used in the preparation (processing) of drilling fluids and cement slurry shall be permitted by the competent authorities in the field of environmental protection and the public health and epidemic prevention agencies. The formation between aquifers should be effectively isolated.

403.When drilling under suction conditions, it is not permitted to bring solutions and materials into the formation containing potable water. At the same time, quick-setting mixtures and equipments and processes to prevent entry should be used.

404.Before well testing, check and ensure that the following conditions are met: the tightness and reliability of monitoring instruments and discharge lines, well test separators (separators), flares, metering devices, storage tanks; Waterproofing of tanks, platforms under separators and surrounding banks.

399. 钻井时，不允许破坏待钻区域以外的植被和土壤层。

400. 为了避免钻井废弃物进入钻井平台、有毒物质扩散至自然环境，应设置一种工程系统，用于有组织地收集和存储废弃物及对工艺场地做防水处理。

401. 不允许将矿产资源利用所产生的废弃物排入地表水体和矿产资源中。

402. 在钻探饮用水源的含水层时，制备（加工）钻井液和水泥浆所使用的化学剂具有的毒性，应通过环保领域主管机构、居民卫生防疫机构的允许。含水层之间的地层应加以有效隔离。

403. 在吸入条件下钻井时，不允许将溶液和物料带入含有饮用水的地层。同时，要使用速凝混合物及防止进入的设备和工艺流程。

404. 在试井之前，要检查并确保满足以下各项情况：监测仪表和排出管线、试井分离装置（分离器）、火炬、计量装置、储罐的密封性和可靠性；油罐、分离器下方的平台和周围堤岸的防水情况。

405.During well testing, produced crude oil, condensate and formation water are collected in storage tanks for subsequent transport. With the exception of cases where exploration wells are tested offshore based on environmental assessment results (appraisal wells), flaring oil and gas is the most environmentally efficient disposal method.

406.In preparation for the development of oil and gas fields, all oil-bearing gas formations are tested to check for the presence of formation water. If water is found during testing, the chemical composition of the water and the gas it contains should be studied to determine its source, and if necessary, the test should be repeated after isolating the water-producing layer.

407.The well can be produced and tested if the height of the cement paste back behind the production casing meets the design requirements and mineral resource conservation.

408.When exposing high-pressure formations at risk of blowout or spontaneous blowout, blowout prevention devices must be installed at the wellhead and wash fluid must be used according to the drilling technology plans.

409.Under the supervision of the production personnel, check and confirm the position of the drilling rig and personnel, check that precautions have been taken to protect workers and residents in the event of an emergency blowout (self-blowout), and start drilling the hydrogen sulphide formation when all steps are in order.

410.In the event of an oil and gas accident, the wellhead should be sealed and the next steps should be carried out according to

405. 在试井过程中，将采出的原油、凝析油、地层水收集在储罐中，以便后续运输。除了基于环境评估结果在海上测试探井（评价井）的情况外，用火炬燃烧油气是对环境最为有效的处置方法。

406. 在准备进行油气田开发时，测试所有的含油气层，检查是否存在地层水。在测试过程中如果发现水，应研究水及其所含气体的化学成分，以查清其来源，必要时在隔离产水层后重复进行测试。

407. 如果生产套管后面的水泥浆上返高度符合设计要求和矿产资源保护的要求，则可以进行井的生产和测试。

408. 揭开有井喷或自喷风险的高压地层时，必须依据钻井技术方案，在井口安装防喷装置，采用洗井液。

409. 在负责生产人员的监督下，检查并确定钻机和人员的到位情况，检查是否采取在发生紧急井喷（自喷）时保护工人和居民的预防措施，一切正常后开始钻探含硫化氢地层。

410. 发生油气事故时，应密封井口，并根据事故处理计划开展下一步的工作。

the accident treatment plans.

411. If the well contains hydrogen sulfide, treat the drilling fluid with a hydrogen sulfide neutralizer.

412. Exploration and production wells are not allowed to be developed and tested without neutralization or sustained combustion if there is no possibility of product recovery.

413. After completion of the well development and fluid dynamics studies, the air in the operating area should be monitored for hydrogen sulphide and the wellhead should be inspected tightness of the valve.

414. When signs of oil and gas appear, workover operations should be stopped immediately and the neutralized drilling fluid should be re-pressed into the well.

415. For uncompleted wells due to technical reasons (accidents or poor wiring quality), isolation operations are performed in the drilled section to prevent interzone flow of oil and gas and liquids, if there is a reservoir containing oil and water.

416. When using hydrocarbon-based drilling fluids (calcareous asphalt, reverse-phase emulsions, etc.), measures should be taken to prevent gas contamination. In order to monitor the pollution degree of harmful gases, the air environment in the rotor, drilling fluid preparation device, vibrating screen and pump room should be measured, and when gas pollution is found, corresponding elimination measures should be taken immediately.

417. In accordance with a plan established at headquarters, the operators of mineral resources carry out clean-up operations in accordance with prescribed procedures.

418. The rig site should be equipped with an exhaust ventilation device, and the sensor

411. 如果井中含有硫化氢，则用硫化氢中和剂处理钻井液。

412. 如果缺少产品回收的可能性，则不允许在未经中和或持续燃烧的情况下开发和测试探井和生产井。

413. 完成井的开发和流体动力学研究后，应监测作业区的空气是否含有硫化氢，并检查井口阀的密封性。

414. 当出现油气迹象时，应立即停止修井作业，将被中和剂处理过的钻井液重复压至井中。

415. 基于技术原因（事故或布线质量差）的未完井，在已钻剖面如果存在含油气水层，则进行隔离作业，以防止油气和液体层间窜流。

416. 当使用碳氢化合物基的钻井液（石灰沥青、反相乳液等）时，应采取措施防止气体发生污染。为了监测有害气体的污染程度，应测量转子、钻井液制备装置、振动筛及泵房中的空气环境，当发现气体污染时，应立即采取相应的消除措施。

417. 根据总部制定的计划，由矿产资源利用者按照规定的流程，开展自喷作业清理。

418. 钻机场所应配备排气通风装置，当硫化氢达到最大允许浓度时，传感器启动。

will be activated when the maximum allowable concentration of hydrogen sulfide is reached.

419. After the completion of drilling, development (testing) of wells and removal of equipments, the land is restored (reclaimed) in accordance with the design resolution.

420. Health protection zones are established according to the production well line and each layer of the oil and gas field, the dimensions of which are determined according to the current health code. For oil and gas field with hydrogen sulfide, the location and size of the health protection area are determined according to the possible accidental discharge and the conditions of hydrogen sulfide leakage.

421. Observation of seismic and geodynamic conditions in oil and gas development zones in order to identify specific sources of seismic activity and study the law of spatial-temporal migration, determine seismic mechanisms and reliably track seismically active zones and possible subsidence areas.

422. The development and operation of production and injection wells are carried out with the appropriate equipment of the wells in order to avoid the release of oil and gas and the possibility of injection water loss.

423. In the event of production casing seal failure, zonal cross-flow, lack of cement block behind the pipe, gap between wellhead flanges, and operation of defective wells, the user of mineral resources shall take necessary measures to restore the integrity of the well and ensure its safe operation. For the above drilling approved special operation mode, develop maintenance and rehabilitation operation plan, and continuously monitor the operation to protect mineral resources and natural environment.

424. Measures were taken to improve

419. 在完成钻井、开发(测试)井和拆除设备后，根据设计决议恢复(复垦)土地。

420. 根据生产井边列及油气田的每个层系，建立卫生防护区，其尺寸依据现行卫生规范确定。针对含硫化氢的油气田，根据可能的意外排放量和硫化氢泄漏的条件，确定卫生防护区的位置及面积。

421. 观察油气田开发区的地震和地球动力状况，以发现地震活动的具体震源和研究空间—时间运移规律，确定地震机制，可靠地追踪地震活跃带及可能的沉陷区。

422. 采油井和注入井的开发和运行是通过井的相应设备进行的，以避免油气释放和自喷，以及注入水损失的可能性。

423. 钻井开发、试井和运行时，如果生产套管密封性失效，存在层间窜流，管后缺少水泥块，井口法兰连接有间隙，以及运行缺陷井时，矿产资源利用者应采取必要的措施恢复井的完整性，保障其安全运行。针对上述钻井批准特殊运行方式，制定维修和修复作业计划，并对作业持续进行监控，以保护矿产资源和自然环境。

424. 采取措施改善钻井总量，包括清除部分套管或导管后的水泥返高不达标的缺陷钻井，钻进构造可靠的副井等。首先改善位于卫生保护区的缺陷井。

the total number of Wells drilled, including defective drilling where the cement return height was not up to standard after the removal of part of the casing or pipe, and drilling of secondary wells with reliable construction. First improve the defective wells located in the health protection areas.

425.Before any oil increase method is actually implemented in each new field, test studies are carried out to demonstrate the main process parameters, compliance with which ensures the safety of the drilling casing and cement ring.

426.The study of the geological structure and hydrogeological conditions of oil and gas reservoirs is a necessary condition for the use of chemical reagents (indicators) in the development of oil and gas fields.

427.When selecting a chemical agent to act on a formation, consideration must be given to its danger level, solubility in water and volatility.

428.Chemical liquid and crude oil spills and leaks should be prevented when drilling and equipment are ready for operation, maintenance and study, when faulty or unchecked gate regulating devices and components are used, when major process procedures are violated, when production casing seals fail.

429.When salt deposition and paraffin deposition scale inhibitors, surfactants, demulsifiers, etc. are injected into the formation, only special equipments can be used to avoid spillage.

430.During well operation after drilling and downhole overhaul, sealing devices should be installed at the wellhead to prevent liquid overflow and gushing.

431.when flooding producing wells

425. 在每个新油田实际实施任意增油方法之前，均要进行试验研究，以论证主要工艺参数，遵守这些参数可确保钻井套管和水泥环的安全。

426. 研究油气藏的地质结构和水文地质条件，是开发油气田使用化学试剂（指示剂）的必要条件。

427. 选择化学剂作用于地层时，必须考虑其危险等级、在水中的溶解度和挥发性。

428. 在钻井和设备准备投入运行、维修和研究，使用有故障的或未经检查的闸门调节装置和部件，违反主要工艺流程，生产套管密封失效等情况下，应防止化学剂和原油发生溢出和泄漏。

429. 将盐沉积和石蜡沉积阻垢剂、表面活性剂、破乳剂等注入地层时，为避免溢出，只能使用专用设备。

430. 在钻井、井下大修后井运行时，应在井口配备密封装置，以防液体溢出和喷涌。

431. 给生产井（采油井）注水时，除了监测油产品的含水率，还应开展专门的地球物理和水文地质研究，以确定水流通过管柱进井的位置、注水来源和深度。

(production wells), in addition to monitoring the water content of the oil product, specialized geophysical and hydrogeological studies should be carried out to determine the location of water flow through the string into the well, the source and depth of the water injection.

432.The user of the mineral resources shall in consultation with the design unit preparing the development plan, decide whether to suspend the operation of the production well.

433.If there are signs of irreversible oil and gas loss, groundwater leakage or interzone cross-flow during field development, the user of mineral resources to identify and remove the cause of the uncontrolled flow of formation fluids.

434.Production wells connected to integrated natural gas treatment units shall be studied using controlled separators and shall not emit and burn natural gas into the atmosphere.

435.In order to protect process plants, downhole equipments, production casings and lifting string operating under the influence of hydrogen sulphide from corrosion, corrosion resistant steels and corrosion inhibitors should be used, as well as stainless anticorrosive steels without corrosion inhibitors, special coatings and technical methods to reduce the corrosion activity of products.

436.Downhole equipments, process equipments, casings and other equipments used in corrosive environments must be resistant to sulphide corrosion cracking.

437.Automatic stationary gas is used in facilities, plants and industrial sites where hydrogen sulphide can be released into the air of the working areas, the detector monitors

432.矿产资源利用者与编制开发方案的设计单位协商，是否采取停止采油井运行的决议。

433.如果在油田开发过程中出现不可挽回的油气损失、水地下泄漏或层间窜流的迹象，则由矿产资源利用者确定和清除地层流体失控流动的原因。

434.连接至天然气综合处理装置的生产井，应使用控制分离器进行研究，且不向大气中排放和燃烧天然气。

435.为了保护在硫化氢影响下运行的工艺装置、井下设备、生产套管和举升管柱免受腐蚀，应使用耐腐蚀钢和腐蚀抑制剂，以及不使用腐蚀抑制剂的不锈防腐钢、特殊涂料和降低产品腐蚀活性的技术方法。

436.在腐蚀性环境中使用的井下装置、工艺设备、套管和其他设备，必须耐硫化物腐蚀开裂。

437.在可以将硫化氢释放到作业区空气中的设施、厂房和工业场所中，利用自动固定式气体检测仪监测空气环境，并定期在可能聚集硫化氢的区域使用便携式气体检测仪或气体分析仪进行监测。

the air environment and periodically uses a portable gas detector or gas analyzer in areas where hydrogen sulfide may accumulate.

438.Take measures to improve formation pressure and maintain system reliability. Replace existing long-used sewage pipes, and implement inhibitor protection and electrochemical protection for all sewage injection pipes.

439.Formation water produced with crude oil shall:

(1) Purify according to the standard content of suspended solids and oil products in the water, and use the purified formation water for formation pressure maintenance system, or inject into the water-absorbing layers for treatment;

(2) Pre-treatment in accordance with environmental regulations and subsequent discharge into artificial evaporation ponds, storage ponds and other wastewater storage ponds designed to prevent soil and groundwater contamination. If necessary, fungicides can be injected into the reservoir to treat the injected water and prevent it from being infected by hydrogen sulfide bacteria, which can lead to the formation of hydrogen sulfide in the oil and water.

440.If necessary, fungicides may be injected into the reservoir to treat the injected water and prevent it from being infected by hydrogen sulphide bacteria, which can lead to the formation of hydrogen sulphide in the oil and water.

441.It is prohibited to discharge formation water into evaporating areas, surface water sources or into underground layers, resulting in groundwater contamination, and to discharge unneutralized liquids containing hydrogen

438. 采取措施，提高地层压力，维护系统的可靠性。更换现有的长期使用的污水管道，并针对所有注入污水的管道进行抑制剂保护及电化学保护。

439. 与原油一同采出的地层水应当：

（1）依照水中悬浮固体和油产品标准含量进行净化，并将净化后的地层水用于地层压力保持系统，或注入吸水层进行处理；

（2）根据环境法规进行预处理，其后排入人工蒸发池、存储池和其他废水存储池，其设计应可防止土壤和地下水污染。如有必要，可向产层注入杀菌剂以处理注入水，防止其被硫化氢细菌感染，从而导致在油和水中形成硫化氢。

440. 如有必要，可向产层注入杀菌剂以处理注入水，防止其被硫化氢细菌感染，从而导致在油和水中形成硫化氢。

441. 禁止将地层水排放到蒸发区、地表水源或注入地下层，导致地下水污染，以及将未经中和的、含有硫化氢的液体排放到露天污水处理系统。

sulphide into open sewage treatment systems.

442.Formation water with high hydrogen sulfide content should be treated and placed in airtight containers.

443.Landfill treatment is carried out by pumping industrial waste into injection wells and by reliably isolating absorptive layers not contain groundwater (used or available for drinking and physiotherapy).

444.The injection of industrial wastewater into absorbent layers is permitted only in exceptional circumstances:

(1)Reservoirs that have not been developed using water drive;

(2)From the initial stage of development to the construction of the water drive system, a small amount of industrial wastewater is generated;

(3)Excessive industrial wastewater compared to the design requirements, and it is not suitable to transport to other areas;

(4)Use formation water as raw material for hydraulic minerals;

(5)The use of unreasonably complex technology to purify some industrial wastewater produced by integrated oil treatment units.

445.In order to bury industrial effluents deep, special facilities (landfills) are created, and surface and underground structures for the collection and disposal of waste, monitoring its condition and transport in pipelines are placed in their areas.

446.Sludge silos are not permitted at the well site and the contents of the sludge collectors should be treated or recovered, followed by land reclamation in the area where the silos have been cleared.

447.Mineral resource users monitor the groundwater situation around the deposit and

442.高硫化氢含量的地层水应进行处理，并将其置于密封的容器中。

443.将工业废水泵入注入井、不含地下水（已用于或可用于饮用和理疗）可靠隔离的吸水层，进行地下掩埋处理。

444.只有在特殊情况下，才允许将工业废水注入吸水层：

（1）未使用水驱开发的油藏；

（2）自开发初期至水驱系统施工前，产生少量工业废水；

（3）与设计需求相比，工业废水过多，并且不适宜将其运输到其他区域；

（4）使用地层水作为水工矿产原料；

（5）采用不合理的复杂技术净化综合油处理装置产生的某些工业废水。

445.为了深埋工业废水，创建专门的设施（垃圾填埋场），在其区域内安置用于收集和处置垃圾、监测其状况和在管路中运移的地面和地下构筑物。

446.不允许在井场设置油泥仓，污泥收集器的内容物应进行处理或回收，其后在已清理仓库的区域复垦土地。

447.矿产资源利用者通过工程井网监测矿床周边及污泥收集器所在区域的地下水状况。

in the area where sludge collectors are located through a network of engineered Wells.

448.The procedures for cleaning up the consequences of oil and gas use, sealing up blocks of oil and gas mineral resources,and sealing up and processing facilities are set out in the regulations for oil and gas exploration and exploitation, and sealing up and cleaning up of uranium exploitation approved by the competent authorities for oil and gas and uranium exploitation.

## 7.3 Russian Regulations on the Development of Oil and Gas Fields (Excerpt)

### 7.3.1 Chapter 6 System Requirements for Oil and Gas field development

6.1 Division of development layers:

6.1.1 The purpose of dividing the development layers of oil and gas fields is to ensure the reasonable development of oil fields and maximize the recovery rate (oil recovery rate, natural gas recovery rate, condensate recovery rate).

6.1.2 The developed layer is defined as having sufficient reserves to ensure the continued production of the well. Separate (primary) development layers and reinjection development layers should be divided and demonstrated.

The main development layer refers to oil (gas) reservoir, a part of oil (gas) reservoir or several oil (gas) reservoirs combined into a development layer and developed by the same development pattern.

6.1.3 For oil and gas reservoirs whose oil and gas reserves are accounted for separately

448. 清理油气使用后果、封存油气矿产资源区块和（或）封存和（或）处理工艺设施的流程，规定于油气和铀开采主管机构批准的油气勘探和开采、铀开采时封存和清理规范中。

## 7.3 俄罗斯的油气田开发法规（节选）

### 7.3.1 第6章 油气田开发系统要求

6.1 开发层系的划分：

6.1.1 划分油气田开发层系的目的是保证油田的合理开发，最大限度地提高采收率（石油采油率、天然气采收率、凝析油采收率）。

6.1.2 划分出的开发层系，其储量足以确保油井持续生产。应划分独立（主要）开发层系和回注开发层系并进行论证。

主要开发层系是指：一个油（气）藏、油（气）藏的一部分或几个油（气）藏合并成一个开发层系，并由同一开发井网开发。

6.1.3 对于油气储量在平衡表中单独核算并在油气藏开发技术设计中合并为一个开发层系的油气层，应分别核算工作剂的注入量及单独计量石油、凝析油、天然气和水的产量。

in the balance sheet and combined into a single development layer in the technical design of reservoir development, the injection of working agents and the production of oil, condensate, natural gas and water should be accounted for separately.

6.1.4 For large multilayer gas fields, the sequence of deployment of development layers shall take into account the dynamic formation pressure, the time of deployment of booster stations or injection stations, and the possibility of using high-pressure gas reservoir energy to transport gas extracted from low pressure gas formations or adjacent gas fields without compression.

6.2 In the technical design of oil and gas field development, the feasibility of oil and gas reservoir consolidation should be demonstrated:

6.2.1 For reinjection formations that have been shown to be uneconomical in the technical design of field development, development must be carried out using wells that have completed the design task and have been transferred from other development formations.

6.2.2 In order to evaluate the deviation of the actual oil and(or) natural gas production from the design values specified in the development plan, the degree of oil and(or) free gas production calculated for the A+ B1 class reserves should be used.

6.3 Development system determination: well distribution diagram and well structure, completion method, number of wells (well pattern density) and use, oil and gas formation operation mode in the development layer, agent to maintain formation pressure and displace hydrocarbons.

6.1.4 对于大型多层天然气田，开发层系投入开发的顺序应考虑到地层动态压力，增压站或注压站投入时间，利用高压气藏能量无压缩运输从低压气层或相邻气田开采出的气体的可能性。

6.2 在油气田开发技术设计中，应论证油气层合并的可行性：

6.2.1 对于油气田开发技术设计中已证明无经济效益的回注层系，必须利用那些已完成设计任务，并从其他开发层系转移的油井进行开发。

6.2.2 为了评价油气田实际石油和（或）天然气产量与开发方案中规定的设计值的偏差，需利用针对A+B1级储量计算的石油和（或）游离气采出程度。

6.3 开发系统确定：井分布示意图和井身结构、完井方式、井的数量（井网密度）和用途、开发层系中的油气层运行模式、保持地层压力并驱替烃的工作剂。

The development system of the development layer is demonstrated in the technical scheme of oilfield development.

6.4 Technical indicators of oil and gas field development:

6.4.1 The oil and gas field development technical indicators calculated in the technical design document for A+B1 oil and gas reserves are used in the current oil and gas production, equipments, drilling number and engineering construction planning.

6.4.2 The technical indicators of oil and gas field development calculated for C2 reserves are used in the prospect planning of oil and gas development, equipment, drilling quantity and engineering construction of A+B1+B2 (C1 in the test production scheme). The field development technical indicators in the technical documents with oil and gas reserve boundaries of A +B1+B2 (C1+C2 in the test production scheme) are used for the planning of prospective oil and gas production, drilling volume and construction works on the basis of:

6.4.3 The development technical index of the development layer is calculated by establishing 3D geological model and fluid dynamics model.

6.5 Selection of working agents for maintaining formation pressure and acting on oil and gas formations:

6.5.1 The working agent pumped into the development layer must:

(a) Ensure chemical compatibility with formation fluids and do not produce secondary sediments that degrade formation properties;

(b) ensure that hydrocarbon properties do not deteriorate in the formation;

(c) Ensure the acceptability of the design.

开发层系的开发系统在油田开发的技术方案中加以论证。

6.4 油气田开发技术指标：

6.4.1 在技术设计文件中针对A+B1级油气储量计算的油气田开发技术指标，被用于当前的油气产量、设备、钻井数量和工程建设规划中。

6.4.2 针对C2级储量计算的油气田开发技术指标被用于A+B1+B2级（在试采方案中为C1级）油气开发、设备、钻井数量和工程建设的前景规划。以A+B1+B2级（在试采方案中为C1+C2级）油气储量边界完成的技术文件中的油田开发技术指标，用于远景油气产量安排、钻井量和建筑工程的规划依据是：

6.4.3 通过建立三维地质模型和流体动力学模型计算开发层系的开发技术指标。

6.5 对保持地层压力和作用于油气层的工作剂的选择：

6.5.1 泵入开发层系的工作剂必须：

（a）确保与地层流体的化学相容性，且不会产生导致地层特性变差的次生沉积物；

（b）确保在地层内油气性质不会变差；

（c）确保设计的可接受性。

6.5.2 The system for maintaining formation pressure shall ensure that:

(a) According to the reservoir development technology design, ensure the working agent injection amount of the development layer and the injection pressure of the wells;

(b) The preparation of the working agent shall meet the requirements of the technical design of the oil and gas reservoir development (composition, physical and chemical properties, mechanical impurities, oxygen, hydrogen sulfide and microbial content, etc.);

(c) The possibility of the system measuring the operating pressure and sensitivity of each well;

(d) Quality control of the working agent in accordance with the period specified in the oil and gas field development technical plans;

(e) Mineral resource conservation requirements.

6.6 When injecting sewage or other corrosive agents into the formation, the wellbore casing (development equipment) must be protected from harmful effects by packer, inhibitor, corrosion resistant coating or other means.

6.7 Chemical reagent solution is injected into the production layer in accordance with the reasonable process in the technical plans to increase production or enhance oil recovery, natural gas recovery and condensate recovery.

6.8 Mineral resource users must ensure that each well's working agent injection and oil and gas production are measured separately in order to accurately assess the increase in production following the use of stimulation or EOR measures.

6.9 If changes in working conditions necessitate adjustments in technical and

6.5.2 保持地层压力系统应确保：

（a）根据油气藏开发技术设计，确保开发层系的工作剂注入量，以及油井注入压力；

（b）工作剂的制备要满足油气藏开发技术设计的条件要求（成分、理化性质、机械杂质、氧、硫化氢和微生物的含量等）；

（c）系统计量每口井工作压力和灵敏度的可能性；

（d）按照油气田开发技术方案规定的周期对工作剂进行质量控制；

（e）矿产资源保护要求。

6.6 在向地层中注入污水或其他腐蚀性工作剂时，必须用封隔装置、抑制剂、耐腐蚀涂层或其他方法保护井筒套管（开发设备）免受有害影响。

6.7 按照技术方案中合理的工艺向生产层中注入化学试剂溶液，以增加产量或提高石油采收率、天然气采收率、凝析油采收率。

6.8 矿产资源使用者必须确保每口井工作剂注入量和油气产量单独计量，以便准确评估使用增产或提高采收率措施后产量增加情况。

6.9 如果由于工作条件的变化，需要对影响矿产安全使用的技术和工艺方案进行调整，则相应的理由应纳入根据俄罗斯联邦法制订的程序编制的油气田开发计划。

technological programmes affecting the safe use of minerals, the corresponding reasons should be included in the plan for the development of oil and gas fields prepared in accordance with procedures established by the law of the Russian Federation.

### 7.3.2　Chapter 7 Uses of Wells

7.1 During the survey, exploration and development phase of a deposit, different classes of Wells are drilled and allocated according to their purpose.

7.2 Reference wells are designed and drilled to study the geological structure and hydrogeological conditions of the entire sedimentary rock formation and to understand the distribution law of formation combinations conducive to oil and gas accumulation.

7.3 Parameter wells are designed and drilled on discovered structures for the purpose of conducting regional mineral resource studies and in combination with other regional research methods; using perfect coring data and comprehensive logging data, the geological structure shall studied in more detail, and the prospect area is divided for further geological exploration.

7.4 Constructional wells are designed and drilled for the purpose of conducting detailed studies of structures revealed by reference and parameter wells in order to deploy census-exploration wells in these structures.

7.5 Census-appraisal Wells are deployed on prospective structures that have been implemented through prior drilling and geological-geophysical studies to explore and discover new oil and gas fields or reservoirs in earlier discovered oil and gas fields.

### 7.3.2　第7章 井的用途

7.1 在矿床的普查、勘探和开发阶段，根据其目的来钻探和分配不同类别的井。

7.2 基准井的设计和钻探是为了研究整个沉积岩层的地质结构和水文地质条件，搞清有利于油气聚集的地层组合的分布规律。

7.3 在已发现构造上设计和钻探参数井是为了进行区域矿产资源研究，并与其他区域研究方法相结合；利用完善的取心资料和全面的测井资料对地质结构进行更详细的研究，划分出前景区域以便进一步开展地质勘探工作。

7.4 构造井的设计和钻探是为了对基准井和参数井所揭示的构造进行详细研究，以便在这些构造部署普查—勘探井。

7.5 在通过前期钻井和地质—地球物理研究成果落实的前景构造上部署普查—评价井是为了在早期发现的油气田中勘探和发现新油气田或油气藏。

7.6 The purpose of designing and drilling exploration wells in areas with industrial hydrocarbon potential is to carry out geological studies and delineate hydrocarbon reservoir boundaries, to obtain raw data for hydrocarbon reserve calculations and to prepare technical development programmes.

7.7 Development wells are designed and drilled during the implementation of test production programmes and industrial development of oil and gas fields:

Production well (oil and gas Well): used to organize development systems and extract oil, gas, condensate, and water from paying formations;

Injection well: The injection of water, gas (a mixture of them) or other working displacing agents to maintain formation pressure and affect the reservoir, and the injection of gas separated from the reservoir or associated secondary useful components to temporarily store and produce oil and gas during production.

7.8 The design and drilling of special wells are used for blasting work using seismic exploration methods during the survey and exploration of oil and gas fields, exploitation of production water (water intake wells), discharge of oil field water to non-productive absorption layers (water intake wells), exploration and exploitation of water, preparation of underground gas storage structures and delivery of natural gas to underground gas storage, elimination of blowouts of oil and gas wells, Environmental monitoring of groundwater (drinking water), transfer of working agents to injection wells and other uses.

7.9 Control monitoring wells are designed and drilled to systematically monitor changes in fluid contact (water-oil, gas-oil, gas-water)

7.6 在具有工业含油气潜力的地区设计和钻探探井是为了开展地质研究和圈定油气藏边界，获取油气储量计算和编制开发技术方案的原始资料。

7.7 开发井是在实施试采方案和油气田工业开发过程中设计和钻探的：

采油井（油井和气井）：用于组织开发系统并从产层中提取石油、天然气、凝析油和水；

注入井：通过注入水、气（它们的混合物）或其他工作驱替剂来维持地层压力并对油气藏产生影响，通过注入从油气藏中分离出的气体或伴生的二类有用成分，用于在开采期间临时储存及开采油气。

7.8 专用井的设计和钻探是用于油气田普查和勘探时用地震勘探方法的爆破工作、生产用水（取水井）开采、向非生产性吸收层（吸水井）排放油田水、水的勘探和开采、地下储气库构造的准备和向地下储气库输送天然气，消除油井和气井井喷，对地下水（饮用水）进行环境监测，将工作剂转注注水井以及其他用途。

7.9 控制监测井的设计和钻探是为了系统监测油藏开发过程中流体间（水—油、气—油、气—水）接触的变化和其他参数（包括地层的油气水饱和度）的变化。控制测压井的设计和钻探是为了控制地层压力和温度的变化。

and other parameters, including formation oil, gas and water saturation, during reservoir development. Controlled wells are designed and drilled to control changes in formation pressure and temperature.

7.10 During the development of the reservoir, the use of the well may change in accordance with the approved technical design or in accordance with paragraph 5.16 (c) of this Regulation.

### 7.3.3 Chapter 8 Well Structure, Drilling Process, Cementing, Zone Perforation and Well Testing

8.1 The drilling of survey, exploration, production and other purpose wells in oil and gas fields should be carried out in strict accordance with the work plan of production drilling operations (single well or combination wells) formulated and approved by the law of the Russian Federation.

8.2 The technical requirements for drilling and exposing the production zones, cementing the well, perforating the production zone and test production proposed in the oil and gas field development technical plans should be clearly stated in the drilling construction design and test production plan in addition to the type of well profile.

8.3 The drilling design of all target wells (single or combination) must ensure reliable well structure, high quality formation opening, cementing and formation separation, completion of logging, hydrodynamic testing and workover, and completion of all requirements of the field development program.

8.4 In accordance with the general recommendations of the technical program for the development of the fields (development

7.10 在油气藏开发过程中，根据批准的技术设计方案或根据本法规第5.16款（c）项，井的用途可能会发生变化。

### 7.3.3 第8章 井身结构、钻井工艺、固井、产层射孔和试井

8.1 在油气田开展普查井、探井、生产井和其他用途井的钻探需严格按照经俄罗斯联邦法制订和批准的生产钻井作业工作规划（单井或组合井）进行。

8.2 油气田开发技术方案中提出的对钻揭产层、固井、用射孔揭开产层和试采等工艺的要求，除井剖面类型外，在钻井施工设计和试采计划中需明确说明。

8.3 所有目的井（单井或组合井）的钻井设计必须确保可靠的井身结构、高质量的产层—开、固井和地层分隔，确保完成测井、流体动力测试和修井工作，以及完成油气田开发方案的所有要求。

8.4 根据油气田（开发层系）开发技术方案的总体建议，矿产资源使用者地质服务部门需要确定生产层（开发层系）的油井剖面，根据测井数据控制进尺，并确定射孔井段或井底结构元件的安装。

layer), the geological services of the users of the mineral resources are required to determine the well profile of the production layer (development layer), control the penetration based on the logging data, and determine the installation of the perforated well interval or the structural element of the bottom hole.

8.5 The perforation interval is controlled by geophysical methods.

Based on the technical design and logging data of each well, the geological service of the mineral resource user may specify the second well interval and downhole equipment type.

8.6 The supporting work of the test production operation including protection, restoration and improvement of the production layer, and the technical means and materials required for the implementation of these works shall be stipulated in advance in the drilling construction design. Test production operations shall be carried out in accordance with a personalized and(or) standardized plan prepared by the user of the mineral resource, which shall provide for measures to ensure the integrity of the cement bond outside the production casing outside the test production interval of the development zone and shall include the following preventive measures:

(a) Production casing deformation;

(b) Formation water (bottom, upper, lower) gushing, gas cap gas channeling;

(c) oil, gas and water well swell;

(d) The formation permeability near the bottom of the hole decreases;

(e) Pollution of mineral resources.

8.7 Responsibility for compliance with the design plan and drilling quality rests with the mineral resource user.

8.5 采用地球物理方法控制射孔井段。

根据每口井的技术设计和测井数据，矿产资源使用者的地质服务部门可以指定二开井段和井下设备类型。

8.6 试采作业的配套工作包括产层保护、恢复和提高，实施这些工作所需的技术手段和材料均应在钻井施工设计中事先予以规定。试采作业需按照矿产资源使用者编制的个性化和（或）标准化计划进行，计划中必须规定相应的措施以确保开发层系试采井段以外生产套管外水泥胶结的完整性，同时还应制订下列预防措施：

（a）生产套管变形；

（b）地层水（底部、上部、下部）涌出、气顶气窜；

（c）油气水井涌；

（d）井底附近地层渗透率降低；

（e）矿产资源污染。

8.7 遵守设计方案和钻井质量的责任归属于矿产资源使用者。

8.8 Drilling shall be deemed completed after all drilling operations specified in the drilling construction design have been completed.

### 7.3.4 Chapter 10 Production of Oil and Gas Wells

10.1 When a well is commissioned and listed as a major producing well, the user of the mineral resource must have the following documents in paper and electronic format:

(a) Drilling engineering design and drilling geological technology construction book (drilling design table);

(b) Spud-in and completion instructions;

(c) Sing mouth and turntable elevation measurement report;

(d) All mine geophysical logging data and their conclusions;

(e) Pipe length measurement (pipe size), section diameter, wall thickness and steel grade information, basic characteristics of non-metal pipe column;

(f) Casing cementing report, laboratory analysis of cementing quality and drilling fluid density measurement results during cementing, wellhead cement return or cement height data (cement bond thickness measurement map), pipe measurement, string combination, drilling fluid density data before cementing;

(g) Tightness test report of all casing, if necessary, wellhead and downhole equipments;

(h) Target zone test or oil test work plans;

(i) Casing perforation report indicating perforation interval, perforation type and mode, and perforation density;

(j) Test or completion test reports for each production layer, accompanied by oil and gas well test data (e.g., individual well production

8.8 钻井施工设计中规定的钻井作业全部完成后,钻井视为完成。

### 7.3.4 第10章 油气井的投产

10.1 当一口井投产并列入主要生产井时,矿产资源使用者必须具备以下纸质和电子格式的文件:

(a)钻井工程设计和钻井地质技术施工书(钻井设计大表);

(b)开钻和完钻指令;

(c)套管口和转盘海拔高度测量报告;

(d)所有的矿场地球物理测井资料及其结论;

(e)管材长度测量(管材尺寸),各段直径、壁厚和钢级信息,非金属管柱的基本特性;

(f)套管固井报告、固井质量实验室分析和固井过程中钻井液密度测量结果、井口水泥返出或水泥返高数据(水泥胶结厚度测量图)、管材测量、管柱组合、固井前井内钻井液密度数据;

(g)所有套管及必要时井口和井下设备的密封性测试报告;

(h)目的层测试或试油工作计划;

(i)套管射孔报告,其中标明射孔段、射孔类型和方式、孔密;

(j)每个生产层系的测试或完井试油报告,随附油气井试井资料(如标注有产液量和含水率的单井产量,地层压力、井底压力、井口压力、套管压力、环空压力,石油、天然气、凝析油和水分析,流体动力学试井数据、矿场地球物理测井资料);

rate marked with fluid production and water content, formation pressure, bottom hole pressure, wellhead pressure, casing pressure, annulus pressure, oil, gas, condensate and water analysis, hydrodynamic well test data, field geophysical logging data);

(k) Formation test results during drilling (report);

(l) Tubing size and type information is indicated on the bottom of the equipments, starting valve installation depth, and a complete diagram of the downhole equipments;

(m) Contains geological logs that describe the entire drilling and completion test process;

(n) Documents on the results of geological engineering monitoring during drilling;

(o) Well specifications containing drilling process, oil(gas、water) display and loss data, and well structure information;

(p) string tension data (if available in the drilling design);

(q) Wellhead equipment documents;

(r) Delivery of geological and technical documents of the oil and gas well by the contractor to the owner.

### 7.3.5 Chapter 11 Requirements for oil production methods in oil and gas Wells

11.1 Production wells can be produced by means of injection and mechanical oil production, lifting formation fluids through tubing, and at the same time, safely carrying out operations related to the utilization of mineral resources in accordance with the prevailing industrial safety codes and regulations in the oil and gas exploitation industry.

The method of oil and gas well production is demonstrated in the technical scheme of oil and gas field development.

（k）随钻中途地层测试结论（报告）；

（l）油管的尺寸和类型信息标注在设备底部，启动阀的安装深度，并附上井下设备的完整示意图；

（m）包含介绍整个钻井和完井试油过程的地质日志；

（n）关于钻井过程中地质工程监控成果的文件；

（o）包含钻井过程、油（气、水）显示和漏失数据及井身结构信息的井说明书；

（p）管柱的张力数据（如果在钻井设计中有）；

（q）井口设备文件；

（r）承包商向业主方交付油气井地质和技术文件的交接单。

### 7.3.5 第11章对油气井采油方式的要求

11.1 生产井采油方式有自喷采油法和机械采油法，通过油管举升地层流体，同时，符合油气开采行业中现行工业安全规范和法规，安全地实施与矿产资源利用有关的作业。

在油气田开发技术方案中论证油气井采油方式。

11.2 According to engineering and process calculations, the installation depth, size and type of downhole equipment should be given in the oil and gas well commissioning (completion and oil testing) scheme or workover plan.

11.3 Under the conditions of reliable and safe production of oil and gas wells, the oil recovery process of production wells should be determined based on the amount of crude oil, natural gas, condensate and liquid determined in the technical plans that can ensure the planned exploitation.

11.4 The production well oil recovery process requires the following main parameters:

(a) Formation pressure, bottom hole pressure, wellhead pressure and wellhead temperature of oil and gas fields containing free gas;

(b) the amount of produced liquid (gas), the moisture content of produced liquid, the gas-oil ratio (condensate output) and the content of mechanical impurities in the product;

(c) The model size of the installed downhole equipment, its mode of operation and time.

11.5 Under the conditions of reliable and safe production of injection wells, the production process should be determined based on the injection volume determined in the technical plans and the working test required for injection allocation within the planned period.

11.6 The injection well production process requires the following main parameters:

(a) Formation pressure, bottom hole pressure and wellhead pressure;

11.2 根据工程和工艺计算，在油气井投产（完井试油）方案或修井计划中给出井下设备的安装深度和尺寸及类型。

11.3 在满足油气井可靠和安全生产条件下，基于技术方案中确定的，可保证按计划开采的原油、天然气、凝析油及液量来确定生产井的采油工艺方式。

11.4 生产井采油工艺方式要求提供以下主要参数：

（a）含有游离气的油气田地层压力、井底压力和井口压力及井口温度；

（b）产液（气）量、采出液的含水率、气油比（凝析油产量）和产品中机械杂质的含量；

（c）已安装井下设备的型号尺寸，其工作方式和时间。

11.5 在满足注入井可靠和安全生产条件下，基于技术方案中确定的配注量，确保在计划期内达到配注要求的工作试剂量来确定生产工艺方式。

11.6 注入井生产工艺方式要求提供以下主要参数：

（a）地层压力、井底压力和井口压力；

(b) Injection capacity of the injection well and the content of mechanical impurities and crude oil in the injection agent;

(c) Temperature of the injector (steam injection well);

(d) The type and size of the installed downhole equipment, its working mode and time.

11.7 In the case of multiple production layers in one well, separate measurement of produced fluid volumes and production logging should be ensured.

11.8 Stratified metering and production logging should be ensured when stratified injection of work reagents into several production layers in a well.

11.9 On the basis of ensuring the design parameters, the user of the mineral resources shall determine and approve the production process system of the oil and gas well at least once a quarter. The determination of the process system shall take into account the approved geological process measure implementation plan.

11.10 During the production of oil and gas wells, it shall be ensured that the production casing engineering status and equipment operation are regularly monitored to obtain the initial data required to optimize the process system.

11.11 All original data (paper, disk and electronic) relating to the monitoring of the operation of oil and gas wells and downhole equipment must be kept in the database of mineral resource users throughout the field development period (excluding daily production reports of crude oil measurements, daily production reports of oil and gas wells, echo measurement curves and automatic force

（b）注入井注入能力及注入剂中的机械杂质和原油含量；

（c）注入剂的温度（注蒸汽井）；

（d）已安装井下设备的型号尺寸，其工作方式和时间。

11.7 在一口井中存在几个生产层系分层开采的情况下，应确保对采出液量的单独计量及开展生产测井。

11.8 在向一口井中的几个生产层系实施分层注入工作试剂时，应确保分层计量和开展生产测井。

11.9 在确保设计参数的基础上，矿产资源使用者至少每季度确定和批准一次油气井生产工艺制度。工艺制度的确定应考虑批准的地质工艺措施实施计划。

11.10 在油气井开采过程中，应确保定期监测生产套管工程状况、设备运行情况，获取优化工艺制度所需的初始数据。

11.11 有关监测油气井和井下设备运行的所有原始资料（纸质资料、磁盘资料和电子版资料），在整个油田开发期间必须保存在矿产资源使用者的资料库中（不包括每日原油计量的生产日报、油气井生产日报、回波测量曲线和自动测力曲线图，其保存应确保符合俄罗斯联邦税法要求期限或是3年内）。

charts). It should be kept in accordance with the requirements of the tax law of the Russian Federation for a period of time or 3 years).

## 7.3.6 Chapter 12 Monitoring and Adjustment of Well Workover, Oil and Gas Field (Oil and Gas Reservoir) Development Process

12.1 Workover operations are divided into major repairs and minor repairs.

12.2 Workover fluids that irreversibly reduce near-wellbore formation permeability are not permitted during major or minor well repairs (except for water isolation or plugging). In cases where workover fluids are used that reduce near-wellbore permeability, subsequent restorative measures must be taken. The density of wellhead and wellbore equipment and workover fluid should avoid the overflow of crude oil, natural gas and water.

12.3 Major and minor repairs of oil and gas Wells shall be carried out in accordance with the current industrial safety norms and regulations of the oil and gas exploitation industry, the requirements of mineral resources and environmental protection, as well as the technical specifications and operational requirements of the operating procedures of the equipments used and the implementation process.

12.4 According to the production status of oil and gas wells, determine the necessity of major and minor repair operations of oil and gas wells, and implement according to the operation plan approved by the user of mineral resources.

12.5 For major and minor repairs to oil and gas wells, job descriptions (major repairs) and job reports (minor repairs) should be

## 7.3.6 第12章 修井、油气田（油气藏）开发过程监测与调整

12.1 修井作业分为大修和小修。

12.2 在油气井大修或小修过程中，不允许使用不可逆转地降低近井地带地层渗透性的修井工作液（隔断或封堵水层作业除外）。在使用会降低近井地带地层渗透率的修井工作液的情况下，后续必须采取恢复性措施。井口和井筒设备及修井工作液的密度应避免出现原油、天然气及水的溢流。

12.3 油气井大修和小修应根据油气开采行业现行工业安全规范和法规、矿产资源和环境保护要求，以及所使用设备操作规程和实施工艺流程的技术规范和作业要求进行。

12.4 根据油气井的生产现状确定油气井进行大修、小修作业的必要性，并按照矿产资源使用者批准的作业方案实施。

12.5 油气井大修和小修工作应当按照规定的格式编制作业说明（大修）和作业报告（小修），并将其记录在井史中。

prepared in the prescribed format and recorded in the well history.

12.6 Monitoring and regulation of oil and gas well production process:

12.6.1 The purposes of development, implementation and monitoring of oil and gas reservoirs are as follows:

(a) Evaluate the effectiveness of the development system and the implementation of the corresponding technological measures adopted by the production line as a whole;

(b) Obtain the information needed to develop the process adjustment and its improvement action plan.

12.6.2 The type, amount and period of testing work during the development of oil and gas fields should be determined through the technical design scheme according to the approved method recommendations, and in some cases according to the work plans.

12.6.3 In order to monitor the development process of oil and gas fields and the production process system of a single well, according to the technical design scheme, a single well must be equipped with a special measuring instrument.

12.6.4 To monitor the production parameters of individual wells, a pressure gauge (device) and a produced fluid sampler are installed at the wellhead. Each well connection manifold shall ensure separate measurement of liquid production and natural gas volume, sampling of produced liquid, wellhead pressure measurement, echo meter measurement, automatic force curve measurement (for wells equipped with rod deep well pumps) and downhole measuring instruments capable of being installed from the gusher.

12.6.5 In order to monitor the working regime of the injection well, a manometer

12.6 油气井生产过程的监测与调节：

12.6.1 油气藏的开发实施监测目的如下：

（a）评估生产层系整体上所采用的开发系统、相应的工艺技术措施实施的有效性；

（b）获取开发过程调节及其改进措施方案所需的信息。

12.6.2 根据批准的方法建议，在某些情况下按照工作计划，通过技术设计方案确定油气田开发过程中测试工作的类型、工作量及周期。

12.6.3 为监测油气田开发过程及单井生产工艺制度，根据技术设计方案，单井必须配备专用的测量仪器。

12.6.4 为监测单井的生产参数，在井口安装压力计（装置）和采出液取样器。各井连接管汇应当保证单独计量产液量和测定天然气量、采出液取样、测量井口压力、回波仪测量、自动测力曲线测量（对于配备有杆式深井泵的井）及能够安装自喷井的井下测量仪器。

12.6.5 为了监测注入井的工作制度，在井口应安装压力计（装置）。各井连接管汇应当保证单独计量井口压力下的注入量（吸液能力）及安装井下仪器。

(device) shall be installed at the wellhead. Each well connection manifold shall ensure that injection volume (suction capacity) at wellhead pressure is measured separately and downhole instruments are installed.

12.6.6 Oil and gas wells that are not equipped with the above measuring and testing equipment are not allowed to produce.

12.7 The following are the main methods and measures for adjusting development methods:

(a) Change the working system of production wells (including increasing or limiting gas production or fluid production, closing high water or free gas breakthrough wells, strengthening fluid production and regularly adjusting production);

(b) Changing the working system of injection wells (including increasing or limiting the injection amount of working reagents, adjusting the injection amount by well, and periodically injecting);

(c) Improving the hydrodynamics of oil and gas wells (e.g. supplementary perforation, near-wellbore formation modification measures, formation hydraulic fracturing);

(d) Blocking or limiting the production of ineffective water (gas) in the well through various processes (cementing, creating artificial non-permeable barriers in the formation, using chemicals liquid);

(e) Use the flow direction adjustment process;

(f) Adjust the perforating interval within the production series;

(g) Implement stratified exploitation and stratified water injection of oil and gas wells in multiple sets of oil and gas fields.

12.7.1 Refine and improve the injection

12.6.6 未配备上述测量和测试装置的油气井不允许生产。

12.7 以下部分属于调整开发方式的主要方法和措施：

（a）改变生产井的工作制度（包括增加或限制采气量或产液量，关闭高含水井或有游离气气窜井、强化采液及定期调整开采量）；

（b）改变注入井的工作制度（包括增加或限制工作试剂的注入量、按井重新调整注入量、周期性注入）；

（c）提高油气井的水动力学完善程度（如补充射孔、近井地带地层改造措施、地层水力压裂）；

（d）通过各种工艺（注水泥、在地层内制造人工非渗透性屏障、使用化学药剂）封堵或限制井内无效的水（气）的产出；

（e）运用渗流方向调整工艺；

（f）调整生产层系内的射孔段；

（g）在多套产层油气田中，对油气井实施分层开采和分层注水。

12.7.1 根据技术设计中确定的当前剩余储量构成，对注水系统进行完善和改进（一种注水系统调整为另一种注水系统、中心式注水、调整注水前缘）。

system based on the current remaining reserve composition as determined in the technical design (adjustment of one injection system to another injection system, central injection, adjustment of injection front).

12.8 Design the development monitoring and adjustment system (taking into account production, injection and adjustment) in the recommended development scheme for specific geo-geophysical conditions and different development stages.

12.8 针对特定地质—地球物理条件及不同开发阶段, 在推荐开发方案中设计其开发监测和调整系统 ( 考虑产量、注入量及其调整 )。